Soils of the Tropics

Soils of the Tropics

Properties and Appraisal

Armand Van Wambeke
Department of Soil, Crop, and Atmospheric Sciences
Cornell University
Ithaca, New York

McGraw-Hill, Inc.
New York   St. Louis   San Francisco   Auckland   Bogotá
Caracas   Lisbon   London   Madrid   Mexico   Milan
Montreal   New Delhi   Paris   San Juan   São Paulo
Singapore   Sydney   Tokyo   Toronto

Library of Congress Cataloging-in-Publication Data

Wambeke, A. Van.
  Soils of the tropics : properties and appraisal / Armand Van
Wambeke.
    p.   cm.
  Includes bibliographical references and index.
  ISBN 0-07-067946-0
  1. Soils—Tropics. 2. Land capability for agriculture—Tropics.
3. Soil management—Tropics.   I. Title.
S599.9.T76W36   1991
631.4'713—dc20   91-30678
                                                     CIP

1 2 3 4 5 6 7 8 9 0   DOC/DOC   9 7 6 5 4 3 2 1

ISBN   0-07-067946-0

The editing supervisor for this book was Joseph Bertuna, and the
production supervisor was Donald F. Schmidt. It was set in Century
Schoolbook by McGraw-Hill's Professional Book Group composition
unit.

Cover Photographer: Donna Nussbaum

Printed and bound by R. R. Donnelley & Sons Company.

This book is dedicated to my wife
Francine

Contents

Preface

There has been in recent years a growing interest by the general public, the media, international institutions, government agencies, multinational corporations, local development organizations, municipalities, and, finally, small farmers in the ways that land resources are utilized in the tropics. Small farmers are the individuals who face the most serious dilemma of supporting themselves, feeding their families, and, at the same time, preserving the environment against degradation.

The public participates in the debates. The media periodically reports on conflicts created by the aggressive expansion of modern agriculture and its consequences with respect to the destruction of rain forests, the burning of savannas, the increasing carbon dioxide levels in the atmosphere, and the issue of global warming.

There are also economic and social reasons for wariness about the use of land. Developing countries become more and more dependent on each other for essential commodities, and their socioeconomic standings are strongly influenced by international trade agreements. The participation of tropical agriculture in commodity production has increased, and certain countries are now powerful competitors in world markets. On the other hand, some nations are unable to upgrade their agricultural technology to become self-sufficient in food production or at least be capable of avoiding periodic famines. Hunger is one of the motivations for the growing interest in agronomy of the soils of the tropics.

Soil science can contribute by clarifying many of the pressing problems in food production and can assist in the conservation of natural resources in the tropics. To achieve these objectives, it is necessary to appraise the potential of land to support people without the deterioration of the environment.

Soils are part of this environment; they are the basis of sustained development. Accurate information on land resources in the tropics is required to obtain objective knowledge on the management properties of tropical soils. This would lead to more reliable predictions about the outcome of development initiatives.

The present text reviews soil conditions in the tropics and describes land types in terms of their qualities and major constraints. The book addresses an audience of individuals who have an understanding of the concepts, definitions, and terms that are used in soil science or in natural resource management. It is written by keeping in mind those readers who are *not* familiar with the tropical environment; therefore, it places some emphasis on the *differences* that exist between soils of the tropics and soils of temperate regions. It calls attention to characteristics that are often overlooked by people who have *not* worked in the tropics. It may for this reason seem somewhat elementary to agronomists who have had long experience in tropical agriculture.

The question has often been raised whether there is a need for a *separate* text on soils of the tropics. Why is a general soil science book not sufficient? It is true that the same physical and chemical principles apply to the tropical as well as to the temperate world. The same laws of chemistry and physics are valid everywhere.

There are differences, however, that justify a publication on tropical soils. They are of two kinds: First, soil formation and soil-plant relationships involve biological processes that are activated by living organisms specific to the tropical environment, and second, soil management practices are essentially technological *packages* that have been tailored to particular ecological conditions. These packages cannot be transferred without their adaptation to the environments in which they are introduced.

Some aspects of soil science are emphasized. One of them relates to the *diversity* of soils: Not all soils were created equally. To understand this diversity it is necessary to know their modes of formation and the genetic pathways that shaped their characteristic attributes. A large part of the text deals with soil genesis.

Part 1 describes the tropical environment and its interactions with soils. However, it is not a comprehensive compilation of the physical geography of the tropics; only points that are relevant to soil formation and management are mentioned. These include climate, parent materials, vegetation, and fauna.

Part 2 describes the major kinds of soils of the tropics. It starts by introducing the classification system that is needed to provide an internationally acceptable terminology. It stresses the properties that relate to soil-forming processes and highlights their importance for soil management. It follows the U.S. Department of Agriculture system given in *Soil Taxonomy* in its latest version of *Keys to Soil Taxonomy*.

Part 2 is further subdivided according to the soil orders of *Soil Taxonomy*. Each order that is discussed is the subject of a separate chapter. This approach may not be perfect, but it facilitates access to information that is specific to each kind of soil. It is based on the belief that all soils have unique properties that dictate their responses to management and determine their suitability for particular land utilization types. There are, of course, no sharp boundaries in the soil universe and many orders share common attributes. Cross-referencing and indexing should help the reader to overcome the disadvantages imposed by the soil classification approach.

Not every defined order is the subject of a separate chapter. A choice had to be made. The mineral soils that cover large areas in the humid or semiarid tropics, that support high population densities, or that present serious management problems when cultivated have been given preference. There is, for example, no chapter on Mollisols because they are not extensive in the tropics and have no major limitations for agricultural use.

The contributions of colleagues who advised in the selection of the subject matter and reviewed the numerous drafts that preceded this publication are gratefully acknowledged. The students of the Agronomy 471 and Agronomy 480 courses that are offered in the College of Agriculture and Life Sciences at Cornell University were constructive by their reactions and suggestions. The teaching assistants who participated for more than 12 years in revising chapters were active contributors to the text.

Many soil scientists and agronomists of the Agricultural Research Organization of the former Belgian Congo and Ruanda-Urundi, and colleagues in the Food and Agriculture Organization of the United Nations (FAO) have influenced the order of the subject matter. There is no doubt a bias toward soil conditions in Africa and South America. Soil scientists of tropical countries contributed considerably in the selection of topics that are emphasized in this book.

Affiliation with the Faculty of Sciences at the State University of Gent, Belgium and teaching activities in the Department of Soil, Crop, and Atmospheric Sciences in the College of Agriculture and Life Sciences at Cornell University stimulated a variety of interests that combined pure science with applied, more practical, agricultural science.

Cooperation with the Soil Conservation Service of the U.S. Department of Agriculture was instrumental in adopting their classification system, *Soil Taxonomy*, as a framework for constructing the outline of the book.

Special thanks are expressed to Dr. Hari Eswaran for providing illustrations and Dr. John Kimble for helping in the classification of the profiles that are included in the appendix. The International Soil Reference and Information Center (ISRIC) at Wageningen, The Netherlands, and Drs. H. Van Baren and L. P. van Reeuwijk gave without delay soil information that is difficult to access otherwise. Dr. M. McBride's assistance in reviewing parts of the sections on weathering is also gratefully acknowledged. The chapter on soil organic matter was constructively reviewed by Dr. J. Duxbury.

A. Van Wambeke

Soils of the Tropics

The Tropical Soil Environment

There are many definitions of the term "tropics." No attempt
is made here to propose a new definition or select the most
appropriate one; the intent is rather to describe the
characteristics of tropical climates as they are defined by
climatologists, placing the emphasis on characteristics that
separate tropical regions from temperate regions.
The climatic attributes of the tropical area derive from its
geographic location between latitudes 23° 27' N and S. This
is the only region where the sun can be directly overhead; the
high elevation of the sun above the horizon allows more
radiation per unit area to reach the outer limits of the
atmosphere than in temperate regions. Another inference
resulting from the geographic location of the tropics is that
the lengths of the days practically remain unchanged during
the whole year.
The high elevation of the sun above the horizon and the
constant length of the day have two consequences. First, at the
time the sun is close to zenith, the strong radiation and
heating causes expansion of the air, rising and cooling of air
masses, condensation of moisture, and formation of clouds,
which bring a maximum of rain at the time of the year when
the sun is directly overhead. Second, the annual seasonal
temperature variations are not pronounced, or, in other
words, there are no seasons determined by marked changes in
temperature.
The tropics are said to be regions of summer rains. In this
expression, summer means the season during which the sun
in a particular hemisphere comes in zenithal position. There
is no connotation that this season is necessarily the warmest

one. In certain cases, because of cloud cover, the
astronomically defined summer may actually be the coolest
season.
The two characteristics, the low seasonal variation in
temperatures and the summer concentration of rain, are
essential to the concept of tropics. Other definitions also
include the absence of frost, more specifically frosts that kill
plants, or low temperatures that cause leaf fall. For example,
the Agroecological Zones Project (FAO, 1981) considers
climates as tropical only if all the mean monthly
temperatures, corrected to sea level by subtracting
0.6°C/100-m elevation, are above 18°C. Those that have one or
more months with corrected temperatures below 5°C are
temperate; the others, between these two extremes, are
subtropical regardless of whether they have summer or winter
(mediterranean climates) rains.
Figure I.1 illustrates some of the major attributes of tropical
climates: three climatic characteristics at two latitudes (0 and
30° N) are shown on a monthly time scale: they exemplify the
major differences between the tropics and temperate regions.
The most striking is the uniform high elevation of the sun at

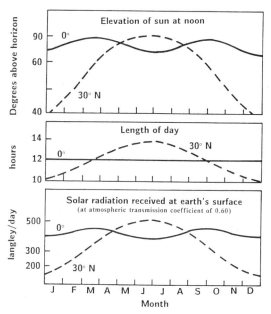

Figure I.1   Comparison of radiation and day length at
the equator and at 30° N latitude. [From Nieuwolt
(1977). John Wiley & Sons, reprinted by permission.]

noon time during the whole year. The second characteristic is the constant day length of close to 12 h; the potential insolation at 0° latitude never reaches 14 h/day as at 30° latitude. The resulting radiation that reaches the earth's surface at 30° latitude varies markedly during the course of the year, as opposed to the rather uniform incoming radiation at the equator. Note that temperate regions may receive more radiation during cloudless dry summers than an equatorial area during the rainy season.

References

FAO, *Report on the Agroecological Zones Project,* vol. 3: *Methodology and Results for South and Central America,* World Soils Resources Report #48/3, Food and Agriculture Organization of the United Nations, Rome, 1981.

Nieuwolt, S., *Tropical Climatology—An Introduction to the Climates of the Low Latitudes,* John Wiley & Sons, London, 1977.

1

Soil Climate

Soil climate is the *only* common attribute that all soils in the tropics share and that separates them from soils in the temperate regions. This common characteristic is important for land use as well as for the correct understanding of soil-forming processes. There are no other soil properties that consistently discriminate tropical soils from the soils of other regions.

This chapter considers soil climate from several viewpoints: one of them relates soil temperature and moisture to plant growth and is of particular interest to agronomists. Another viewpoint is that of soil formation; this perspective contributes to the understanding of soil genesis and the geographic distribution of the major kinds of soils.

When the second objective of relating soil properties with climate is pursued, both the present and past climates are important. Many soils started their formation during earlier geological times and acquired properties that are characteristic for climatic conditions different from the present ones. Many soils of tropical regions are relics. They remained on the landscapes because no glaciers scraped them away, as happened in many temperate regions, and because the soil constituents that were formed resist alteration by the present climatic conditions.

1.1 Soil Temperature

1.1.1 Effect of radiation on soil temperature

Because of the low *angle of incidence* of sunshine, unit soil areas in the tropics receive more radiation than at high latitudes. At the same time, the thickness of the atmosphere through which it has to pass before it reaches the ground is much thinner; consequently the intensity

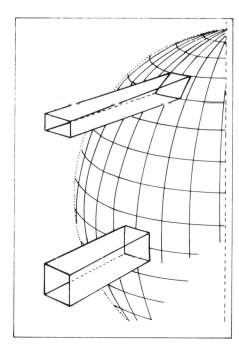

Figure 1.1 Unit areas at the equator receive more radiation that travels through less atmosphere than at high latitudes. [*From Trewartha et al. (1967). McGraw-Hill, redrawn with permission.*]

of the radiation that enters the lower atmosphere is greater than in temperate regions (Fig. 1.1).

The high rise of the sun above the horizon makes the high radiation intensity one of the most critical factors for plant growth, because the radiation is capable of increasing soil temperatures to values[1] that may be lethal for germinating crop seeds and young seedlings. It also maximizes the differences between day and night soil temperatures. The surface soil temperature amplitudes in the tropics are generally wide, with up to approximately 20°C difference between night and day temperatures for uncovered surface soil. The pronounced changes in soil temperature that often coincide with moisture fluctuations cause organic matter to decompose faster than in soils with uniform weather conditions.

The angle of incidence of sunshine also depends on the aspect of slopes. Topography may cause some slight variations in soil temperatures. However, given the low latitude of the tropics, the north or south aspect of hill slopes does not cause the large differences in soil temperature that exist in temperate climates. For example, at 14° N latitude, differences of only 3°C were observed in the soil at 50-cm depth between north- and south-facing slopes (Alexander, 1976).

---

[1]Some critical temperatures for plant growth are given in sections 1.1.2 and 1.1.4.

There may be small (less than 1°C at 50-cm depth) but significant differences in soil temperature on east-facing slopes, which are generally slightly warmer than west-facing slopes during the rainy season. These variations are apparently related to morning sunshine and afternoon cloudiness (Alexander, 1976).

Soil properties influence the dimension of the temperature changes that takes place in the soil. The dimensions vary according to the specific heat of the soil and its thermal conductivity. To set orders of magnitude, the specific heat of soil minerals is approximately one-fifth of that of water (0.2 cal/g/°C or 838 J/kg/K) and the specific heat of organic matter is roughly one-half of that of water (2000 J/kg/K). At field capacity the specific heat of a medium-textured soil is more or less 50 percent higher than its specific heat at wilting point.

The thermal conductivity of soils varies with their moisture content, bulk density, organic matter, and quartz content. Water increases the thermal conductivity of sands at low water contents (< 10 percent) but only at high water contents in clayey soils. Structure also influences heat transfer. An open, rough seed bed has a lower thermal conductivity than a fine, slumped seed bed; a loose structure is said to help the emergence of seedlings by allowing them to avoid too close contact with the hot soil as they approach the surface (Sinclair, 1987).

The physical constants that were mentioned imply that dry topsoils that contain little organic matter are the most susceptible to extremes of temperatures. Sandy surface layers suffer most from incoming radiation and often need to be protected against direct sunshine.

Figure 1 2 provides an overall view of daily and monthly soil temperature variations at 1 cm depth in Allahabad, India (25° 27′ N, 81° 50′ E), a weather station with a hyperthermic soil temperature regime. Figure 1.3 shows the annual cycle of monthly means of soil temperatures measured at 3:00 p.m. at different depths in soils of Ibadan, Nigeria (Harrison-Murray and Lal, 1979). The soils of Ibadan have an isohyperthermic temperature regime. There are other examples of maxima in soil temperatures: 50°C at 5 cm depth in bare sandy soils of Brazil and in red soils of Guyana. The hottest temperatures most frequently occur at planting time, which is therefore the most critical period of the growing season.

There are several *management techniques* available to avoid the high soil temperatures and at the same time to reduce the amplitude of the daily variations. Cover crops and mulches are the most common. In the examples given above temperatures may drop from 50°C to approximately 30°C by mulching. The daily amplitude is also reduced to 5°C.

Living cover crops or natural vegetation depress soil temperatures more than dead mulches because part of the incoming radiation is transformed into latent heat that is used for transpiration. In dry cli-

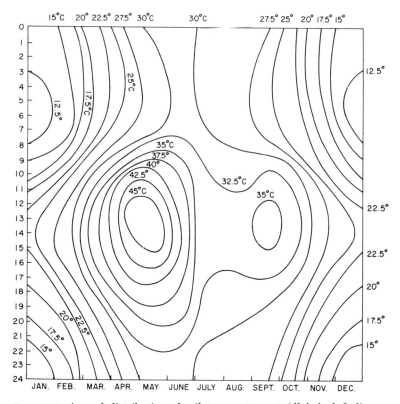

Figure 1.2 Annual distribution of soil temperature at Allahabad, India, recorded at 1-cm depth. [*From Monteith (1979). John Wiley & Sons, reprinted by permission.*]

mates the living cover crops may not be appropriate because they compete for water with the main crop.

When the soil surface is completely protected against radiation, for example, by a dense vegetative cover, the surface may never reach temperatures that are higher than the temperatures of the air above it. For example, observations in the Guyana rain forest at 10-cm depths indicate uniform temperatures around 24°C.

1.1.2 Effect of air temperature on soil temperature

Soils that are not directly influenced by radiation tend to adjust their temperatures to the temperatures of the air above the ground. In the tropics the *seasonal* fluctuations at a given site are small, and therefore temperature is not a major criterion for the selection of planting times. Moisture conditions instead are more important.

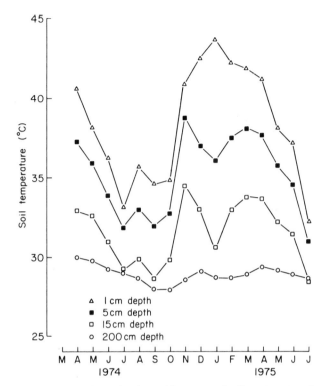

Figure 1.3   Annual cycle of monthly means of soil temperature in two bare fallow soils at Ibadan, Nigeria. [*From Harrison-Murray and Lal (1979). John Wiley & Sons, reprinted by permission.*]

Annual air temperatures vary according to the elevation above sea level. They define altitudinal climatic belts in mountainous areas of the tropics to which vegetation, agriculture, and people adapt.

Land clearing methods also modify air temperatures by burning vegetation. The high temperatures that are generated affect soil properties.

Relation between air and soil temperatures.   Correlations between soil and air temperatures are used by soil taxonomists for classification purposes. They use annual and monthly means to distinguish classes in soil temperature regimes.

Annual soil temperatures in the tropics are always higher than the annual air temperatures but the differences between the two vary with the moisture regimes of the soils. The largest differences occur at the driest sites that usually have an aridic moisture regime and can reach more than 5°C. In soils with udic moisture regimes the difference is approximately 2°C. The higher specific heat and better ther-

mal conductivity of wet soils, together with the lower proportion of solar radiation that reaches the ground in rainy climates, explain the
different behaviors (Nullet et al., 1990). There are at present no computational models that allow taxonomists to estimate with sufficient
precision soil temperatures at any given time and depth from atmospheric data.

The computation of the differences between winter and summer soil
temperatures suffers from the same uncertainties as the computation
of the annual means. Interest in these figures results from the intent
of the authors of *Soil Taxonomy* (Soil Survey Staff, 1975) to distinguish tropical soils on the basis of the small seasonal variations in
temperatures that characterize the low-latitude regions. The criterion
that is used for this purpose is the difference between the 50-cm depth
average soil temperatures of 3 summer and 3 winter months; the difference should be less than 5°C for a soil temperature regime to be
considered as an *isotemperature regime,* a term that is intended to be
synonymous with tropical temperature regime.

Despite the computational shortcomings, the definition of
isotemperature that is given above is adhered to in this text. In most
cases it succeeds in distinguishing the summer rain areas in the world
that have only small yearly seasonal temperature variations. However, it does not include all summer rain regions, particularly those
that have monsoon climates, such as the northern hyperthermic part
of the Indian peninsula.

Vertical zonation.   Vertical zonation is the term used to designate the
variations in soil properties that are linked to changes in altitude. Air
temperatures generally drop by 0.6°C for each hundred meters increase in elevation; soil temperatures follow the same trend.

The major effect of decreasing temperatures on soil properties is
due to the slower decomposition of soil organic matter in cold climates. If the precipitation is adequate to support vegetation and
produce biomass, soils at high elevation in the tropics contain more
organic matter than in tropical lowlands. Mountain soils are generally darker than their low-elevation counterparts. The organic
matter and soil temperature relationship will be discussed in more
detail in Chap. 3.

The location of temperature zones in tropical mountains has
changed considerably during the quaternary period. Figure 1.4 illustrates how dramatically the vegetation belts have shifted their position along the slopes of the Andes Cordillera. The snowline was more
than 1000 m lower 14,000 years ago, and all mountain climates in the
tropics were much colder than they are presently (Flenley, 1981).

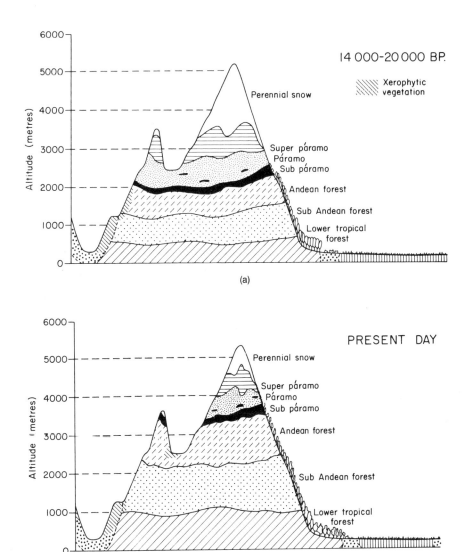

Figure 1.4 Changes in the position of vegetation belts in the South American Andes. [*From Flenley (1981). Butterworth, reprinted by permission.*]

Crops have their specific temperature requirements and the geographic distribution of crops is linked to the altitudinal belts. Figure 1.5 illustrates this association for east Africa as determined by Acland (1971). Table 1.1 adapted from Voorhees (1981) gives the optimum soil temperatures for maximum yields. Similar climatic toposequences exist in all tropical mountains.

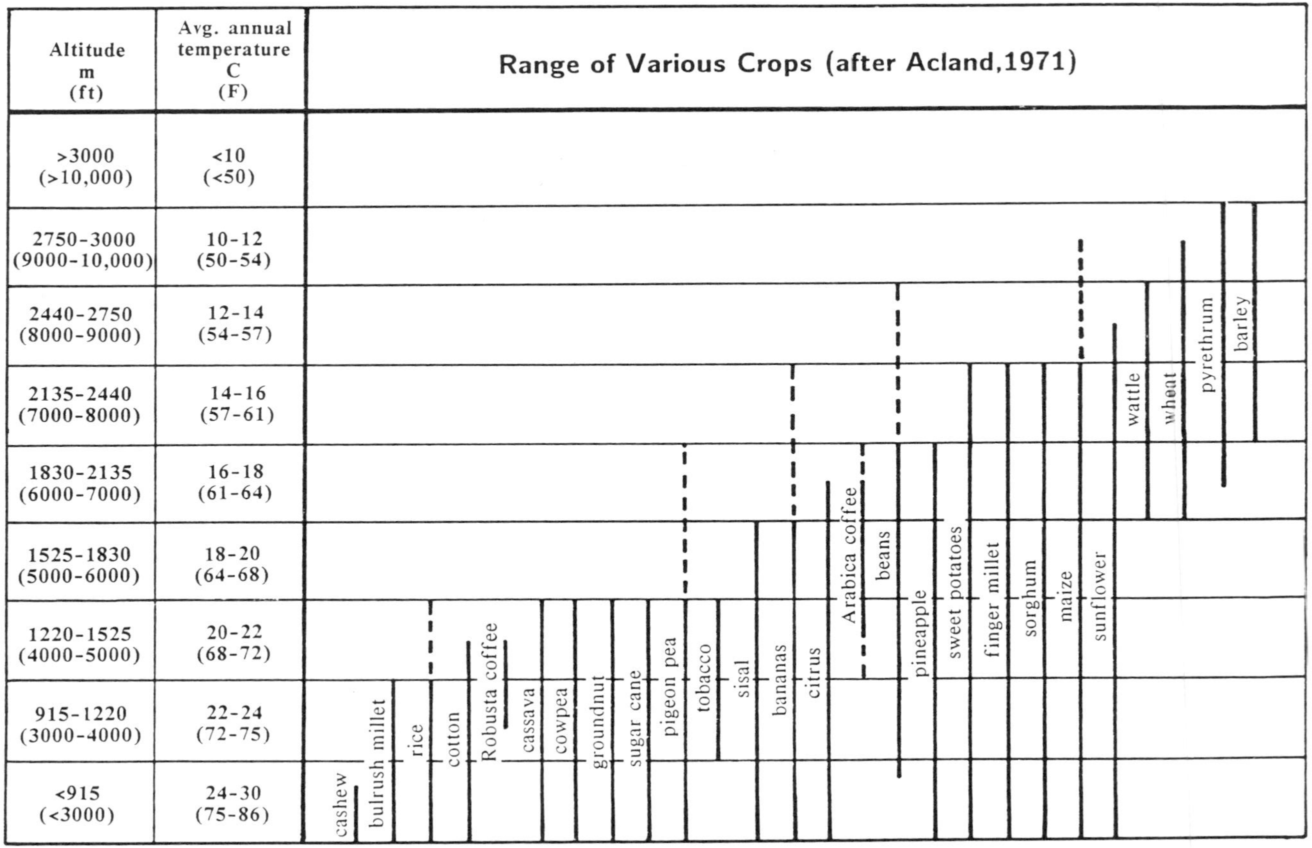

Range of Various Crops (after Acland, 1971)

Altitude m (ft)	Avg. annual temperature C (F)
>3000 (>10,000)	<10 (<50)
2750–3000 (9000–10,000)	10–12 (50–54)
2440–2750 (8000–9000)	12–14 (54–57)
2135–2440 (7000–8000)	14–16 (57–61)
1830–2135 (6000–7000)	16–18 (61–64)
1525–1830 (5000–6000)	18–20 (64–68)
1220–1525 (4000–5000)	20–22 (68–72)
915–1220 (3000–4000)	22–24 (72–75)
<915 (<3000)	24–30 (75–86)

Crops: barley, pyrethrum, wheat, wattle, sunflower, maize, sorghum, finger millet, sweet potatoes, pineapple, beans, Arabica coffee, citrus, bananas, sisal, tobacco, pigeon pea, sugar cane, groundnut, cowpea, cassava, Robusta coffee, cotton, rice, bulrush millet, cashew

Figure 1.5   Relationship between crops, altitude, and annual air temperatures in east

12

TABLE 1.1   Optimum Soil Temperatures for Maximum Crop Yields

Plant species		Temperature, °C
Common name	Scientific name	
Alfalfa	*Medicago sativa L.*	21–27
Apple	*Malus sp.*	25
Barley	*Hordeum vulgare L.*	18
Beans	*Phaseolus vulgaris*	28
Brome grass	*Bromus inermis Leyss.*	19–27
Coffee	*Coffea arabica*	20 (night)–26 (day)
Corn	*Zea mays L.*	25–30
Cotton	*Gossypium hirsutum L.*	28–30
Gardenia	*Gardenia grandiflora*	23
Grape	*Vitis sp.*	28
Guayule	*Parthenium argentatum*	28
Jack pine	*Pinus banksiana*	27
Kentucky bluegrass	*Poa pratensis L.*	15
Lucerne	*Medicago sativa L.*	28
Oats	*Avena sativa*	15–20
Onion	*Allium sp.*	18–22
Orange	*Citrus sp.*	25
Orchard grass	*Dactylis glomerata L.*	20
Pea	*Pisum sativum L.*	18–22
Potato	*Solanum tuberosum L.*	20–23
Rice	*Oryza sativa L.*	25–30
Rye grass	*Lolium perenne L.*	20
Snapdragon	*Antirrhinum L.*	20
Soybean	*Glycine max L.*	22–27
Soybean (inoculated)	*Glycine max L.*	19
Soybean (not inoculated)	*Glycine max L.*	30
Squash	*Cucurbita sp.*	27
Strawberry	*Fragaria sp.*	18–24
Sugar beet	*Beta vulgaris*	20–24
Sugar cane	*Saccharum officinarum L.*	25–30
Sunflower	*Helianthus sp.*	23
Tobacco	*Nicotiana tabacum sp.*	22–26
Tomatoes	*Lycopersicon esculentum Mill.*	26–34
Wheat	*Triticum aestivam L.*	20

SOURCE: *Voorhees (1981). American Society of Agricultural Engineers, copied by permission.*

The temperature belts and the specific temperature requirements of plant species have to be taken into account when selecting crops and planning the use of land. Many annual crops of temperate regions can be grown at high elevation in the tropics provided that the air and soil temperatures during their growing season are sufficiently cool.

The introduction of tree crops or other perennials may be more difficult. The absence of a cold season may preclude their satisfactory adaptation to the tropical environment. For example, apple trees need chilling at some period for optimum production, and the simulation of this event by other physiological processes does not always lead to the expected results.

Burning vegetation

Effect on soil temperature. Burning vegetation is a common practice in the tropics. It is for many farmers the most economic and practical way to remove vegetation from the fields when opening land for agriculture. It is also used to stimulate regrowth in savannas and for many other purposes. The effects of burning on the environment and the consequences it has on soils is a much debated subject of controversy.

The changes in soil temperature that are caused by burning vegetation depend on the intensity of the fire and its duration. The temperature reached in the burning vegetation itself is a function of the amount of organic matter, its water content, and the combustion rate. The latter is influenced by wind, which supplies oxygen to the fire and determines the speed of spread of the fire front.

The flame front in a typical tropical savanna fire usually progresses at a rate that varies between 75 to 300 m/h and occasionally reaches 900 m/h. The temperatures in the burning vegetation range from 300°C at the soil surface, to approximately 200°C at 1.5 m above the ground (Fig. 1.6).

The effects of fire on soil temperatures depend on the length of time of exposure of the soil to the fire. In savanna fires soil exposure is only a few minutes. Temperatures immediately below the soil surface seldom exceed 100°C; in fact, as long as free water is present in the soil, the heat is used to boil off the water.

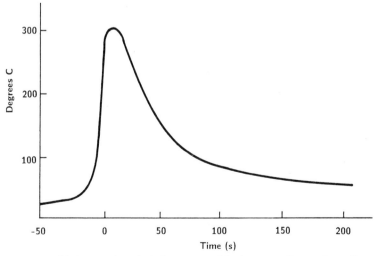

Figure 1.6   Air temperature just above ground during grass fire on clay soil near Port Moresby, Papua. Time in seconds; time zero is arrival of fire front. [*From Scotter (1970). Australian Journal of Soil Research, redrawn by permission.*]

The soil temperatures during a fire that is spreading at a normal rate in savanna vegetation decrease rapidly with depth. The length of time that soil temperatures are above 50°C at 1 or 2 cm below the surface usually does not exceed 1 or 2 minutes. Pitot and Masson (1951) found an increase of 2°C at 5-cm depth during the 5 minutes that followed the passage of the fire front in grassland vegetation.

Effects on soil properties. Only the effects of the high temperatures are discussed here; the influence of ashes that are produced by the fire on soil fertility will be dealt with in later sections.

It is generally accepted that the short-lived increase in soil temperature during savanna fires has little direct effect on the properties of the soil. However, indirect effects, which result from the removal of the plant cover during the fire, are important because (1) the soil has reduced infiltration rates, which favor erosion due to increased runoff; (2) the removal of the vegetation eliminates transpiration (the fire may be a soil water conservation practice to retain the little amount of moisture that may still be present in the subsoil); and (3) the absence of vegetation affects the exposure of the soil to raindrop impact and direct insolation (this is the most critical indirect effect). The timing of the fires is frequently such that soils are left unprotected against erosion at the beginning of the rainy season.

The time at which the burning takes place in the dry season indirectly influences the long-term effects of savanna fires on soil properties. Moore (1960) compared plots on flat topography at Ibadan, Nigeria, which for 30 years had been subjected to either early (light) burning or late (fierce) burning or were completely protected against burning. He observed that:

1. The organic matter content of the soil in the early-burned plot was 30 percent higher than the organic matter in the late-burned plot. The early-burned plot also had a higher organic matter content than the protected plot.

2. Nitrogen levels were highest in the early-burned plot and lowest in the protected plot.

Moore considered that the differences between the treatments were more related to the botanical composition of the plant cover induced by the treatments than to the direct effects of fire. He concluded that early burning produces a somewhat higher soil fertility level than complete protection.

Trapnell (1959) reported that after 23 years of fire treatments, early burning maintains a closed canopy woodland. Late burning on the contrary results in hot fires that encourage invasion by grasses.

Forest fires, or the burning of wood heaps after clearing, release a large amount of heat from glowing combustion and may last for several hours. These fires may locally produce high temperatures in the soil. Sanchez (1976) reported temperatures of 450 to 650°C at 2 cm above the surface, which decrease in the top 5 cm of the soil at a rate of 100°C/cm. The high temperatures may locally activate the destruction of organic substances in the soil and have a direct effect on the microflora and fauna. Little is known quantitatively about the losses of specific soil constituents that occur during the burning of wood when rain forests are cleared. They are probably well compensated by the release of plant nutrients that are contained in the ashes.

High temperatures produced by fire may have specific objectives. Sertsu and Sanchez (1978) studied the *guie* (soil-burning) practice used by Ethiopian farmers to restore soil fertility when crop productivity is declining. Soil clods are heated in heaps mixed with dry grass and cow manure for 10 days or more. To simulate the effects of temperature, Sertsu and Sanchez heated surface samples of a Paleudult derived from volcanic ash, and a Vertisol for 48 h at 100, 200, 400, and 600°C. Temperatures of 400°C and higher changed textures from clayey to sandy. Organic matter started to decompose between 100 and 200°C and was essentially eliminated at 400°C. Total nitrogen was reduced by half at 400°C. Nitrogen in the form of ammonium increased sharply at 200°C but decreased at higher temperatures. The most pronounced chemical change was a marked enrichment in available phosphorus in all soils at 200°C or above. The authors conclude that the beneficial effect in the volcanic ash soil is obtained by heating at temperatures that are low enough to prevent important changes in organic matter content but are high enough to increase the availability of phosphorus and the amount of ammonium nitrogen.

1.1.3   Effect of temperature on soil formation

It is difficult to single out soil properties that are exclusively and directly attributable to temperature. Most changes in temperature coincide with moisture changes, i.e., warming often causes drying. Many temperature effects on soil properties are indirect in that they involve specific vegetation types that leave their specific marks in the soils. It is not possible to dissociate these interactions from the primary cause, which may be the soil temperature.

Temperature governs the kinetics of chemical processes. The rate of most inorganic reactions increases exponentially with temperature, approximately two- or threefold for each 10°C (referred to as the "Q_{10}"). This general principle is valid as long as water is available to maintain the reactants in solution. Given this relationship, it is obvious that the warm summer rains in the lowland tropics accelerate chemical decomposition. In the case of minerals the transformations are usually referred to as *weathering*; in the case of organic matter they are often referred to as *mineralization*. If the high temperatures act together with surplus rains, leaching of the reaction products further increases the decomposition rates. Weathering and mineralization are discussed in detail in separate chapters.

The most obvious consequences of the temperature regimes in the tropics are related to the *absence* of low-temperature seasons that include freezing or that are characterized by air that is cold enough to induce leaf fall. Freezing of wet soil often improves the structure of plow layers in temperate regions; this mechanism does not operate to restore the structure of compacted light-textured topsoils that have deteriorated under hot, dry weather conditions in the tropics.

In tropical climates plants do not shed their leaves because of a sudden drop in air temperature as they do in temperate regions; there is consequently no concentration of leaf fall at the beginning of a cold winter. At high latitudes, in autumn and winter, leaf litter is soaked by rain and snow and produces organic extracts that penetrate into the soil. They often mobilize sesquioxides and clays in the topsoil and transport them in the lower parts of the profiles where they form illuvial horizons.

In the wet, dry tropics leaves fall on dry soil, and the microfauna is the first to consume the plant residues. The transformations follow metabolic pathways that are different from those that prevail in regions with cold winter rain. There is no immediate production of leachates, but rather a comminution of dry plant materials by insects. Therefore, the decomposition products of organic matter in the tropics do not lead to the formation of strongly contrasting eluvial and illuvial horizon sequences.

There are other soil properties associated with warm climates: Color is one of them. Many soils of regions with a hot, dry season have strong reddish or yellowish colors that are characterized by high chromas. There are several iron oxide minerals that act as pigments to produce these colors: Goethite and hematite are the most important ones. The first is yellowish-brown, while the second is red. According to Schwertmann (1985), hematite forms preferentially at high temperatures and crystallizes from precipitates, and goethite crystals tend to form in dilute solutions that are not subjected to high temperatures and drying. Organic matter also plays a role in coloring soils; as a rule it attenuates bright colors. There is little consistent quantitative evidence, however, on the exact influence of soil organic matter on the color of soils of the tropics.

Fires that reach temperatures of about 400 to 500°C convert aluminum-substituted goethite or hematite into maghemites. This often occurs in the upper horizons of tropical soils (Schwertmann and Fechter, 1984).

1.1.4 Effects of soil temperature on plant growth

The optimum temperature for *seed germination* or the temperature where the highest percentage germination is obtained in the shortest time varies from species to species. Cotton, for example, will best germinate at 34°C, corn at 27°C, and sorghum at 16°C. The germination of most crops is severely reduced at temperatures higher than 38°C. The lethal temperature for germinating sorghum is 45°C (Sinclair, 1987). Minimum temperatures for germination of these crops, although less critical in the tropics, are as follows: Sorghum starts germinating at temperatures between 7° and 10°C, corn at 10°C, and cotton will only germinate at temperatures above 17°C.

High temperatures produced by grassland fires may stimulate germination of *Themeda triandra* (Troloppe, 1984). The germination of *Acacia* spp. is enhanced by heat treatment as experienced during a fire.

Root growth is also affected by temperature. The optimum for cotton is reported to be 27°C, and for corn it is somewhere between 20 to 30°C. The maximum temperatures for most crops are approximately 37°C.

Most response curves of root dry weight to soil temperatures have been obtained by imposing constant temperatures to roots in a controlled pot experiment. In the field however, particularly in the tropics, the diurnal changes may be extremely high, and reactions of roots are difficult to predict. Optimum temperatures for root growth seem to differ between night and day. For coffee the optimum growth was obtained at 20°C during the night, while 26°C was optimal during the daytime.

The activity of *microorganisms* is strongly affected by soil temperatures. *Rhizobium* has been studied most intensively. Nodulation with legumes seems to be optimal at 25°C. It is reduced to 70 percent of its maximum at 17 and 30°C. No growth occurs at temperatures less than 12°C or more than 33°C. Approximately the same values apply to *Azotobacter paspali.*

At temperatures above 40°C, the ability of *Nitrobacter* to oxidize nitrite to nitrate is severely reduced. *Nitrobacter* dies at 57°C. The effect of temperature on these microorganisms may cause toxic nitrites to accumulate in the soil.

1.2 Soil Moisture

Most of the climatic variability of soils in the tropics is due to seasonal variations in moisture. There is a large diversity of moisture regimes that ranges from deserts to perhumid climates. The growing seasons of most crops in the tropics are defined by the duration of the rainy season rather than by temperature changes.

The moisture that is stored in the soil is, *in addition* to atmospheric precipitation, an important supplier of water to plants. It helps crops to overcome periods without rain, often allowing them to survive dry spells during the growing season. The water-holding capacity of soils is therefore often the most critical soil quality for crop production in the tropics.

Direct precipitation on the surface and groundwater are the two main natural suppliers of soil water. There are some additional sources: One of them is water from runoff that flows into depressions and increases the amount of water that enters the soil. It is particularly important in semiarid regions. This redistribution by surface flows of water on the topography adds to the local soil variability. The concave parts of the landscape usually have more favorable soil moisture regimes than the convex parts. It is often stated that without runoff there would probably be no agriculture in the dry tropics.

Groundwater is the second source of soil moisture and is often an essential component of the soil climate. In the tropics its level may fluctuate several meters during a year. It transports elements in solution, both within profiles and between landscape facets. It may cut off oxygen from entering soil horizons and prevent root growth. It may protect soil constituents against oxidation.

The amount of water that is stored in the soil is also a function of losses by evapotranspiration and percolation. The potential rate of evapotranspiration in tropical areas depends mostly on cloudiness because there are only small seasonal variations in radiation and day lengths. During summertime (rainy season), when the sun is directly overhead and clouds are formed, the potential rate of evapotranspiration as well as the temperature are usually lower than during the dry season. However, the amplitude of the monthly fluctuations is not as large as those of temperate regions.

1.2.1 Soil moisture and atmospheric climate

Geographic distribution of rain. The tropical area is a region of summer rains. The optimum time for cloud formation and maximum rains takes place when, in a particular location, the sun is in an overhead position at noon time. The course of the zenithal position of the sun during a year is shown in Fig. 1.7. It illustrates the relation between the seasonal maxima of rainfall, the time of the year, and latitude. It also provides a practical means to identify moisture regimes and growing seasons.

The continuous rain zone at the equator is an area of permanent moisture and is called the equatorial zone. At higher latitudes it grades into climates with bimodal rain patterns that include two maxima and two minima. As the tropical parallels are approached, this regime changes into a monomodal rain distribution with one rainy season and one dry season (Fig. 1.7). Finally, at about 30° N and S, desert conditions prevail.

The vegetation types that are associated with the rain patterns grade from rain forests at the equator to woodlands and savannas, and to shrub vegetation that changes into deserts. The moisture regime is the most important land quality that determines whether rain forests will return after land is cleared and cultivated for several years, and whether trees will succeed in dominating the regrowth instead of grasses. If the dry season does not exceed 2 months (with less than 50 mm rain), there is a very high probability that the pioneer plant as-

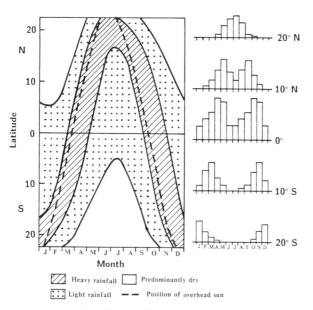

Figure 1.7   A simplified model of seasonal rainfall distribution over tropical continents. [From Nieuwolt (1977). John Wiley & Sons, reprinted by permission.]

sociations in the fallows after cropland will include many rain forest tree species.

The regular rainfall distribution related to latitude is often modified by monsoon circulations. It is also changed by orographic rains, produced by the cooling of air that is raised along mountain slopes. In the rain shadow of winds that cross mountains, extremely dry regions may alternate with humid ones; very contrasting climates, from perhumid to desertic, may occur over very short distances, not more than 10 km.

The durations of rainy and dry seasons are the basis for several climatic and soil classification systems. Only the moisture regime classes of *Soil Taxonomy* (Soil Survey Staff, 1975) are used in this text. They are described in Part 2. Other classifications that are internationally accepted follow the agroecological zone concepts of FAO (1981). None of these systems takes into account the actual water-retention characteristics of the soils. They are therefore of limited usefulness when *detailed* land resources data are needed. They nevertheless provide general information on land units that are shown on small-scale maps.

Variability of rainfall. Annual rainfall may vary considerably from year to year in the tropics. Nieuwolt (1977) considered that variations between 50 and 150 percent of the annual mean over a period of 30 years are representative, the negative departures from the mean being the most frequent. He also pointed to the persistence of dryness over several consecutive years.

Rains may be very irregular during the wet season, and several days in a row may not receive any precipitation. These dry spells are most critical at the onset of the growing period when annual crops have not fully developed their root system and no stored water is available in the subsoil after the dry season.

The variability of rainfall is particularly damaging in the semiarid tropics. When selecting crop land in these areas, its position in the landscape should be taken into account: i.e., will it receive inflow from upland areas or, on the contrary, lose water by runoff.

Table 1.2 gives the frequency of dry spells during the rainy season in an area around Brasilia and its effect on the depth of drying in two soils.

Intensity of rain. Rain intensity, or the amount of rain per unit time, is an important attribute of precipitation because it affects the proportion of water that is lost by runoff and because it characterizes its ero-

TABLE 1.2 Depth of Soil (Initially at Field Capacity) Required to Furnish 6 mm of Evapotranspiration for Selected Periods of Consecutive Days without Rain

		Depth of soil, cm*	
Consecutive days without rain	Frequency at Brasilia	Latossolo Vermelho Escuro (Brazil)†	Catalina clay (Puerto Rico)‡
8	3 per year	40	26
10	2 per year	50	33
13	1 per year	65	43
18	2 years in 7	90	60
22	1 year in 7	110	78

*Depth of soil, initially at field capacity, that is reduced to the wilting point.
†Soil is Anionic Acrustox.
‡Soil is Typic Hapludox.
SOURCE: *From Wolf (1975).*

sive power. Nieuwolt (1977) reported that tropical rainstorms, defined as short periods of uninterrupted and intense precipitation, contribute up to 90 percent of the total rainfall. They produce a spotty rainfall distribution and are highly variable from one day to another.

Observations in Kenya (Ahnert, 1982) indicate that for a *given* station, the average amount of rainfall per rain day increases when the annual precipitation increases. However, when *several weather stations* are compared, the average amount per rain day is higher in dry regions than in wet regions. As a rule of thumb, the drier the climate, the higher the erosivity of the rain and the higher the risks for erosion of land that is usually less protected against raindrop impact by a sparser vegetation.

Periodic *catastrophic* rains are another characteristic of tropical climates. For example, the recurrence time of a 100-mm rain in 1 day is 5 years at Machakos, Kenya (1° 32′ S, 27° 1′ E) (Fig. 1.8). Most rain showers take between 1.5 and 2 h. Heavy rains of about 10 mm in 15 min occur approximately 6 times a year. Rains of 15 mm in 15 min happen once a year, and 20-mm rains once in 3 years (Ahnert, 1982). If one accepts a 25-mm rain per day as an intensity beyond which average soils cannot absorb water fast enough to avoid runoff, then there would be approximately 12 days in a year that there is a very high risk of erosion of the soils that are left unprotected. Roose (1977) reports that, in the Sahel region of west Africa, catastrophic rains with a 50- to 100-year recurrence time are the most severe land-destructive forces.

Erosivity. Hudson (1981) defined erosivity as the potential ability of rain to cause erosion. Erosivity depends on the physical characteristics of rainfall: the size of the drops, their number per unit time, their

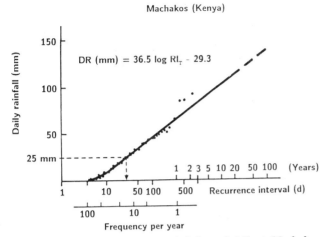

Figure 1.8 Recurrence interval of daily rainfall at Machakos, Kenya. [*From Ahnert (1982). Catena Verlag, reprinted by permission.*]

velocity, and the resulting kinetic energy they carry. Limited data indicate that tropical rains have larger drops than temperate rains. For example, at Samaru, Nigeria, the predominant diameter during a rainstorm is 2.34 mm (Kowal and Kassam, 1976). In general, the greater the rain intensity, the greater the proportion of larger drops and the faster their terminal fall velocities. For these reasons the energy load of tropical storms is very high (Lal, 1990) as is the force by which the drops hit the ground.

Most observations indicate that the best estimators of erosivity are the intensity of rainfall (millimeters of rain per hour) and the kinetic energy it carries. Rain falling at low intensities is usually nonerosive.

An intensity of 25 mm/h is accepted by many as a critical value above which erosion hazards increase markedly. Following this norm, Hudson (1981) considered that only 5 percent of the rain in temperate regions is erosive versus 40 percent in the tropics. Based on other considerations, he concluded that the erosive power of tropical rain is about 16 times greater than that of temperate rain.

The product of the 30-min intensity (the largest amount of rain, in mm, during a 30-min period of a rainstorm) times its energy[2] is called the EI_{30} index (Wischmeier, 1976). It has been accepted as a useful estimator of the erosivity of rain in the tropics. For example, Lo et al.

---

[2]There are several formulas to compute the kinetic energy, E. One of them, proposed by Lal (1982) for Nigeria, is $E = 24.5\ R + 27.6$, where R is the amount of rain in mm.

(1985) found that it is linearly related to soil loss in Hawaii and that it is suitable for quantifying the erosivity of Hawaiian rainstorms. The erosivity of the rain also depends on the time of the peak intensity of the precipitation during a storm. If the highest density occurs at the end of the storm, after a period of light rains that saturates the topsoil, the infiltration rate of the soil is easily exceeded and runoff is more likely to occur than when heavy rain initiates the event. Wilkinson (1975) reported that in the rain forest area of Western Nigeria the usual pattern of maximum intensity occurs early in the storms and that two storms with high erosivity are usually more than 2 days apart, thus allowing the soil to drain sufficiently and restore its water acceptance capacity between storms. These favorable precipitation characteristics, however, do not diminish the high risks of erosion that catastrophic rains with short recurrence intervals impose on the land.

Pauwelijn et al. (1988) derived an equation to estimate the EI_{30} index from daily rain data on the basis of a statistical analysis of precipitation records in Zambia. For daily rains of less than 80 mm, EI_{30} was equal to $2.26 \times N_{tot}^{1.93}$, where N_{tot} is the daily rain in cm.

There are only a few reliable estimates of the amount of erosion that soils of the tropics can tolerate. In certain African countries 9.0 and 11.2 Mg/ha/year are considered acceptable for sandy and clayey soils, respectively (Hudson, 1981). It is not possible to generalize these figures; many tropical soils have their land qualities concentrated at the surface and are therefore more vulnerable than soils that have them evenly distributed with depth.

1.2.2 Soil moisture and groundwater

Effects on soil properties. The effects of groundwater on soil characteristics are essentially the same in tropical and temperate regions. Water tables create special environments in soil profiles and landscapes, different from conditions that prevail in materials that are constantly in contact with air. For example, groundwater that contains organic matter that is soluble enough to function as an electron donor reduces Fe^{3+} into Fe^{2+}. The groundwater cuts off air that penetrates into the soil, and in the case of potential acid sulfate soils, the groundwater protects pyrite against oxidation. Groundwater carries many elements in solution that may move with its flow from one landscape facet to another; the composition of the solutes often defines the type of clays that are synthesized. When seepage water from groundwater emerges at the surface and evaporates, it deposits salts that may lead to salinization of crop land.

The reason to discuss groundwater in a description of the tropical soil environment is that in many areas the levels of the water tables have changed considerably during the time of soil formation. On the old continental shields, many of which date back to tertiary age and cover large areas in the tropics, the water tables dropped as a result of the dissection of the shields by river systems. Many marks that the groundwater imprinted in the soil profiles are still present, although groundwater does not actually saturate the soil in any season. The marks are said to be fossil evidences of past conditions. Examples of such occurrences are some rotten rock (saprolite) zones and the mottled or pallid zones of deep-weathering profiles.

The gleying and mottling phenomena that are often used to define internal drainage classes in soil surveys may also be fossil. To correctly identify these classes, periodic observations of the level of the actual water table over several years are often necessary.

Fluctuations in the level of the actual water tables are greater in monsoon climates than in other climates. Certain soils that are inundated by rain in the monsoon season may be completely dry down to several meters during the dry season.

Soil sections that experience fluctuations in groundwater rich in Fe^{2+} are particularly suited to accumulate precipitable iron hydroxides because of the oxidation of Fe^{2+} into insoluble Fe^{3+}. In the rainy season groundwater brings in Fe^{2+}; in the dry season the water table drops, and air is introduced that oxidizes the Fe^{2+}, which is almost irreversibly precipitated as Fe^{3+}. When this mechanism is repeated year after year, it results in the accumulation of iron oxides that may lead to the formation of *plinthite,* which will be defined later.

Effects on plant growth. There are only a few studies that measure the direct effects of the depths of groundwater on crop performance. They either relate to drainage studies of poorly drained soils or to the yield variations that correlate with differences in the depths of water tables in semiarid regions. For example, Worou (1989) reported that in deep sandy soils of Togo maximum yields of corn are obtained during the dry season in areas where the groundwater occurs at 50- to 60-cm depth; the yields are reduced almost seven-fold in sites where the water level is only 10 cm deeper. These extreme differences are probably enhanced by the sandy nature of the soil materials that only have a very low hydraulic conductivity. The observations nevertheless demonstrate the dependence of crop performance on the depths of groundwater levels that very often correlate with small variations in topography. Concave landscape facets that collect runoff water offer more chances of success than convex parts that present the highest risks of crop failure.

1.3   Soil Climate

The combined effects of soil temperature and moisture create the soil climate that influences soil formation, soil behavior, and plant growth. The objective of this section is to summarize the characteristics of the tropical soil climate as it affects these three aspects of agronomy.

1.3.1 Soil climate and soil formation

The major attributes of the tropical climate that were highlighted in the previous sections with respect to soil formation were:

1. Leaf fall in the wet and dry tropics does not occur when the soil is moist. Litter is not soaked by rain or melting snow to produce organic leachates that may have a marked effect on the mobility of soil constituents. For these reasons the potential for producing soil horizons by the translocation of clay seems to be lower in the tropics than in temperate regions.

2. Summer rains, which moisten the soil when temperatures are high, significantly increase weathering rates because of faster chemical decomposition and promote leaching of soluble weathering products. In the warm tropical lowlands the decomposition rates of organic matter are greater than in temperate regions.

3. The interpretation of soil characteristics in terms of climatic factors needs to consider past climates as well as the present one. In the lowland tropics the climatic fluctuations mainly involved changes in precipitation from arid to perhumid; at high elevations the temperature variations were large enough to influence the rates of organic matter decomposition.

1.3.2 Soil climate and soil management

Moisture conditions during growing season. Cropping seasons in the tropics are essentially determined by the actual rainfall with only small amounts of storage water in the soil as carryover from previous seasons. In regions with dry months the growing season starts with a dry soil with little available water. Therefore, annual crops with shallow rooting systems depend mainly on actual rain rather than on water stored in the soil during a time when no water was extracted by the root system (e.g., during a cold winter).

Annual crops should be planted under optimum initial growing conditions so that their root system can develop as fast as possible in deep layers where water may be partly protected against direct evaporation from the warm surface soil. Depth and time of planting are crucial for the successful growing of annual crops. Water acceptance and water retention are very important land qualities. Root penetration into deep layers should be stimulated by eliminating toxic aluminum compounds or calcium deficiencies in the subsoil. Generally perennial crops are not exposed to the same drought risks at the beginning of the rainy season because they have permanent rooting systems in deep soil layers.

Erosion hazards. Dry seasons usually end with bare soils. Rain that falls on soils not covered by vegetation is most erosive. Soils with alternating wet and dry seasons should be protected as much as possible against the direct impact of rain. Direct exposure to sunshine increases the risks of crusting and topsoil compaction that reduce seed emergence and root penetration. Mulches and cover crops are recommended practices that will diminish these risks.

Soil management should maximize the water acceptance of soils and increase their water-holding capacity in order to reduce runoff. Certain topsoils that overlie slowly permeable subsoils may become saturated after a few millimeters of rain. Once these topsoils are saturated, runoff becomes the dominant process leading to severe soil erosion. "Duplex" soils, which have very strong textural differences in their profiles, are common in many dry tropical areas. They are the most difficult soils to protect against erosion.

Nitrogen flux. The dry season does not necessarily stop the decomposition of soil organic matter. During that period the plant residues are divided into smaller particles by termites and other insects. Some ammonification and nitrification may take place. At the end of the dry season easily decomposable organic matter and mineral forms of nitrogen may have accumulated in the soil.

The first rainfall at the beginning of the rainy season triggers mineralization and causes a strong nitrogen flux. At that time the vegetation looks very green. The nitrates that are produced, however, are easily leached out of the profile and lost. A well-developed root system to intercept the nitrates is recommended for maximum yields. The onset of the growing season, therefore, is the most critical phase in the overall production system.

Salinization. Land that mainly depends on irrigation water to grow crops needs to be drained more intensively in the tropics than in temperate regions, where winter rains partly leach salts out of the root zone. Many irrigation systems in the tropics deteriorate because of the accumulation of salts from the irrigation water. Soils in mediterranean climates (which have winter rains) do not require the same drainage precautions.

References

Acland, J. D., *East African Crops*, FAO-Longmans, Rome, 1971.

Alexander, E., "Soil temperatures and slope aspect around hill 998," *CEIBA*, vol. 20(1), Tegucigalpa, Honduras, 1976.

Ahnert, F., "Untersuchungen über das morphoklima und die morphologie des inselberggebietes von Machakos, Kenia," in *Contributions to Tropical Geomorphology*. Catena Verlag, Germany, 1982, pp. 1–72.

De Martonne, E., *Traité de Géographie Physique*, Armand Colin, Paris, 1957.

FAO, *Report on the Agroecological Zones Project*, vol. 3; *Methodology and Results for South and Central America*, World Soils Resources Report #48/3, Food and Agriculture Organization of the United Nations, Rome, 1981.

FAO, *Land, Food and People*, Food and Agriculture Organization of the United Nations, Rome, 1984.

Flenley, J. R., *The Equatorial Rain Forest*, Butterworth, London, 1981.

Harrison-Murray, R. S., and R. Lal, "High soil temperature and the response of maize to mulching in the lowland humid tropics," in R. Lal and D. J. Greenland (eds.), *Soil Physical Properties and Crop Production in the Tropics*, John Wiley & Sons, Chichester, 1979, pp. 285–304.

Hudson, N., *Soil Conservation*, 2d ed, Cornell University Press, Ithaca, NY, 1981.

Kowal, J. M., and A. H. Kassam, "Energy load and instantaneous intensity of rainstorms at Samaru, northern Nigeria," *Trop. Agric.* (Trinidad) 53(3):185–197 (1976).

Lal, R., "Temperature profile of soil during infiltration," *Niger. J. Soil Sci.* 2:87–100 (1982).

Lal, R., *Soil Erosion in the Tropics: Principles and Management*, McGraw-Hill, New York, 1990.

Lo, A., S. A. El-Swaify, E. W. Dangler, and L. Shinshiro, "Effectiveness of EI_{30} as an erosivity index in Hawaii," in S. A. El-Swaify et al. (eds), *Soil Erosion and Conservation*, Soil Conservation Society of America, Ankeny, IA, 1985, pp. 384–393.

Miller, A. A., *Climatology*, 9th ed., Methuen, London, 1971.

Moore, A. W., "The influence of annual burning on a soil in the derived savanna zone of Nigeria," *Trans. 7th Int. Cong. Soil. Sci.* 4:257–264 (1960).

Nieuwolt, S., *Tropical Climatology—An Introduction to the Climates of the Low Latitudes*, John Wiley & Sons, London, 1977.

Nullet, D., H. Ikawa, and P. Kilham, "Local differences in soil temperature and soil moisture regimes on a mountain slope, Hawaii," *Geoderma* 47:171–184 (1990).

Pauwelijn, P. L. L., J. S. Lenvain, and W. K. Sakala, "Isoerodent map of Zambia. Part I: The calculation of erosivity indices from rainfall data bank," *Soil Technology* 1:235–250 (1988).

Pitot, A., and H. Masson, "Quelques données sur la température au cours des feux de brousse aux environs de Dakar," *Bull. IFAN* 6:47–62 (1951).

Roose, E. J., "Use of the universal soil loss equation to predict erosion in West Africa," *Soil Erosion: Prediction and Control*, Proceedings of a National Conference on Soil Erosion, Soil Conservation Society of America, Ankeny, IA, 1977, pp. 61–74.

Sanchez, P., *Properties and Management of Soils in the Tropics*, John Wiley & Sons, New York, 1976.

Schwertmann, U., "Transformation of hematite to goethite in soils," *Nature* 232:624–625 (1971).

Schwertmann, U., "The effect of pedogenetic environments on iron oxide minerals," *Advances Soil Science* 1:171–200 (1985).

Schwertmann, U., and H. Fechter, "The influence of aluminum on iron oxides. XI. Aluminum-substituted maghemite in soils and its formation," *Soil Sci. Soc. Amer. J.* 48:1462–1463 (1984).

Scotter, D. R., "Soil temperatures under grass fires," *Aust. J. Soil Research* 8:273–279 (1970).

Sertsu, S. M., and P. A. Sanchez, "Effects of heating on some changes in soil properties in relation to an Ethiopian land management practice," *Soil Sci. Soc. Amer. J.* 42(6): 940–944 (1978).

Sinclair, J., "Soil properties and management in Botswana," In *Alfisols in the Semiarid Tropics*, ICRISAT, Patancheru, India, 1987, pp. 49–58.

Soil Survey Staff, *Soil Taxonomy. A Basic System of Soil Classification for Making and Interpreting Soil Surveys*, Agriculture Handbook No. 436, Soil Conservation Service, U.S.D.A., Washington, D.C., 1975.

Soil Survey Staff, *Keys to Soil Taxonomy*, SMSS Technical Monograph #19, Blacksburg, VA, 1990.

Sombroek, W. G., H. M. H. Braun, and B. J. A. van der Pouw, *Exploratory Soil Map and Agroclimatic Zone Map of Kenya*, 1980, Exploratory Soil Survey Report No. E1, Kenya Soil Survey, Nairobi, Kenya, 1982.

Trewartha, G. L., A. H. Robinson, and E. H. Hammond, *Physical Elements of Geography*, McGraw-Hill, New York, 1967.

Trapnell, C. G., "Ecological results of woodland burning experiments in northern Rhodesia," *J. Ecology* 47(1):129–168.

Troloppe, W. S. W., "Fire in savanna," in P. de V. Booysen and N. M. Tainton (eds.), *Ecological Effects of Fire in S. African Ecosystems*, Springer-Verlag, New York, 1984, pp. 149–176.

Voorhees, W. B., "Alleviating temperature stress," in G. F. Arkin and H. M. Taylor (eds.), *Modifying the Root Environment to Reduce Crop Stress*, ASEA Monograph #4, American Society of Agricultural Engineers, St. Joseph, MI, 1981, pp. 217–268.

Wilkinson, G. E., "Rainfall characteristics and soil erosion in the rainforest area of western Nigeria," *Expl. Agric.* 11:247–255 (1975).

Wischmeier, W. H., "Use and misuse of the universal soil loss equation," *J. Soil Water Conservation* 31(1):5–9 (1976).

Wolf, J. M., "Water constraints to corn production in central Brazil," Ph.D. Thesis, Cornell University, Ithaca, NY, 1975.

Worou, S., "L'organisation des sols et la dynamique des nappes perchées semipermanentes: un lien fonctionnel et un atout agronomique pour les cultures de contresaison," *Soltrop 89*, pp. 179–190 (editions de l'ORSTOM, collection colloques et séminaires, Paris) (1989).

Soil Parent Materials

Soils are often considered as being the products of the interactions between climate and the geologic formations of the earth's crust. The processes involve chemical reactions and physical work and are influenced by biological factors such as vegetation and fauna that act on the surface materials to produce soil horizons. The materials in which the horizons form are designated as the *parent materials* of the soils. They may have their own history, or prehistory, because they may have undergone several transformations under the influence of climates before they were deposited at their present location.

Despite their sometimes long history, many attributes of soils in the tropics still derive from the properties of the rocks from which they developed. As indicated before, the attributes of the soil also depend on conditions the soil parent materials experienced before being deposited at their present location. Large areas in the tropics belong to old stable continental shields that are covered by surface mantles carrying the marks of processes that took place during previous geologic periods. These surface mantles are still present on the land surfaces because no glaciers have removed them as has happened in temperate regions during pleistocene times.

Weathering is considered one of the major soil-forming processes in the tropics. It is argued that the moisture and temperature conditions during part of the year favor the chemical decomposition of minerals and promote the leaching of soluble weathering products out of the soil mantle. In addition, many soil parent materials are very old and at least once underwent strong weathering and leaching.

Some definitions are necessary. The mantle of unconsolidated soil parent materials at the earth's surface is called *regolith*. Its thickness over solid unaltered rock depends on the rate of weathering as well as on the rate of removal by erosion or addition by sedimentation. Time also controls the thickness of regoliths. Regolith formation is called *geogenesis* in this text. The development of horizons in the regolith is named *pedogenesis,* and it includes, among others, eluviation and illuviation processes of soil constituents, accumulation of organic matter, enrichments of iron hydroxide minerals, etc.

The present chapter considers two aspects of parent material formation. The first is weathering that results in the production of surface mantles. The second is the redistribution of regoliths by erosion and sedimentation as a function of the topography. The chapter should lead to a better understanding of the geography of parent materials and soils in the tropics and serves as an introduction to the rationale of soil classification.

2.1   Weathering

Weathering includes the physical disintegration and the chemical decomposition of solid rocks or other earth materials upon exposure to the atmosphere. Weathering leaves residues when decomposition is *incongruent,* not proceeding at same rate for all constituents, and releases electrolytes in the soil solution. The solutes may eventually recombine to form secondary minerals, or they may be removed by leaching.

This section is divided into three parts: (1) the breakdown of primary minerals and rocks, (2) the formation of secondary minerals, and (3) the stability of secondary minerals. Most descriptions relate to processes occurring in warm humid environments and under free-drainage conditions.

2.1.1   Breakdown of primary minerals and rocks

Physical disintegration.   Physical disintegration of rocks by volume expansion due to changes in temperature is not a major weathering process in humid environments. It is rather common in deserts and semiarid climates. It may have contributed significantly to parent material formation in the dry tropics bordering arid regions.

Physical disintegration, however, has been recognized as a possible mechanism that participates with chemical decomposition in the formation of clay in humid environments. Some minerals enter the clay fraction by comminution of sand and silt size particles. For example, Rebertus et al. (1986) mentioned that the clay increase in certain Hapludults during saprolite-to-soil transformations may be largely due to the addition of clay produced by the breakdown of sand size kaolinite pseudomorphs. The same process is mentioned by Harris et al. (1985) for kaolinization in soils of Virginia.

TABLE 2.1   Clay and Silt Contents in Deep Weathering Profiles of Puerto Rico

Depth, cm	Horizon	Clay, %*	Silt, %†
18–33	B_{21}	76.8	9.3
33–53	B_{22}	79.1	9.2
53–81	B_{23}	78.9	9.4
81–112	B_{24}	76.8	14.1
112–150	B_{25}	69.2	19.8
150–183	B_{26}	68.2	21.3
183–213	C_1	59.4	28.7
213–275	C_2	51.7	32.7
275–320	C_3	49.1	34.6
320–335	C_4	39.0	42.0

* $< 2 \mu m$.
† 2 to 20 μm
SOURCE: *Soil Conservation Service (1967), pp. 172–173.*

The reduction in particle size by comminution also contributes to the production of silt. When primary minerals decompose and cleavage planes are weakened, some grains break apart and join smaller particle size fractions. It is commonly accepted that during chemical decomposition such mechanisms supply a large part of the fine silt (2 to 20 μm) fraction present in soil materials. Fine silt is thus an indicator of actively weathering soil horizons. When all weatherable minerals are decomposed, this source ceases to exist and the silt fraction percentages either remain constant or decrease because of dissolution. The more the weathering of a horizon approaches the ultimate stage, the less the fine silt fraction becomes. The decline in silt content is illustrated in Table 2.1 by analytical data of a deep Typic Hapludox profile in Puerto Rico. It should be noted in this example that as weathering proceeds, the clay percentages increase because of the formation of secondary minerals in that fraction.

The decrease in fine silt content during weathering is used in classification systems to characterize weathering stages in soils of the tropics. For example, the silt-to-clay ratio is used in the FAO-Unesco legend of the soil map of the world as a criterion to identify extremely weathered soil horizons. A ratio of 0.2 or less is considered indicative of strong weathering (FAO, 1988). Earlier investigations found that parent materials that have reached the ultimate degree of chemical decomposition have silt-to-clay ratios of less than 0.15 (Van Wambeke, 1959, 1962).

Chemical decomposition of rocks and minerals.   Rocks are solid aggregates of primary minerals. The weatherability of rocks for the most part is defined by the kinds of minerals that they contain.

The decomposition rates of primary minerals depend on *intrinsic* factors such as their chemical composition, their crystal structure, and the size of the surface area that is exposed to air or to water. These factors determine the *weatherability* of the minerals, which is the same both in tropical and temperate regions. The *extrinsic* factors are soil climate (temperature, moisture regime, length of wet season, etc.), the composition of the soil water, and the presence or absence of air.

Intrinsic factors. The chemical composition of *primary minerals* is discussed here first. It is beyond the scope of this text to provide a comprehensive treatise on the mineralogy of rocks and sediments. Only general principles of their resistance to weathering will be discussed. Readers who need detailed information on the kinds of rocks that form the earth's crust are referred to petrography textbooks. The chemical composition of the most common rock types is given in Table 2.4.

The major building stone of minerals is the silica tetrahedron. The resistance to weathering of primary minerals *increases* with the number of direct Si-O-Si bonds that link the Si tetrahedra together in the crystal structure. Table 2.2 gives the number of oxygens that are shared between adjacent Si tetrahedra for a number of silicate minerals.

The resistance to weathering *decreases* as the fraction of tetrahedra substituted by Al is increased. The Al bond is weaker because the Al ion does not fit as well as the Si ion in a four-fold coordination; the greater the fraction of Al tetrahedra, the weaker the mineral structure. Feldspars are examples of minerals containing Al tetrahedra.

The weakest bonds in the structure of minerals are the oxygen-*common cation* bonds. They are the sites where the initial breakdown of minerals takes place; if many of these bonds are exposed to the soil solution, the minerals decompose faster than when the bonds are sandwiched between sheets of silica tetrahedra. Minerals with the lowest stability usually contain the highest amounts of Fe, Ca, and Mg and release these elements at a faster rate than the more stable minerals.

Volcanic glasses, which are seldom considered in stability series,

TABLE 2.2   Number of Direct Oxygen Bonds between Si Tetrahedra in Silicate Mineral Structures

Number of direct Si-O-Si bonds between Si tetrahedra	Structure*
None	Nesosilicate[1]
2	Inosilicate, single chain[2a]
2 and 3	Inosilicate, double chain[2b]
3	Phyllosilicate, sheet[3]
4 (all)	Tectosilicate, network of tetrahedra[4]

*The superscripts are used to indicate the kind of structure to which the minerals that are listed in Table 2.3 belong.

are usually the least resistant minerals. There is by definition no crystal structure, and consequently no arrangements exist whereby weak bonds are systematically protected by silica sheets.

The Goldich stability series of primary minerals (1938) follows the general rules given above. It includes quartz at the most stable end and olivine at the most weatherable end (see Table 2.3).

The mineral's area that is exposed to water and air may in many instances modify its weatherability. For example, the solubility of quartz is highly surface area dependent. According to Jackson et al. (1948), clay-sized quartz is more weatherable than clay-sized illite. Quartz grains are known to corrode severely under tropical environments (Eswaran and Stoops, 1979). Claisse (1975) found that the solubility of quartz increases markedly when the particle size is smaller than 40 μm. Particles larger than 200 μm on the other hand dissolve very slowly. The influence of surface area on the solubility of quartz increases considerably when the temperature exceeds 20°C.

The solubility of small size quartz at high temperatures may explain why strongly weathered soil materials contain practically no fine silt, even when they developed from rocks that are rich in quartz. As mentioned before (Table 2.1), most weathering profiles show a decrease in silt as soil materials grade into more advanced weathering stages.

According to a dissolution equation proposed by Henin, weathering would take 60,000 years to reduce the diameter of 1-mm quartz grain to 0.8 mm in water at pH 7 and at 25°C.

The *weatherability of rocks and sediments* will now be considered.

TABLE 2.3   Ranking of Stability of Common Rock-Forming Minerals and Their Structural Classification*

Ferromagnesian (mafic) minerals[†]	Nonferromagnesian (felsic) minerals[†]
	Ca^{2+}-plagioclase[4]
Olivine[1]	An_{80-100}
Fe-rich	An_{65-80}
Mg-rich	
Hypersthene[2a]	
Augite[2a]	An_{50-65}
Hornblende[2b]	
Biotite[3]	Na^+-plagioclase[4]
	K^+-feldspar[4]
	Muscovite[3]
	Quartz[4]

*Ranking of stability is from lowest to highest.
[†]Superscripts with mineral name are keyed to the numbered aluminosilicate structures of Table 2.2.
SOURCE: *Copied from Birkeland (1974), p. 136.*

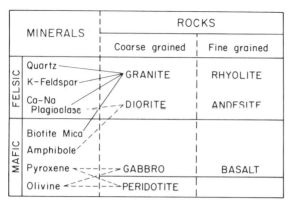

Figure 2.1 Simplified chart of silicate minerals and igneous rocks. [*From Strahler and Strahler (1983).*]

The stability series of the minerals is used to evaluate the resistance of rocks to weathering. Figure 2.1 is a simplified chart of silicate minerals that form the most common igneous rocks. If feldspars and quartz (Si) dominate, they are called *felsic* or acid rocks; if minerals rich in magnesium and iron are abundant, they are named *mafic* or basic rocks. Table 2.4 gives information on the chemical composition of these rocks; nonigneous rock types have been added to the table. The rocks are listed in order of decreasing SiO_2 content. Table 2.5 pro-

TABLE 2.4   Chemical Composition of the Most Common Rocks

	Percentage							
	SiO_2	Al_2O_3	Fe_2O_3	FeO	CaO	MgO	K_2O	P_2O_5
Quartzite	95	1	0.4	0.2	1.6	0.1	0.2	—
Granite*	70	14	1.0	1.9	2.0	0.8	4.9	0.2
Granodiorite*	67	16	1.3	2.6	3.6	1.6	3.0	0.2
Dacite*	63	17	2.3	3.0	6.1	2.4	1.2	0.2
Syenite*	59	17	2.2	2.8	4.1	2.0	6.5	0.4
Shale	58	15	4.0	2.4	3.1	2.4	3.2	0.2
Trachyte*	58	18	2.5	2.0	4.2	2.0	7.4	0.2
Andesite*	55	18	2.4	5.6	7.9	4.9	1.1	0.3
Diorite*	52	16	2.7	7.0	8.4	6.1	1.3	0.3
Basalt*	51	14	2.9	9.0	10	6.3	0.8	0.2
Gabbro*	48	17	2.5	7.9	11	8.0	0.6	0.2
Basalt*	46	15	2.7	9.0	11	9.0	0.9	0.4
Peridotite*	44	4	2.5	10	3.5	34	0.2	0.1
Limestone	5	1	(	0.5	)	42	7.9	0.3	0.1
Dolomite	0.3	0.1	(	0.1	)	30	21	0.1	0.01

*Igneous rocks are marked by asterisk.
SOURCE: *Examples from Nockolds et al. (1978) and other sources.*

TABLE 2.5   Norms or Calculated Mineralogical Composition of Most Common Igneous Rocks

	Quartz	Ortho-clase	Albite	Anorthite	Hypers-thene	Olivine	Magne-tite	Apatite
Granite	27	29	27	8.3	2.2	—*	1.4	0.5
Granodiorite	28	28	29	8.3	1.9	—	2.3	0.3
Dacite	19	7	32	26	7.6	—	3.3	0.3
Syenite	2	38	33	10	4.2	—	3.3	0.9
Trachyte	—	44	29	10	—	—	3.7	0.5
Andesite	4	7	31	28	15	—	3.5	0.7
Diorite	1	8	28	26	18	—	3.9	0.8
Basalt	3	5	19	26	17	—	4.2	0.5
Gabbro	—	3	19	34	14	6.6	3.7	0.6
Basalt	—	6	19	26	—	17	3.9	0.9
Peridotite	—	2	5	7	14	59	3.7	0.1

*Below detectable level.
SOURCE: *Examples are from Nockolds et al. (1978).*

vides averages of the actual mineralogical composition of the most common igneous rocks.

It can be seen from Table 2.4 that as the SiO_2 content decreases, the Fe and CaO percentages increase. As a rule the resistance to weathering diminishes accordingly. For example, under laboratory conditions basalt decomposes 11 times faster than granite (Pedro, 1964). Mafic rocks release per unit time more calcium and iron than felsic rocks. The rates at which electrolytes are supplied to the soil solution often determine the kind of secondary mineral that is formed. For example, mafic rocks release iron hydroxides in such quantities that they precipitate as amorphous materials or ferrihydrite rather than slowly crystallize into goethite. Ferrihydrite is a precursor of hematite, which gives well-drained soils that are developed from mafic rocks their typical dusky red colors.

The rocks that contain only small amounts of iron weather into yellow materials that are colored by goethite. More information on iron oxides is given in Sec. 2.1.2.

Parent rocks leave an imprint on soils by the residues that are left unchanged at the site of their formation, even after intensive weathering: coarse-grained quartz is one of the most resistant minerals and its quantity and particle size characteristics often allow the identification of the origin of parent materials. The texture of many strongly weathered soils that developed in situ is well correlated with their parent rocks. Table 2.6 illustrates this point; the particle size analyses reported in this table refer to extreme weathering stages: The clay contents are related to the amounts of weatherable minerals that were present in the original rock, the size distribution in the

TABLE 2.6   Relation between Parent Rock, Color, and Texture of Strongly Weathered Soils in Central Africa

Rock type	Color	Clay, %	Fine silt, %	Coarse silt, %	Sands* Fine, %	Sands* Coarse, %
Basalt	Dusky red	88	2.8	3.0	3.6	2.7
Diorite	Dusky red	88	3.8	2.8	3.8	1.0
Gabbro	Dusky red	77	4.5	2.2	8.6	7.7
Shales	Yellow	72	6.6	10.2	9.9	2.4
Limestone	Red	64	4.9	11.0	12.7	7.4
Granite	Yellow	58	5.8	3.6	10.7	21.7
Gneiss	Red	53	2.5	3.3	9.0	32.2
Schists	Red	52	3.8	7.8	21.1	15.3
Granite	Yellow	48	4.0	4.3	11.6	32.1
Quartzite	Yellow	36	2.3	3.1	21.6	16.0
Sandstone	Yellow	33	2.0	5.6	31.1	28.3
Psammite	Yellow	30	2.2	7.3	59.9	0.6
Sandstone	Yellow	12	1.3	4.4	30.9	51.4

*Fine and coarse sands have diameters of 50–250 and 250–2000 μm, respectively.
SOURCE: Sys (1972) and Van Wambeke (1961).

sands reflects the coarseness of the rock formation, and the color is indicative of the iron content.

There are many variations to the correlation between rocks and soil materials. For a given area, however, the field characteristics derived from parent rocks provide valuable information on the potential of the land to support rural communities.

Mineral composition is not the only factor that determines the weatherability of rocks. Other properties such as porosity, cleavage, surface area, etc. are equally important. Coherent impermeable rocks, with no cleavage or fissures, only weather at their surface, usually producing iron oxide crusts that protect the fresh rock against further decomposition. This weathering pattern strongly contrasts with the processes that occur in permeable formations into which water penetrates, invading the rock or the sediment and increasing the surface area of minerals exposed to water and acids. Granites and gneisses are examples of such rocks. The soaked rocks transform into saprolites or "rotten" rock by in situ chemical decomposition. The quartz grains in the felsic rocks form a mesh structure that acts as a sieve and facilitates the movement of groundwater (Carroll, 1970).

Volcanic ashes are examples of materials that weather rapidly. They release large quantities of $Si(OH)_4$ and aluminum hydroxides that saturate or supersaturate the soil solution in a short time. They precipitate as amorphous gels or as allophane before they transform into crystallized clays. Allophane, therefore, is a typical clay mineral of young volcanic ash soils.

The weatherability of rocks is an important determinant of the level of erosion that soils can tolerate. As long as nutrients lost by erosion are replaced by those released by weathering, soils are considered at steady state with little risk for deterioration. However, there is little data on this subject. Owens (1974) calculated for granites in Zimbabwe that weathering in a 900-mm rainfall area decomposed 150 kg rocks/ha/year. When precipitation was 1200 mm, this amount increased to 400 kg/ha/year. For west Africa, the time to weather 1 m of granite to an advanced stage was estimated by Leneuf and Aubert (1960) to be between 22,000 and 77,000 years.

Extrinsic factors. Temperature and rainfall are the most important extrinsic factors that determine weathering intensities. The rate of chemical reactions increases approximately two- or three-fold for each 10°C increase in temperature. Because decomposition is only active in the presence of water, the yearly weathering rates are also dependent on the length of the rainy seasons. The intensity of rain further controls leaching. A seasonally "flushing" regime, which constantly removes weathering products from the soil, drives reactions to complete decomposition. This trend is enhanced by the warm temperatures of summer rains in the tropics.

The weathering rates of soil materials and minerals at a given location also depend on drainage conditions. This is important because no transformations take place when chemical reactions without phase modifications reach an equilibrium in closed environments. Weathering rates of minerals, therefore, are dependent on the removal of the products from the solution phase either by rain water and *drainage,* or by precipitation or crystallization into solids. This is the reason why the intensity of weathering may differ very strongly between well-drained and poorly drained sites in the same climatic region. It also explains why the fastest decompositions occur in freely draining soils or in the parts of them that are in contact with percolating water and oxygen.

The oxidation of metals such as iron has several consequences: First, iron hydroxides are precipitated, removing iron from the solution phase. Second, oxidation of iron is an acid-producing reaction that intensifies weathering. Both processes favor the extraction of iron out of primary minerals and, thus, the acceleration of their disintegration.

Weathering rates also depend on the organic matter content of the soils. Huang and Keller (1970) and Huang and Kiang (1972) showed that feldspars weather faster in the presence of organic matter. Organic matter also acts as an electron donor in oxidation-reduction reactions such as $Fe^{3+} + 1e \rightarrow Fe^{2+}$, with the Fe^{2+} being more soluble. This process may remove or prevent the formation of protective iron oxide coatings around primary minerals. Schalscha et al. (1967) indicated that organic acids increase the efficiency of water to extract Fe from minerals.

2.1.2   Formation of secondary minerals

There are several mechanisms by which minerals enter the clay fraction. Comminution has already been discussed in section 2.1.1. The other modes include the formation of *residual clays,* which are clays that have retained all or part of the original crystal structure of the primary minerals from which they are derived. Residual clays have been altered by isomorphic substitution or loss of some but not all of their original constituents. The process is often called *alteration* in the literature (Brady, 1984).

The second mechanism involves complete crystallization of minerals from electrolytes dissolved in the soil water and is called *clay synthesis* or *neoformation* in this text. The kinds of minerals that form depend on the conditions prevailing in the soil solution. They are described below as clay formation environments.

Residual clays.   Residual clays include minerals that enter the clay fraction by mechanical diminution without chemical modification. Residual clays also include minerals that are formed by extraction or substitution of elements in the crystal lattice of primary layer silicates without complete destruction of the original structure.

Muscovite is one of the primary aluminosilicates that best resists chemical decomposition because the weakest bonds in its structure are protected by the silica sheets that form the 2:1 layers in its lattice. Muscovite is often found in the clay fractions of soil materials that were recently exposed to weathering. In many cases muscovite transforms directly into chlorite-like minerals; Herbillon (1981) reported that weathering muscovite in dilute solutions becomes readily coated by aluminum hydroxide polymers similar to the interlayers found in chlorites. The mechanism of transformation would involve hydrolysis of the exposed potassium, attack by protons of the octahedral sheets, followed by desilication of the tetrahedral sheets. The decomposition results in the relative accumulation of aluminum.

The presence of residual clays in the fine particle fractions is often considered a sign of incomplete weathering. It is used as a criterion in several classification systems to differentiate young soils or horizons from older ones that have reached the extreme stage of chemical decomposition.

Clay synthesis.   Synthesis of clays implies the construction of entirely new clay mineral lattices from electrolytes in the soil water. It also includes the formation of *complete* pseudomorphs. It is assumed that most of the building elements of the synthesized clays have passed through a solution phase.

There are numerous examples of the formation of complete pseudomorphs. Muller and Bocquier (1986) describe macrocrystalline kaolinite (10 to 50 μm) as pseudomorphs of muscovite in saprolites. Harris et al. (1985) found that kaolinitic pseudomorphs of biotite form without a significant intermediate vermiculite phase in the kaolinization process. Soubies and Gout (1987) mentioned that many kaolinite pseudomorphs actually consist of a mixture of disordered kaolinite and large amounts of x-ray amorphous materials having the same chemical composition as kaolinite.

The kinds of minerals that synthesize depend on the activities of the elements present in the soil solution where the clays form. The rates at which the clays grow are a function of the kinetics of crystal formation: Some minerals nucleate and enlarge at slower rates than others. The actual mineral composition of the clay fraction is, therefore, not always in equilibrium with the conditions at the point of formation.

This section introduces the concept of *clay formation environments*. They are defined by the kinds of elements that are present in the soil solution, their activities, and their supply rate. The dimensions of the formation environments may vary from microfeatures such as microvoids and proximity of roots to soil fragments as large as horizons or even landscape segments. On a worldwide scale, there is a relationship between dominant clay minerals in soils and climatic zones.

Microvariability may be the result of short-distance changes in drainage and leaching conditions. For example, in soils derived from basalts and at sites under stones (*leaching shadows*), soil solutions are usually more concentrated than elsewhere and nontronite forms instead of kaolinite (Sherman and Uehara, 1956). The microvariability makes theories on clay mineral sequences a controversial subject. The approach followed here is that there is no specific clay mineral chain for the humid tropics. The minerals that synthesize are those that respond to the conditions that prevail in the soil solution at the point of formation. These may change over very short distances even within the same soil horizon. Details of the kinds of clays that synthesize are given below.

Formation of silicate clay minerals and aluminum oxides. Many theories have been suggested to explain the synthesis of kaolinite or gibbsite during weathering under humid conditions in the tropics. Tsuzuki (1967) applied solubility diagrams to show how either gibbsite or kaolinite may form during the decomposition of microcline, a potassium feldspar, given constant H^+ and K^+ activities in the soil solution.

The diagrams (Fig. 2.2) use the solubility product of gibbsite (10^{-34}) to set the upper limit of Al^{3+} activity and of amorphous silica ($10^{-2.74}$)

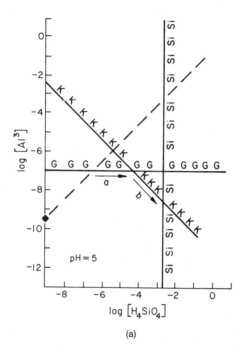

(a)

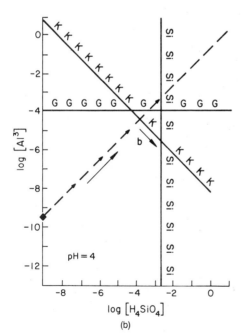

(b)

Figure 2.2 Dissolution of microcline at 25°C; $[K^+] = 0.001$ and (a) pH = 5 and (b) pH = 4. [*From Tsuzuki (1976). The Clay Mineral Society, adapted by permission.*]

to control the H_4SiO_4 concentration in the soil solution. As soon as the activities of Al^{3+} or Si exceed the levels set by these equilibrium constants (shown by lines G or Si in the diagrams), these ions will be removed from the soil solution by the formation of gibbsite or amorphous silica. The G lines change position in the diagrams of Fig. 2.2 because of a decrease in pH. The model also uses the equilibrium constant of $10^{7.69}$ for the hydrolysis of kaolinite to construct the dissolution or formation line of this mineral (line K in Fig. 2.2).

$$Al_2Si_2O_5(OH)_4 + 6H^+ \rightleftarrows 2Al^{3+} + H_2O + 2H_4SiO_4$$
kaolinite

The driving force that supplies Si and Al in constant relative molar proportions of 3:1 to the soil solution is the congruent dissolution of microcline.

$$KAlSi_3O_8 + 4H^+ + 4H_2O \rightleftarrows K^+ + Al^{3+} + 3H_4SiO_4$$

The initial increases in concentration of Al and Si in the soil solution are shown by the arrows along the lines that start at the $\diamond$ marks on the diagrams of Fig. 2.2. When the initial pathway on the pH 5 diagram (Fig. 2.2a) reaches the G line, gibbsite is formed and is the first to be synthesized (line segment a). Once the Si activity arrives at the kaolinite line, gibbsite dissolves and kaolinite is formed; when all the gibbsite is consumed, the clay synthesis continues along the line segment b.

The diagram constructed for pH 4 (Fig. 2.2b) does not show the formation of gibbsite; under the assumed conditions kaolinite is the only mineral to be synthesized (line segment b). At this pH the higher solubility of gibbsite and the constant supply of $Si(OH)_4$ by the weathering microcline favors the direct formation of kaolinite. At higher concentrations of Si and in the presence of other cations such as Mg, not kaolinite but 2:1 clay minerals may crystallize.

The equilibrium constants can be used to draw stability diagrams such as Fig. 2.3 that show the ion activity domains in which given clay minerals are stable. The $Si(OH)_4$ activity is usually plotted on the x axis; the H^+ activity or pH and the electrolyte concentration (M^+) in the soil solution are set out on the y axis, usually as pH – pM^+ (in the instance of potassium, it is pH – pK^+; for magnesium it is pH – 1/$2pMg^{2+}$). Acidity is usually inversely related to cation concentrations. Figure 2.3 thus indicates that the stability of gibbsite increases with a decrease of the cation activities, i.e., the pH and the $Si(OH)_4$ concentration. Kaolinite, 2:1 clay minerals, and feldspars are, in contrast, stable at high values of these variables.

The stability diagram for magnesium shows that Mg-beidellite (a

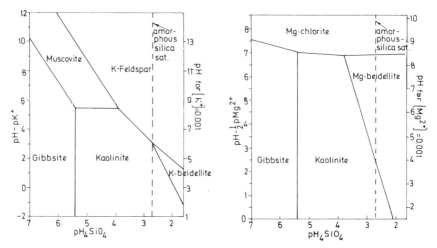

Figure 2.3   Stability diagrams of minerals at 25°C and 10 kPa. [*From Van Breemen and Brinkman (1976) Elsevier Science Publishers, reprinted by permission.*]

2:1 clay) can form at H_4SiO_4 activities that are lower than those found in equilibrium with amorphous SiO_2, provided that either the pH or the Mg activity is high enough.

Amorphous clays.   The crystallization of clays is an extremely slow process, and there are many circumstances where there is not sufficient time for crystals to nucleate and grow before solutes precipitate as amorphous materials. The best known example is the rapid release by volcanic ash of aluminum and silica that precipitate as *allophane* or *amorphous silica* before regular crystal structures can be built. A detailed discussion of the formation of amorphous clay minerals is given in Chap. 10 on Andisols.

Iron hydroxides and oxides.   The synthesis of iron oxide clay minerals follows pathways different from those of silicate clays. The dimensions of Fe^{2+} and Fe^{3+} ions do not favor a tetrahedral coordination with oxygen, and the very low solubility of $Fe(OH)_3$ does not allow Fe^{3+} to participate in the building of the octahedral sheets in phyllosilicates. Iron is incorporated in the clay lattice structure only in the case of nontronite.

Iron hydroxides in permanently well-aerated and oxidizing conditions usually occur as coatings on clay, silt, or sand particles. Iron hydroxides may also precipitate on themselves.

The kinds of iron hydroxides that form depend on the rates at which their ion activity products in the soil solution increase beyond their solubility. These solubilities in turn change according to the protection that chelation by organic matter imparts on these hydroxides and according to the oxidation-reduction potential of the soil water that solubilizes the iron as Fe^{2+}.

According to Schwertmann and Taylor (1977), *goethite* forms when the hydrolysis of iron during weathering proceeds *slowly* and the ion activity product of $Fe(OH)_3$ exceeds the goethite solubility product (but remains below the slightly higher solubility product of ferrihydrite). *Fast* hydrolysis of Fe^{3+} on the other hand results in the precipitation of *ferrihydrite* in soil solutions that are supersaturated with respect to ferrihydrite and goethite. Ferrihydrite is considered a necessary precursor of *hematite*, which forms by dehydration and re-arrangement into a crystal lattice, without passing through a dissolution phase. The same authors contend that at normal soil temperatures, goethite and hematite cannot be interconverted by simple solid state reactions but that these transformations always involve dissolution and reprecipitation processes.

There are other factors that control the types of iron oxides that form. According to Schwertmann and Murad (1983), the formation of goethite is favored when the relative concentration of monovalent $Fe(OH)_2^+$ is at its maximum, which occurs at approximately pH 4. Hematite on the other hand shows preferential formation when the $Fe(OH)_2^+$ content is at its lowest, which occurs at a pH of 8. These values hold when the pH values of the soil solutions are 4 or higher. For pH values well below 4, hematite is formed because the activity of $Fe(OH)^{2+}$ starts to control its formation.

Ferrihydrite and hematite color soils to reddish hues. Hematite is a very effective pigmenting agent that may mask the yellowish colors produced by goethite when both minerals are present in the same material. Mixtures may also produce intermediate colors between red and yellow.

Cool temperatures, high moisture, and high organic matter contents favor the formation of goethite. These attributes either lower the concentrations of iron in the soil solution or increase the solubility of its hydroxides. The soil colors in such environments tend to be yellow. Under warmer or drier conditions red becomes the dominant color.

Parent rocks may have similar effects on the colors of the soils that are derived from them. Felsic igneous rocks, or any other rocks that are low in iron, tend to produce yellow soils, and mafic rocks generally form dusky red soils, containing large amounts of hematite.

Clay formation environments. The clay formation environments that are important for soil genesis are shown in Fig. 2.4. The left-hand shaded side of the graph lists the major *primary minerals* arranged in a sequence from silica-poor to silica-rich. The top shaded part of the figure represents the unweatherable minerals, essentially quartz.

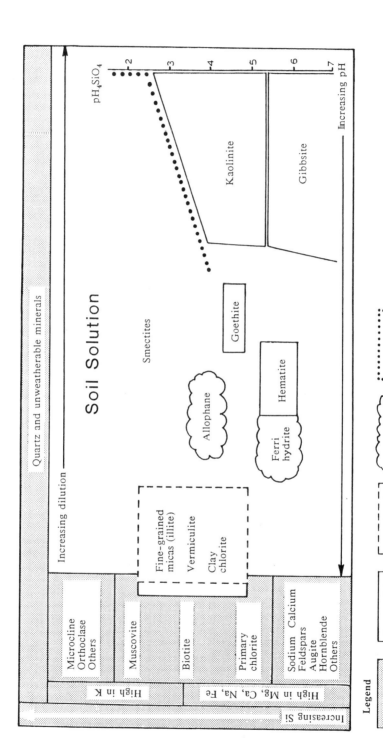

Figure 2.4 Conceptual representation of clay formation environments.

The rest of Fig. 2.4 (the unshaded portion) pictures the soil solution with the high cation activities on the left-hand side and the most dilute solutions on the right-hand side (pH is decreasing from left to right). In the highly diluted part, the activity of H_4SiO_4 decreases from top to bottom. The stability domains of kaolinite and gibbsite from Fig. 2.3 are pictured in the right, lower corner. There is a special box between the primary mineral part and the soil solution part for the residual clays.

The lower left-hand corner of the soil solution part of the figure is an area where amorphous iron oxide clays (ferrihydrite) precipitate, because the supply of Fe by weathering exceeds the rate of crystallization of goethite. Allophanes are placed at comparable concentrations for solutions containing dominantly Si and Al produced by rapidly decomposing volcanic ashes.

Goethite is a common mineral that is formed by crystallization of dilute Fe solutions, generally in the presence of organic matter. Goethite mostly occurs as coatings on clay minerals, although it may occur as discrete bodies such as nodules, microaggregates, etc.

The right-hand side of the graph represents dilute concentrations in which clay minerals have sufficient time to crystallize. It is divided horizontally on the basis of $Si(OH)_4$ activity into three fields: The upper field with high Si activities in which smectites are most likely to form; an intermediate field, which is favorable for kaolinite formation; and a lower portion where gibbsite is the most stable.

Figure 2.4 places the most common clay minerals as separate fields at locations that qualitatively relate them to their formation requirements. The way the figure is drawn suggests that for given conditions only one mineral would form. In real soil solutions it is obvious that the chemical characteristics constantly change due to temporary variations in soil moisture, and that mixtures of clay minerals are synthesized. Their composition, however, reflects the most frequent set of soil solution attributes that are designated as clay formation environments. Four kinds of environments are recognized:

1. *Residual clays.*   These were discussed previously. They include illites, vermiculites, and chlorites. In the tropics they characterize young sediments.

2. *Smectites.*   These form in soil solutions rich in Si and Mg. The Si activity is close to or exceeds the activity in equilibrium with amorphous silica. Smectites are either found in semiarid regions in depressions into which water seepage brings silica, or they form close to weathering basic rocks that are protected against leaching.

3. *Amorphous clays.*   These form in solutions oversaturated with aluminum hydroxides and $Si(OH)_4$. Allophane is the dominant clay

mineral often associated with imogolite and halloysites. Amorphous aluminum hydroxides and ferrihydrite may also be present.

4. *Low activity clays.* These form in soils in which the soil solution is very dilute and has a low silica content. Gibbsite may be an intermediate stage to the synthesis of kaolinite. As long as the solubility product of ferrihydrite is not exceeded, goethite is synthesized directly from the soil solution. Organic matter facilitates goethite formation. Otherwise iron oxides precipitate as ferrihydrite, which subsequently may crystallize by dehydration and rearrangement into hematite.

2.1.3 Stability of secondary minerals

The clay formation environments are not permanent and they may change during the time that parent materials are built and soils develop. Clays that form under one set of conditions and that are subsequently exposed to new chemical environments slowly react and decompose, and the decomposition products that enter the soil solution may recombine into other types of clays. The time frame is pedologic or geologic. For example, no lattice clay mineral transformations are to be expected as a result of soil management practices.

Climatic modifications from dry to wet conditions are more likely to accelerate chemical decomposition than changes toward arid environments. The latter changes rather tend to stop chemical reactions and freeze an acquired situation. There are soils with the typical mineralogy of humid soil moisture regimes in tropical deserts, but the attributes of aridic conditions are seldom found under presently wet climates. An example of a stable clay mineral assemblage is the *low activity clays* that characterize soils in large areas of the continental shields. Figure 2.5 shows where they are dominant in soil associations of Africa, as compiled in the FAO-Unesco *Soil Map of the World* (FAO-Unesco, 1977).

Roose (1981) analyzed surface, ground, and well waters at eight isohyperthermic stations in west Africa that received between 800 and 2100 mm of rain per year. The rainy season lasts 11 months in the south and 4 months in the north. The bedrock in the region is dominated by granites and gneisses or quartz-rich sediments derived from them. All analyses show $Si(OH)_4$, H^+, and Na^+ concentrations that fall within the stability field of kaolinite. Roose concluded that in the pH range of the soils in the region (6 to 7), kaolinite is stable and persists.

In the same study (Roose, 1981), some of the surface waters were found to be undersaturated with respect to quartz and would thus allow its dissolution. On the other hand, groundwaters that are in con-

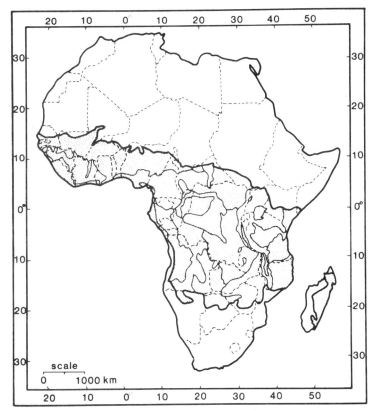

Figure 2.5 Distribution in Africa of soil associations in which low activity clay soils dominate.

tact with weathering rocks have Si contents indicating stable conditions for quartz.

The 2:1 clays, more particularly the smectites, are pedologically unstable when the clay formation environments move toward more dilute soil solutions that occur either because of climatic changes or as a consequence of time. Their disintegration releases large amounts of Al^{3+} that produce hydrolysis and partly restore the acidity that was used to consume them. For this reason, there are only a few soils dominated by smectites that have pH values in the strongly acid range; when they occur, their soil solutions contain toxic amounts of exchangeable aluminum.

The most drastic changes in clay mineralogy are caused by changes in oxidation-reduction potential. The oxidation of Fe^{2+} produces H^+ that often contributes to the dissolution of silicate clay minerals. For example, Brinkman (1979) called *ferrolysis* the destruction of clays during the oxidation of Fe^{2+} in soils that are seasonally waterlogged. There are many reports in the literature that relate the disintegration of clays to waterlogging at microsites or in soil horizons. Ambrosi et al. (1986) postulated that the dissolution of kaolinite during the precipitation of hematite in plinthites is due to the action of H^+ released during the oxidation of Fe^{2+}.

Soil horizons that have flooded or are water-saturated and that contain organic matter that acts as an electron donor are effective reducers of iron hydroxides, which are solubilized by this process. Many iron oxide coatings on particles can be removed or altered in this way. Hematite and ferrihydrite are the first to be dissolved, thus, causing changes in color from red to yellow, when goethite persists, or to the color of the bare mineral grains, when all coatings are removed.

2.2 Deep Weathering Profiles

2.2.1 Horizon sequence

Many soil profiles in the tropics reach depths that are unusual to pedologists who gained most of their experience in temperate, glaciated areas. These extraordinary deep tropical soils are the products of extremely long periods of intensive weathering without interruptions by massive erosion caused by glaciers.

The deep profiles are often used to illustrate soil and parent material formations in the humid tropics. The complete horizon or layer sequence is most commonly found on old geomorphic surfaces at freely drained sites, although groundwater may be present in the lower parts of the profiles or, if absent, may have influenced the deep horizons in the past. The depth of these profiles on average may reach 30 m, but regoliths more than 100 m deep have been reported (Thomas, 1974).

The profiles that are considered the most typical for soil formation in the tropics all include a lower portion that has retained the structure of the original rock; in situ weathering and leaching, however, have deeply altered the mineral composition and softened its consistency. This "rotten" rock or *saprolite* has suffered severe desilication, probably at the time it was saturated by groundwater. Saprolite may include several "zones" that have reached different degrees of decomposition; the upper one is often called the *mottled* zone that overlies a *pallid* zone that grades into fresh solid rock. The upper part of saprolite may contain plinthite (defined later).

In profiles that are considered complete, the saprolite is overlain by materials that have lost the original rock structure. They may have been deposited on top of the saprolite by erosion and sedimentation

Figure 2.6 Exposure of a profile that contains a surface mantle, a stone line, and saprolite in Burundi.

processes. More likely they are the products of mixing of soil materials by animals, mainly rodents and insects, or uprooting by vegetation. These processes are designated as *pedoturbation, bioturbation,* or *pedoplasmation.*

The transition between saprolite and the pedoturbated surface mantle may be marked by a gravelly stone line or by an ironstone crust (see Fig. 2.6). When a stone line separates the saprolite from the top horizons, it is currently assumed that the saprolite was once exposed at the surface and that erosion produced a gravelly lag deposit that was later covered by the surface mantle.

Figure 2.7 shows a sequence of horizons commonly found in deep weathering profiles. The lowest portion, developed in the soft saprolitic material, is called the *pallid* zone. It has not been influenced significantly by oxidation and probably weathered under the lowest level of a fluctuating water table. The groundwater could freely drain and evacuate soluble weathering products.

The pallid zone grades upwards into a *mottled* zone in which the number of iron hydroxide mottles and iron concentrations increase. The mottled zone probably corresponds to the major zone of seasonal water table fluctuations. The amount and the kinds of iron hydroxides may be such that the horizon has the properties of soft *plinthite* or *ironstone* derived from it by irreversible induration. Birot (1968) esti-

Deep weathering profile

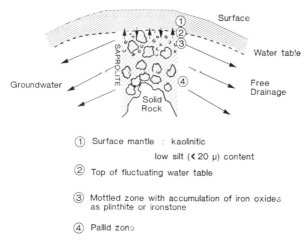

① Surface mantle : kaolinitic

low silt (< 20 μ) content

② Top of fluctuating water table

③ Mottled zone with accumulation of iron oxides
as plinthite or ironstone

④ Pallid zone

Figure 2.7   Conceptual representation of a deep weathering
profile.

mated that the rate of weathering in the zone of alternate wetting and
drying is at least 20 times faster than in the permanently saturated
part of the profile.

The layer or horizon sequence that is described above is often labeled by Greek symbols (Stoops, 1989) where α designates the surface
mantle, β the gravel layer or stone line, and γ the saprolitic layers
that rest on the solid unweathered rock (δ).

2.2.2   Stone lines

There are several theories on the origin of stone lines, which are thin
(usually a few decimeters) layers of angular gravel that separate
saprolites from surface mantles.

The hypothesis that was presented in Sec. 2.2.1 assumes that stone
lines probably form at a time when the saprolite is at the surface and
is exposed to strong erosion, which removes all fine soil materials and
leaves behind the gravel. Similar surface accumulations of coarse materials presently take place in many arid and semiarid regions that
are not protected against raindrop impact and runoff. The gravel layers are commonly called *desert pavements*. The mineralogy of the resistant minerals in most stone lines is similar to that of saprolites.

The theories for the deposit of the surface mantle on top of the stone
line invoke sedimentation from surface waters on top of the pediment
surface as the backslopes retreat, deposition by wind, or transporta-

tion by termites or by other organisms (Fig. 2.8). None of these theories is mutually exclusive. The most plausible is nest building by insects, more specifically termites, using materials extracted from the saprolitic layers. This activity constructs a surface mantle made of particles that are well sorted, without gravel, and mineralogically comparable to the underlying rock. This hypothesis does not require that the stone line was once exposed at the surface, because termites,

Figure 2.8 Exposure of a quartz vein that is at the origin of the gravel that forms the stone line that underlies the surface mantle.

by removing small particles and clay from the subsoil, undermine the gravels and stones that sink to the level where the insects penetrate the sediments. The presence of artifacts, patines on gravel, the sharp boundaries between surface mantles and stone lines, as well as the evidence of dry periods during Holocene and late Pleistocene, however, favor the exposure theory.

2.2.3   The solum

The present climate and vegetation that now influence soil formation on old landscapes act mainly on surface mantle materials. They are often considered the parent materials in which the genetic A and B horizons, or the *solum,* forms. It is therefore generally accepted that many soil parent materials in the tropics have been *preweathered,* a situation not often found in the glaciated areas of temperate regions.

Not all deep weathering profiles include the complete sequence of weathering zones described above; some may be missing because of their removal by erosion or by a lack of time to develop them. On the other hand, erosion and sedimentation may distribute the materials of each zone horizontally over the landscapes when valley slopes are formed following incision by rivers. The deep weathering profiles are then transformed into weathering toposequences, or catenas, which vary according to the depths of the river incision into the weathering zones.

2.3   Erosion and Sedimentation

The diversity of soils in the tropics is not only due to differences in climate and rock types. The weathering products that form the regolith are also subjected to forces that create variability. Most surface mantles have been sorted and redistributed over the land by the incision of rivers, erosion during the shaping of valley slopes, colluviation by surface waters, transportation by wind, etc. The soil patterns that result from these processes are at least as important for the understanding of soil geography as those that have been shaped by climate and rock types.

The thickness of the regoliths that cover landscapes and the soils that develop in them depends on the combined actions of weathering, erosion, and sedimentation. In this text this variability is examined from two viewpoints. The first one is global and relates soils to the structure of whole continents. It places geography in a perspective that subdivides land into a few broad regions, each of which is dominated by a small set of typifying soils. It emphasizes the "long distance" similarities between natural regions even if they are spread over several continents. The division of land into broad geographical areas is a useful tool for comparing agricultural systems and evaluating the feasibility of transferring them.

The second viewpoint from which soil variability is looked upon is local and considers the reworking of parent materials by surface water. The most frequent examples are valley slopes that form as a consequence of the dissection of landscapes by rivers. These geomorphic processes create soil toposequences that are often referred to as *catenas*. Catenas contain the soil associations that farmers are well aware of and usually deal with when making decisions.

Section 2.3 anticipates somewhat the second part of this book because it links landscape features to soils that have not yet been defined. Since only broad concepts are used, it should serve as an introduction to Part 2, which describes these soils in more detail.

2.3.1 The structure of continents

Physical geography subdivides continents into structural regions: *continental platforms* and *orogenic belts*. Each of them has a distinct history of mountain building, erosion, and sedimentation. The soil patterns that characterize them have typical proportions of dominant soils that create unique soil landscapes. These "soilscapes" could easily be considered different *soil worlds*. The soilscapes have been better preserved in the tropics than elsewhere because they have not been erased by quaternary glaciations. The continental platforms and orogenic belts are schematically shown in Fig. 2.9 with their major associated parent material attributes.

OROGENIC BELT			CONTINENTAL PLATFORM			
MOUNTAIN RANGE		SUBSIDING ZONE	SWELLS			BASINS
			SHIELD	RIFT VALLEY	SHIELD	
VOL–CANOES	YOUNG DEPOSITS / FRESH ROCK	YOUNG ALLUVIUM		L A V A / G R A B E N / H O R S T		OLD SEDIMENTS

Figure 2.9 Schematic cross-section of world structural regions.

The continental platforms. Continental platforms consist of stable, massive blocks of sialic crust that have not been subject to orogeny since the Cambrian period. They are dominantly made of crystalline rocks. Their major landforms include swells known as *continental shields* and *basins*. The shields are often broken by *rift valleys*.

The major continental shields in the tropics belong to the Gondwana shield (Murphy, 1967). They are the Deccan shield in India, the African continental shield, which spreads almost over the entire continent, and the Brazilian and Guiana shields in South America (Fig. 2.12). The oldest erosion surfaces that are most commonly referred to in the literature are Jurassic-Cretaceous (approximately 150 million years old) and Miocene (about 15 million years old). In locations that have not been dissected by river systems, the shields are covered by deep, strongly weathered, residual materials. Their soils are mostly developed in sediments with low activity clays that underwent intensive leaching during one of the rainy, pluvial periods that characterized Pleistocene times. The soils that dominate the landscapes belong to Oxisols and *kandic* great groups of Ultisols and Alfisols. The major soil limitation for agricultural development in these areas is the extremely low nutrient content of the parent materials. This deficiency generally does not allow sustained types of agriculture without substantial use of fertilizers. The shields are usually sparsely populated and shifting cultivation is the most common farming system. Figure 2.10 shows a typical landscape of a shield area in Uganda.

Some parts of the shields have been tectonically broken by rift valleys. These rifts include the *grabens* that sank and the associated raised *horsts* and volcanic systems. The African Rift is the most typical; the San Francisco valley in Brazil is another example. The uplifted areas that border the sunken grabens are mineralogically not different from the shield regions described in the previous paragraph. If the horst was raised several hundred meters, the soils will contain more organic matter because of the lower temperatures of the high altitudes. The cooler climate also attracts more people to move into the area, but the agricultural potential of the old landscape parts is not significantly superior to that of the lowland shields.

The young sediments that filled the grabens sharply contrast with the deeply weathered soil materials that cover the noneroded parts of the horsts. They still contain weatherable minerals, and volcanoes may have rejuvenated many sediments. There are climatic differences also. Parts of the grabens are situated in the rain shadows of the mountains, and at these sites leaching is minimal. Certain areas in the grabens may be saline or alkaline. The parent materials dominantly contain residual clays or smectites. Lava flows and volcanic ashes are the third component of the rifts. The third component is

Figure 2.10 Landscape in a continental shield area of Africa.

characterized by soils that contain amorphous clays such as allophane. Volcanoes often transform landscapes into productive, densely populated regions of the cool tropics.

The orogenic belts. The orogenic belts correspond to weak areas in the earth's crust that have been subjected to folding and mountain building. They include the *mountain ranges* that have been lifted and the *subsiding zones* that usually border them (Fig. 2.9).

The mountain ranges. The mountain ranges that are important for soil formation in the tropics belong to the Alpine system that started during the Jurassic period and is still active. Volcanic eruptions and frequent earthquakes characterize these regions. The Andes Cordillera, Central America and the Caribbean islands, the Himalayas, and almost all of southeast Asia belong to the Alpine system.

The most striking characteristic of the orogenic belts is the pronounced relief that actually continues to develop and for that reason does not allow soil materials to reach advanced stages of weathering. Erosion can usually remove earthy materials faster than weathering can produce them. Clays are either residual or amorphous, depending on the importance of the activity of volcanoes. There is also a wide climatic diversity. There are many instances of short-distance changes from cool perhumid conditions on top of the mountains to hot aridic regimes in neighboring valleys. In the dry parts of the broad valleys,

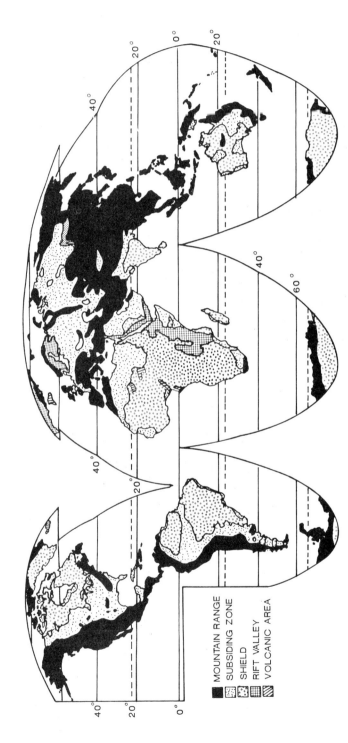

Figure 2.12 World structural regions. [*From Murphy (1968). Association of American Geographers, reprinted by permission.*]

MOUNTAIN RANGE
SUBSIDING ZONE
SHIELD
RIFT VALLEY
VOLCANIC AREA

smectites may be the dominant clay minerals. The low temperatures of the high elevations preserve organic matter more efficiently than the lowlands, and, depending on the amount of precipitation, thick, humus-rich surface horizons that are either acid or neutral may form. Umbric and mollic epipedons frequently occur where the temperature regime is either isothermic or cooler. The orogenic belts also include areas of unstable slopes that are often affected by catastrophic landslides, particularly in the humid parts where soils are dominated by amorphous clays.

People have adapted to the excessive relief situations, mainly by building terraces on the steep slopes or by developing the broad valleys, such as the altiplanos of South America. If the temperatures are not too low nor the climate too dry for rain-fed agriculture, then the population density is high and often exceeds the carrying capacity of the sloping land.

The subsiding zones. Most mountain chains are bordered by subsiding zones. They are very flat huge plains at the foot of the mountains that slowly sink and are gradually filled with alluvium transported by rivers that cut into the mountain ranges. Subsiding zones belong to aggradational geomorphic surfaces. They are often flooded by overflow of the rivers or inundated by intensive rains, particularly in monsoon climates. The subsiding zones differ from the basins on the continental platforms in that their sediments originate from the erosion of fresh

Figure 2.11   Landscape in the subsiding zone east of the Andes Cordillera in Colombia.

rocks that may be rich in weatherable minerals. If this is the case, their soils offer a high potential for agriculture; if not, the alluvium is poor and little agricultural production is possible without adequate fertilization. The Orinoco-Parana subandean depression east of the Andes in South America (Fig. 2.11) and the Ganges valley at the foot of the Himalayas in India are examples of subsiding zones (Fig. 2.12).

The extremely flat topography of the subsiding zones, the low permeability of many of their sediments, and the concentration of precipitation in a short but intensive rainy season often make periodic flooding one of the major impediments for agricultural development. In the monsoon climates the water table may fluctuate over several meters and bring soil moisture conditions from complete water saturation in the rainy season to extreme dryness in the dry season. These conditions are favorable for the formation of plinthite.

2.3.2 Slope evolution

Most continental shields have been dissected by rivers. Several mechanisms are involved. *Downcutting* and *headward* erosion are the major processes at the beginning stages of the dissection. During these early phases V-shaped valleys are formed (Fig. 2.13). Their steepness depends on the angle of repose of the underlying materials. In the case of landscapes that are well protected by vegetation, such as a rain forest, downcutting is probably the most active geomorphic process that determines the shape of the slopes. Some widening of the alluvial flat may occur as a result of undercutting by the meandering rivers, but slope angles remain constant and shapes are linear in recently developed parts of the dissected topography.

The second mechanism to consider for understanding the most common soil patterns in tropical environments is *slope retreat* or *backwearing.* During this process the retreating slope, called *backslope,* keeps a constant angle. Backwearing is invoked by many soil scientists to explain both the landforms and the nature of the parent materials that cover the valley slopes. It is particularly active in areas that are poorly protected by vegetation, such as semiarid regions. By the mechanism (Fig. 2.13) known as *pedimentation,* a concave slope, the *footslope,* forms at the expense of the retreating backslope. The footslope contains an erosion surface, the *pediment,* that is often marked by an erosion pavement. When this erosion pavement is covered by a soil mantle, the *pedisediment,* it corresponds to a stone line (Ruhe, 1974).

Pediment formation very often provides satisfactory explanations for age and weathering relationships among sediments in toposequences or catenas. One important factor in the understanding

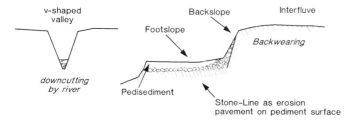

Figure 2.13   Schematic representation of pedimentation processes.

of these catenas is the depth to which rivers cut into the deep weathering profiles. If they only penetrate into the strongly weathered surface mantle, not much variability can be expected in the mineralogy of the soils that cover the valley slopes. If, on the other hand, the rivers cut into fresh or weathering rocks, most likely the soil toposequences will include units with striking differences in weathering stages, mineralogy, and horizon development.

Not all relief forms relate to erosional processes; many features result from structural factors, such as resistance of rocks to weathering and tectonic movements. Other landscapes are shaped by aggradational processes such as deposition of sediments by rivers. The reasons for discussing pedimentation in more detail are the large areal extent that pediments cover on the continental shields and the multiple erosion cycles that cut several nearly horizontal surfaces as steps in the slopes. Each of these steps has a different age and is characterized by parent materials that have reached distinct weathering stages. The occurrence of dry climatic periods during the Pleistocene in almost all regions in the tropics made it possible for pedimentation to leave its marks on many landscapes, even in regions that are presently humid.

2.3.3   Soil catenas and toposequences

The concept of catenas or chains of soils that repeat themselves over the landscapes originated from work by Milne (1935) on soils of Tanzania. Since that time many toposequences have been described, and several ways of grouping them have been suggested (Moss, 1968; Gerrard, 1981). The diversity of soils in a catena can be the result of differences in clay formation environments, weathering stages, or parent materials.

Texture gradients.   Textural toposequences most commonly occur on valley slopes of rivers that cut into uniform, deeply weathered sediments. Raindrop impact disrupts the aggregates in the topsoils and separates clay particles from coarser sand grains. Running water picks up the finest particles and transports clay in suspension downslope into the rivers. When the rivers flow freely out of the watersheds into streams, this sorting process leaves the nonclay particles and the resistant aggregates behind as colluvium in the lower parts of the topography.

The mechanism described above creates sequences of soils that become sandier downslope. In many cases it also explains the occurrence of lighter textured surface layers in soils of sloping land. The textural differentiation in the profiles is often called *appauvrissement* by French pedologists (Roose, 1981). The decrease in clay content in the soils or in the horizons usually coincides with color changes from reddish to yellow in the sandier materials. It is assumed that organic matter contributes to the solubilization of iron oxides during the disruption of the aggregates by raindrop impact and that parts of them are evacuated in the rivers, leaving yellower deposits behind on the lower valley slopes or in the upper part of the profiles. In some cases, however, poorer drainage conditions in depressions may be responsible for the color changes.

An example of textural catenas (Fig. 2.14) is the Yangambi catena in central Africa (De Leenheer et al., 1952). However not all textural sequences follow the trend illustrated by this case. When the incision by the rivers cuts into materials that become considerably heavier with depth, the textural gradient may be reversed or partly obliterated.

The sorting of parent materials during transportation on valley slopes should normally result in sand particles becoming finer downslope. This is usually the case when the grains travel as discrete particles. Fine sands, however, may form stronger aggregates than coarse sands and may be trapped in granules that resist disruption and are less mobile than the individual grains. The upper slope

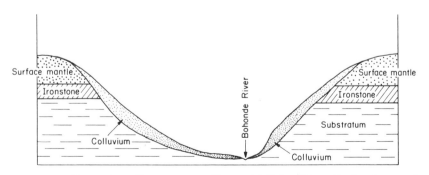

Figure 2.14   Textural gradient catena at Yangambi, Zaire. [*From De Leenheer et al.* (*1952*).]

colluvia for this reason may contain higher proportions of fine sand than the lower slopes that collected the discrete coarse sand particles.

Color catenas. One of the most typical soil catenas of dissected old geomorphic surfaces in the tropics is the sequence of red upland soils that change into yellow soils on adjacent slopes, which in turn may grade into gray bottomlands. These color catenas were first described by Milne (1935) and have since been observed in many tropical regions. The color changes usually correspond to various hematite-to-goethite ratios and may coincide with changes in texture as mentioned in the previous section.

The most valid explanations for the color changes either consider different clay formation environments associated with poorer drainage conditions in the lowest parts of the topography or they assume preferential dissolution of hematite from existing hematite and goethite mixtures. The first hypothesis associates lower Fe concentrations in the soil solution when drainage is retarded. The dilution effect then favors the crystallization of goethite (yellow) rather than the precipitation of ferrihydrite.

The second theory involves the reduction or chelation of iron hydroxides in the presence of organic matter in the surface horizons, and the subsequent transportation and deposition of discolored topsoil as colluvium downslope. Macedo and Bryant (1989) described the preferential microbial reduction and removal of hematite over aluminum-substituted goethite in an Anionic Acrustox of Brazil. By this process color changes from red to yellow are achieved, provided a source of energy is available for the microbial population, which is often the case in surface horizons that are periodically saturated with water.

When the catena is formed by the transportation of discolored topsoil, the red to yellow toposequence is associated with a parallel textural horizon differentiation that results in a lighter textured yellowish topsoil overlying a more reddish subsoil. The two mechanisms that are invoked in this section obviously are not mutually exclusive and may act simultaneously to form a color catena.

The most typical color toposequences occur when the sediments that form the catena originate from the same parent rock and have comparable, preferably advanced, weathering stages. If not, differences in mineralogy may mask the catenary relationship.

Weathering catenas. Weathering catenas develop when valleys cut into incompletely weathered rocks that underlie old erosion surfaces. They also occur when several erosion cycles follow each other and create stepwise sequences of pediments between the river and the interfluve. Differences in the duration of weathering and in the thick-

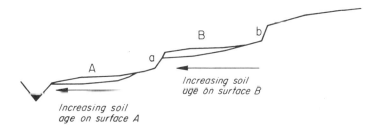

a: backslope of A   b: backslope of B

Figure 2.15   Cross section of landscape that contains weathering catenas.
[*From Gerrard (1981). Allen and Unwin, adapted by permission.*]

ness of the regoliths are usually at the origin of very complex
toposequences; soil horizon formation in the pedisediments may further complicate matters.

A hypothetical cross section (Fig. 2.15) adapted from Gerrard (1981)
summarizes the most common time-dependent soil sequences on mul-
tilevel erosion surfaces. Since erosion surface A was built after surface
B, most pedisediments on B are older than on A and, as a rule, are

Figure 2.16   Land surface under savannah affected by inundation erosion in the Llanos
Orientales of Colombia.

more weathered. Because pedisediments form by slope retreat, soil formation on each surface is more advanced away from its backslope than close to it. In Fig. 2.15, the backslope of pediment A is labeled "a."

The complexity of slope formation and the interactions between (1) parent rocks, (2) length of erosion cycles, and (3) climatic changes make typical tropical weathering catenas difficult to describe. They may vary markedly from one region to another. In regions with comparable history, however, they are valuable tools to predict the kinds of soils on the basis of their position in the landscape.

2.3.4 Inundation erosion microrelief

There are extensive flat areas in the tropics, generally located in the subsiding zones of the orogenic belts, that are inundated every year by monsoon rains. The slow movement of the ponding water, which is usually not deeper than a few decimeters, causes differential removal of soil materials in those parts that are not protected against detachment by individual plants. The slope gradients of the terrains are usually very small, less than 1 percent, and uniform over large areas. There are no concentrations of runoff water that cut deep gullies evacuating water abruptly after heavy showers. Since direct heavy precipitation is the main cause of flooding, practically no sediments are added to the inundated land, and as a result of the slow overall movement of the water, ditches between tussocks are gradually formed in grasslands (Fig. 2.16). In forests, the ditches may separate large trees and isolate them as if they were placed on pedestals. The areas affected by the inundation erosion microrelief may cover several thousands of hectares as, for example, in the Llanos regions of South America (Fig. 2.17).

Flood-eroded land occurs both under grassland and rain forests. It makes land inaccessible for most uses and its formation cannot be stopped as long as flooding occurs and water moves over the surface.

Figure 2.17   Aerial view of land surface under savannah affected by flood erosion in the Llanos Orientales of Colombia.

References

Ambrosi, J. P., D. Nahon, and A. J. Herbillon, "The epigenic replacement of kaolinite by hematite in laterite—petrographic evidence and the mechanisms involved," *Geoderma* 37:283–294 (1986).

Birkeland, P. W., *Pedology, Weathering, and Geomorphological Research,* Oxford University Press, New York, 1974.

Birot, P., *The Cycle of Erosion in Different Climates,* Batsford, London, 1968. (Translated from: Les cycles d'érosion sous différents climats, University of Brazil, 1960.)

Brady, N., *The Nature and Properties of Soils,* 9th ed. Macmillan, New York, 1984.

Brinkman, R., *Ferrolysis, a Soil Forming Process in Hydromorphic Conditions,* Center for Agricultural Publishing and Documentation, Wageningen, The Netherlands, 1979.

Carroll, D., *Rock Weathering,* Plenum Press, New York, 1970.

Claisse, G., "Etude de la solubilisation différentielle du quartz pur en laboratoire," *Cahiers ORSTOM serie Pédologie,* 13(1):61–88 (1975).

De Leenheer, L., J. D'hoore, and K. Sys, *Cartographie et Caractérisation Pédologique de la Catena de Yangambi,* Publication Inéac, Série Scientifique #55, Bruxelles, Belgium, 1952.

Eswaran, H., and G. Stoops, "Surface textures of quartz in tropical soils," *Soil Sci. Soc. Amer. J.* 43:420–424 (1979).

FAO-Unesco, *Soil Map of the World,* vol. 6: Africa, Unesco, Paris, 1977.

FAO, *FAO-Unesco Soil Map of the World. Revised Legend,* World Soil Resources Report 60, Food and Agriculture Organization of the United Nations, Rome, 1988.

Gerrard, A. G., *Soils and Landforms,* George Allen & Unwin, London, UK, 1981.

Goldich, S. S., "A study in rock-weathering," *J. Geol.* 46:17–58.

Harris, W. G., L. W. Zelazny, J. C. Baker, and D. C. Martens, "Biotite kaolinization in Virginia Piedmont soils. I. Extent, profile trends, and grain morphological effects," *Soil Sci. Soc. Amer. J.* 49:1290–1297 (1985).

Herbillon, A. J., "Degree of weathering and surface properties of clays," in D. J. Greenland (ed.), *Characterization of Soils,* Clarendon Press, Oxford, 1981, pp. 80–96.

Huang, W. H., and W. D. Keller, "Dissolution of rock-forming silicate minerals in organic acids. Simulated first-state weathering of fresh mineral surfaces," *Amer. Mineralogist* 55:2076–2094 (1970).

Huang, W. H., and W. C. Kiang, "Laboratory dissolution of plagioclase feldspars in water and organic acids at room temperature," *Amer. Mineralogist* 56:1082–1095 (1972).

Jackson, M. L., S. A. Taylor, A. L. Willis, G. A. Bourbeau, and R. P. Pennington, "Weathering sequence of clay-size minerals in soils and sediments—fundamental generalizations," *J. Phys. Colloid Chem.* 52:1237–1260 (1948).

Kampf, N., and U. Schwertmann, "Goethite and hematite in a climosequence in southern Brazil and their application in classification of kaolinitic soils," *Geoderma* 29:27–40 (1983).

Leneuf, N., and G. Aubert, "Essai d'évaluation de la vitesse de ferrallitisation," *Proc. 7th Int. Congr. Soil Sci.* 225–228 (1960).

Macedo, J., and R. Bryant, "Preferential microbial reduction of hematite over goethite in a Brazilian Oxisol," *Soil Sci. Soc. Amer. J.* 53:1114–1118 (1989).

Milne, G., "A soil reconnaissance journey through parts of Tanganyika territory," *J. Ecology* 35:192–269 (1935).

Moss, R. P., "Soils, slopes and surfaces in tropical Africa," in R. P. Moss (ed.), *The Soil Resources of Tropical Africa,* Cambridge University Press, London, 1968, pp. 29–60.

Muller, J., and G. Bocquier, "Dissolution of kaolinites and accumulation of iron oxides in lateritic-ferruginous nodules: mineralogical and microstructural transformations," *Geoderma* 37:113–136 (1986).

Murphy, R. E., "Spatial classification of landforms based on both genetic and empirical factors—a revision," *Annals Assoc. Amer. Geographers* 57(1):185–186 (1967).

Murphy, R. E., "Landforms of the world," *Annals Assoc. Amer. Geographers* 58(1): Annals map supplement #9 (1968).

Nockolds, S. R., R. W. O'B. Knox, and G. A. Chinner, *Petrology for Students,* Cambridge University Press, Cambridge, 1978.

Owens, L. B., *Rates of Weathering and Soil Formation on Granite in Rhodesia (Zimbabwe),* Department of Agriculture, University of Zimbabwe, 1974.

Parfitt, R. L., C. W. Childs, and D. N. Eden, "Ferrihydrite and allophane in four Andepts from Hawaii and implications for their classification," *Geoderma* 41:223–241 (1988).

Pedro, G., "Contribution à l'étude expérimentale de l'altération géochimique des roches crystallines," *Annales Agronomiques* (Paris) 15:85–191, 243–333, 339–456 (1964).

Rebertus, R. A., S. B. Weed, and S. W. Buol, "Transformations of biotite to kaolinite during saprolite-soil weathering," *Soil Sci. Soc. Amer. J.* 50:810–819 (1986).

Roose, E., *Dynamique Actuelle des Sols Ferrallitiques et Ferrugineux Tropicaux d'Afrique Occidentale,* Travaux et documents de l'ORSTOM #130, Paris, 1981.

Ruhe, R., *Geomorphology,* Houghton Mifflin Co., Boston, MA, 1974.

Schwertmann, U., and M. Murad, "The effect of pH on the formation of goethite and hematite from ferrihydrite," *Clays Clay Min.* 31:277–284 (1983).

Schwertmann, U., and R. M. Taylor, "Iron oxides," in R. C. Dinauer (ed.), *Minerals in Soil Environments,* Soil Science Society of America, Madison, WI, 1977, pp. 145–180.

Schalscha, E. B., H. Appelt, and A. Schatz, "Chelation as a weathering mechanism. I. Effect of complexing agents on the solubilization of iron from minerals and granodiorite," *Geochim. Cosmochim. Acta* 31:587–596 (1967).

Sherman, G. D., and G. Uehara, "The weathering of olivine basalt in Hawaii and its pedogenic significance," *Soil Sci. Soc. Amer. Proc.* 20:337–340 (1956).

Soil Conservation Service, *Soil Survey Laboratory Data and Descriptions for Some Soils of Puerto Rico and the Virgin Islands,* Soil Survey Investigations Report #12, U.S.D.A., Washington, D.C., 1967.

Soil Survey Staff, *Keys to Soil Taxonomy,* 4th ed., SMSS Technical Monograph No. 19, Blacksburg, VA, 1990.

Soubies, F., and R. Gout, "Sur la cristallinité des biotites kaolinitisées des sols ferrallitiques de la région d'Ambalavao," *Cahiers ORSTOM, serie Pédologie* 23(2): 111–121 (1987).

Stoops, G., "Contribution of in situ transformations to the formation of stone-layer complexes in central Africa," *Geo-Eco-Trop,* 11:139–149 (1989).

Strahler, A. N., and A. H. Strahler, *Modern Physical Geography,* John Wiley & Sons, London, 1983.

Sys, C., *Caractérisation Morphologique et Physico-Chimique de Profils Types de l'Afrique Centrale,* Publications I.N.E.A.C., hors série, Bruxelles, Belgium, 1972.

Thomas, M. F., *Tropical Geomorphology,* The Macmillan Press Ltd., London, 1974.

Tsuzuki, Y., "Solubility diagrams for explaining zone sequences in bauxite, kaolin and pyrophyllite-diaspore deposits," *Clays Clay Miner.* 24:297–302 (1967).

Van Breemen, N., and R. Brinkman, "Chemical equilibria and soil formation," in G. H. Bolt and M. G. M. Bruggenwert (eds.), *Soil Chemistry, A. Basic Elements,* Elsevier, Amsterdam, The Netherlands, 1976.

Van Wambeke, A., "Le rapport limon/argile, mesure approximative du stade d'altération des matériaux originels des sols tropicaux," *Comptes-Rendus III° Conférence Interafricaine des Sols.* (Publication CCTA) 50(1):161–167 (1959).

Van Wambeke, A., "Les sols du Rwanda-Burundi," *Pédologie* 11:289–353 (1961).

Van Wambeke, A., "Criteria for classifying soils by age," *J. Soil Sci.* 13(1):124–132 (1962).

Vegetation, Soil Organic Matter, and Crops

Soil scientists have examined soil organic matter from many different perspectives, two of which are the most frequently used in investigations related to agriculture and natural resources management. The first perspective is a *dynamic,* conceptual approach that subdivides organic substances in soils on the basis of their stability and distinguishes several *organic pools* or *compartments,* variously labeled as labile or stable but without much concern about their actual chemical composition. The second perspective is *analytical* and is largely adhered to by chemists who attempt to identify the chemical composition of soil organic matter using various fractionating procedures.

The two viewpoints only partly intersect. On the one hand, the dynamic approach lacks satisfactory chemical identification of the pools. On the other hand, the fractionating procedures are seldom linked to practical land use or soil genesis interpretations that receive general acceptance.

This chapter has two parts. The first is a discussion of the *dynamics* of organic matter contents in tropical soils. The second looks into fractionation and the *chemical* properties of organic substances that have been identified in soils of the tropics.

3.1 Soil Organic Matter Dynamics

Soil organic matter contents are usually expressed as *total organic carbon* (C) or *total nitrogen* (N) percentages of fine earth (less than 2 mm in diameter), which are obtained by specific analytical methods. These are rough estimates that ignore the complex nature of soil organic matter; they include carbon and nitrogen, which are part of finely divided fresh plant residues, charcoal, and living microorganisms. In spite of these limitations, studies of the total organic carbon and nitrogen contents have contributed significantly to the understanding of the processes that govern soil organic matter changes in soils of the tropics.

Interpretations of the soil organic C and N data have often been made without regard for the original reasons for collecting the data, without consistent application of the definitions of the terms that were originally used when the data were reported, or without adequate information on the origin or nature of the analyzed samples. To avoid further confusion, the definitions of the terms that are used in this chapter are briefly explained in the following sections.

Living organisms, such as plants and animals, are the primary sources of soil organic N and C. They produce dead residues, litter or manure, which can be considered a separate organic pool that decomposes into *soil organic matter,* (SOM) and other substances. The SOM is a distinct pool, frequently called *humus,* that eventually may be subdivided into several compartments. It is generally accepted that the SOM is firmly linked to the mineral fraction of the soil and that it no longer has the original cell structure of the primary organic sources.

All dead organic matter (SOM, residues, etc.) is intimately mixed with living microorganisms that actively participate with the mesofauna in various transformation processes, such as the decomposition or the formation of organic pools. In some models the living organisms in the soil are considered to be a separate pool.

The term "addition" in this text is used to designate the amounts of C or N that actually enter one of the organic matter pools. The pool or compartment may be one that is distinguished in the dynamic models or it may be the total of all the organic substances. For example, in the case of SOM the additions (A in the equations that follow) are the quantities of C or N that enter the SOM pool, *not* the total amount of plant residues that are added to the soil; indeed, some parts of these residues may be oxidized and may actually never reach the SOM stage.

The removal of organic substances from the pools is either called *decomposition* or *mineralization.* The losses may also be the result of *erosion* by runoff water. Erosion is seldom adequately taken into account in investigations regarding changes in soil organic matter contents of the tropics. For this reason, many conclusions based exclusively on chemical or biological transformations of organic matter have reduced applicability for real field conditions. The pathways that organic substances may follow in the soil environment are illustrated in Fig. 3.1.

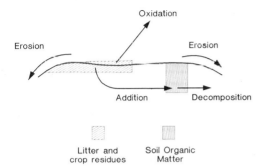

Figure 3.1   Processes affecting the level of soil organic matter.

3.1.1   Basic equations

The changes in soil organic matter contents, expressed here as total nitrogen, have been described mathematically by several authors (Greenland and Nye, 1959; Nye and Greenland, 1960; Laudelout, 1961). Their approaches are summarized in the following equations in which N could be substituted by C without modifying their meaning. The equations state that the change (dN/dt) is simply the difference between additions (A) and removals (kN); the equations may be applied to any of the pools or to the total of organic substances in the domain that is considered in the models.

$$\frac{dN}{dt} = A - kN \qquad (3.1)$$

where dN/dt = change in soil nitrogen per unit time
A = additions of nitrogen to the soil nitrogen pool per unit time
k = fraction of total soil nitrogen that is removed per unit time
N = amount of soil nitrogen

At equilibrium, when no changes occur, A equals kN and

$$N_E = \frac{A}{k} \qquad (3.2)$$

where N_E is the steady-state nitrogen content or its equilibrium level. This equation can be used to predict the amounts of organic matter in soils that are kept under stable ecological conditions such as in mature rain forests, climax savannas, long-term cropping systems, and permanent tree crops.

The integration of Eq. (3.1), with N_0 as initial N content at time 0, results in the following relationship (see Fig. 3.2):

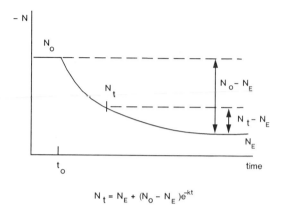

$$N_t = N_E + (N_0 - N_E)e^{-kt}$$

Figure 3.2   Soil organic matter decomposition following a general decay series. [*From Laudelot (1961).*]

$$N_t = N_E + (N_0 - N_E)e^{-kt} \qquad (3.3)$$

where N_t = soil nitrogen content at time t
  N_0 = soil nitrogen content at time zero
  e = basis of natural logarithms
  k = decomposition constant of soil nitrogen, assuming that no removal by erosion takes place

The rearrangement of the Eq. (3.3) permits the calculation of the half-life $(t_{1/2})$ of the $(N_0 - N_E)$ fraction of organic matter. The half-life is the time it takes for one-half of the $(N_0 - N_E)$ organic matter to decay, which is given by:

$$t_{1/2} = \frac{ln\ 2}{k} \qquad \text{or} \qquad t_{1/2} = \frac{0.693}{k} \qquad (3.4)$$

These equations have been used to explain the soil organic matter changes that occur when environmental conditions are modified, e.g., when rain forests are cleared for cropping. As stated previously, the equations may be applied to certain fractions (pools), to all dead organic matter in the soil, or only to the soil organic matter or humus. In this section on dynamics, only two pools are considered: fresh organic matter, mainly plant residues or leaf fall, and SOM. The transformation from the first to the second is called *humification* in Fig. 3.1. The time steps that are considered in the equations applied for tropical soil conditions usually equal one year.

3.1.2   Additions, equilibrium levels, and decomposition constants

The terms of Eqs. (3.1)–(3.4) are used to organize the following discussion of soil organic matter dynamics. The plant biomass production of ecosystems that are at their equilibrium levels (N_E) provides an estimate of the maximum value of the organic matter additions (A). The equilibrium levels themselves give target values for natural resources conservation and can be used to set standards for judging the effects of different types of land use on the sustainability of agricultural systems based on soil resources. The equations are also useful for measuring the consequences of interventions by people when they clear forests and open land for agriculture and livestock and for evaluating the effectiveness of the agronomic practices that they use to maintain or restore adequate organic matter levels in agricultural soils.

Biomass and biomass production.   There is an ample diversity of vegetation types in the tropics, ranging from dense rain forests in the most humid zones to desert shrubs in the arid zones. All of them produce biomass that either returns to the atmosphere by respiration, accumulates in plant tissue, or leaves residue on or in the soil as litter or decaying roots. The physical factors that control the potential to produce biomass and set an upper limit to photosynthesis are solar radiation, the availability of water, i.e., the rainfall and the length of the rainy season, and the supply of plant nutrients.

Only two vegetation types are discussed in this section. The first type, the rain forests, are dominated by trees that close the canopy in such a way that grasses do not receive enough light to survive in the understory. The second type, the savannas, are essentially made of *gramineae* that cover the ground with scattered trees and that may develop into open woodlands, such as *miombos* in Africa or *cerrados* in South America. Roughly speaking, rain forests are the natural vegetation types in the humid tropics where the length of the dry season (months with less than 50 mm of rainfall) does not exceed 2 months; the savannas may occupy regions that have between 3 and 9 dry months.

Tropical rain forests.   The standing biomass of lowland rain forests with little water stress averages 425 Mg/ha dry matter. The lowest values are found in areas with the shortest rainy season, e.g., 239 Mg/ha for a deciduous forest in India (Medina and Klinge, 1983). The figures given by Brown and Lugo (1982) are in the same range. They report that dry matter weights of tropical rain forests vary between 210 and 540 Mg/ha, with values of below-ground biomass carbon between 14 to 132 Mg/ha.

The biomasses of forests in the humid tropics also depend on the age of the vegetation. Saldarriaga et al. (1988) measured the above-ground and root biomasses of slash and burn forest successions in the Amazon basin that ranged from 10-year-old stands to mature forests, i.e., more than 80 years since disturbance. The above-ground living biomass in the sequence increased from 44 to 326 Mg/ha; the root biomass was about one-fifth of these amounts during the whole succession. Leaf biomasses, not counting herbaceous vegetation, did not show significant differences during development and varied between 6 and 10 Mg/ha. The weight of the *physiologically active part* of the biomass, i.e., the leaves, in lowland and montane rain forests averages 8.5 ± 0.9 Mg/ha (Medina and Klinge, 1983).

The annual *net primary production* of biomass dry matter varies between 11 and 20.8 Mg/ha/year (Brown and Lugo, 1982). According to Fournier and Sasson (1983), the average net dry matter production of tropical forests is 20 Mg/ha/year with variations between 9 and 32 Mg/ha/year.

According to these investigations, it appears that there is wide diversity in the characteristics of the tropical rain forests. It is true that the total biomasses of mature stands always exceed the amounts ever found in any other type of vegetation worldwide; however, the dry matter weights of the parts that are active in photosynthesis and respiration do not always reach the maximum values observed elsewhere. The same is true for the net biomass production. For example, a 14-year-old plantation of oil palms *Elaeis guineensis* has a net dry matter production of 37 Mg/ha/year (Westlake, 1963).

The average *litter production* of tropical rain forests is approximately 10 Mg/ha/year dry matter, with values that vary between 5 and 15 Mg/ha/year. There is a relationship (Brown and Lugo, 1982) between the total biomass of a system and its litter production:

$$\text{litter fall} = \frac{\text{biomass} - 23.8}{33.7}$$

The average *life span* of leaves in the lowland rain forests is generally 12 months; in drier environments the life span may be shorter. Litter fall in rain forests is seasonal but does not always coincide with climatic parameters (Medina and Klinge, 1983); most leaves die during the driest season. If there is no rainless period, the leaves fall as soon as they die. If there is a marked dry period (e.g., less than 50 mm of rain per month), the leaves drop at the beginning of the following wet season, being washed down by the heavy rains. Although living leaves may only make up 5 percent of the living biomass, they may contain up to 23.5 percent of the nutrients stored in it. Dead leaves contain much less nutrients (Enright, 1979).

Mature rain forests and the pioneer vegetations that precede them are extremely powerful extractors of plant nutrients from the soil. They store these nutrients in their biomass, most of which is environmentally inactive, and are thus immobilized in stems, branches, and large roots. Table 3.1 sets an order of magnitude of the amounts that accumulate in various mature rain forests; the differences between the data relate to the diversity in soils and the kinds of parent rocks. In the case of Bajo Calima, Colombia, that receives about 8000 mm of rain per year, only traces of calcium and magnesium are left in the top 55 cm of the soils. These data are an illustration of the fragility of most rain forest ecosystems: If for some reason the rain forests are cleared and the ashes that were produced by burning are leached, then most likely the nutrient-deficient soils would be unable to support the growth of the pioneer trees needed to restore the fertility of the soil. This is even more true if erosion, drought, invasion by grasses from savannas, and bush fires make growing conditions more adverse.

The rain forests play a significant role in the conservation of natural resources, stability of climates, and control of the quality of the environment. From a soil science perspective, rain forests provide maximum protection against erosion and therefore conserve watersheds in an ideal way. However, rain forests provide per unit area very few commodities that would justify the acceptance of rain forests as economically valuable assets. In a *short-term* vision, rain forest soils are to be preferred to savanna soils only for the ashes that are released by burning the vegetation and the concentration of organic matter and nutrients in their topsoils.

From a *sustainable* agricultural development perspective, rain forests are no more than enormous masses of weeds that have to be removed before any resourceful land use system can be implemented. This cannot be done without severe risks of environmental deteriora-

TABLE 3.1   Nutrient Storage in Mature Rain Forests

Location	Above-ground biomass, kg/ha					Root biomass, kg/ha					Source
	N	P	K	Ca	Mg	N	P	K	Ca	Mg	
Manaos, Brazil	1428	59	434	424	202	581	7	52	83	55	(1)
Kade, Ghana	1568	106	774	1959	289	214	11	88	146	44	(2)
El Verde, Puerto Rico	1021	59	926	1129	253	27	6	21	29	8	(3)
Bajo Calima, Colombia	1210	36	505	337	386	526	1	128	144	63	(4)

SOURCE: (1) *Klinge et al.* (1975), (2) *Greenland and Kowal* (1960), (3) *Golley* (1978), and (4) *Rodriguez* (1989).

tion. The key issue in the rural development of rain forest areas lies in the selection of sites for each land use alternative, discriminating between locations that are suitable for development and those that are not. The criteria to be used are topography and soil properties on the one hand, and the requirements of the land utilization types on the other hand; the characteristics of the forest itself is not a criterion.

The compromise between the need to produce food and fibers and to protect the rain forest environment may be reached by using only those areas where the chances that the rain forest will return spontaneously if the land is abandoned are maximal. The diversity of the land qualities in the rain forests is such that often some choices are possible. The areas where the prospects for the trees to come back are minimal are better kept under their original forest cover or replaced by tree crops. As a rule of thumb, the avoidance of sandy soils with little water-holding capacity, elimination of convex slopes for annual crops, design of a road system that is adapted to the topography, and selection of soils that in the past have shown a real potential of recovery are some initial, affordable steps in the right direction.

Savannas and woodlands. The *Andropogoneae* Guinean savannas (Lamto savannas) that receive 1300 mm of rain per year and border the semideciduous dense humid forests of Ivory Coast have between 7 and 11 Mg/ha above-ground herbaceous biomass and 10 to 21 Mg/ha roots. The density of the woody species ranges from treeless savannas to stands that cover 45 percent of the surface; their biomass varies accordingly between 7 and 58 Mg/ha above-ground and 3 and 27 Mg/ha below-ground dry matter. Thus, the total biomass of the whole vegetation fluctuates between 31 and 110 Mg/ha (Menaut and Cesar, 1979).

The *Andropogoneae* Lamto savannas produce between 27 and 36 Mg/ha/year total dry matter, of which at least 80 percent is supplied by the herbaceous vegetation. The roots of the herbaceous vegetation each year add between 10 and 19 Mg/ha dry matter to the soil, which corresponds to a minimum of 40 percent of their total net biomass production. The contribution of the above-ground herbaceous vegetation to net biomass production is comparable to the 8 to 16 Mg/ha/year obtained in similar types of tropical savannas elsewhere. Much of the dry matter produced by savannas is lost by fires: Ahn (1970) estimated that in West Africa savannas do not add more than 2.5 Mg/ha/ year.

According to these data, the grasses in the savannas contribute to soil organic matter buildup mostly through their root systems. The trees on the contrary have the highest proportion of their biomass above the ground. In the example of the Lamto savannas the tree roots produce not more than 0.4 Mg/ha/year dry matter. The leaves that fall after the bush fires, however, contribute to the protection of the soil against erosion and direct solar radiation.

The productivity of savannas depends on rainfall. Ohiagu and Wood (1979) found a positive linear relationship between annual rainfall and maximum standing grass biomass in central and west Africa. The regression equation is $MB = 0.0057P - 1.675$, where MB is the dry weight in Mg/ha and P is the annual rainfall in mm.

Soil organic matter contents: Climatic factors. Jordan (1985) reviewed the literature on total nitrogen contents of forests and soils of the tropics. His findings are summarized in Fig. 3.3 where each bar refers to a specific site in a tropical forest. Soils store significantly more nitrogen than the vegetation they support. No meaningful differences are detected between the nitrogen contents of soils of the tropical lowlands and the soils of temperate regions taken as a whole. The variability within these geographic areas is too large to arrive at statistically valid conclusions. Jordan's graph indicates, however, that tropical montane forest soils contain much more organic matter than temperate and lowland tropical forest soils.

Sanchez et al. (1982) compared the organic matter contents of several pedons of the tropical and temperate regions. The mean N content of the top 15 cm of the pedons sampled in tropical soils is significantly higher (0.153 percent) than the mean calculated for pedons from temperate regions (0.123 percent). The same trend exists between the mean N content in the 0- to 100-cm depth section (0.078 versus 0.060 percent). In their set of samples, the Oxisol pedons of the tropics and the Mollisol pedons of high latitudes had the largest average amounts of soil organic matter. The conclusion that can be drawn from this

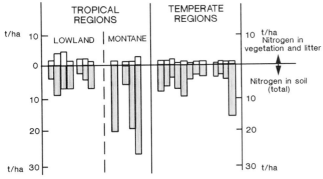

Figure 3.3 Nitrogen contents in biomass and soils of forest ecosystems. [*From Jordan (1985). John Wiley & Sons, Ltd., reprinted by permission.*]

study is that some tropical environments have the potential to produce soils with organic matter contents that are as high as soils of
temperate climates.

The results of this statistical analysis should not be used to compare
the *geographic* extent of soils with high or with low organic matter
contents in each of the climatic zones. To reach conclusions on that
subject, soil maps would be necessary that indicate the area covered
by each kind of soil as well as the degree of erosion that they have
suffered. Eroded soils with low organic matter contents may actually
be very extensive in the tropics, particularly in semiarid regions. The
important issue is whether the acreage of soils with inadequate
amounts of organic matter in some regions is such that food production would be threatened.

Sanchez et al. (1982) compared organic matter contents in grassland
and forest pedons sampled in the lowland tropics. Forest topsoils (0–15
cm) contain significantly more organic C and N than grassland soils
(C: 1.89 versus 1.35 percent; N: 0.182 versus 0.106 percent). The protection that the forest vegetation offers against erosion is certainly
one of the key factors to explain these differences.

The C/N ratios of organic matter in the tropical savanna soils that
were included in the study of Sanchez et al. (1982) are higher than in
rain forest soils (16.0 versus 11.3 in the 0- to 15-cm layer). The opposite relation is true for soils from temperate regions. The grasslands of
high latitudes mostly cover soils developed in dry environments on
loess rich in bases. The bases, particularly calcium, have not been
leached out of the profiles as they have in the more humid forested
parts. Calcium-rich environments produce humus with low C/N ratios.
The majority of the tropical savannas on the contrary occur in acidic
and nutrient-deficient soils, while the topsoils in tropical rain forests
are usually well saturated with bases, having benefited from the action of tree roots to pump up nutrients from the subsoils.

Soils of the humid tropics with a udic moisture regime contain more
organic carbon and nitrogen than soils with an ustic moisture regime
(Sanchez et al., 1982). This difference is probably due to differences in
vegetation, stronger erosion, and more frequent burning of the savannas in regions with a dry season.

Soil organic matter contents: Edaphic factors.   Soil properties influence
organic matter contents. For lowland tropical forests, Jordan (1985)
reported the lowest N contents in a nutrient-poor, coarse, sandy
Spodosol (approximately 800 kg N/ha) and the highest amounts in a
volcanic ash soil containing allophane (20,000 kg N/ha).

Some soil constituents protect organic matter. The influence of clay
is well illustrated in Fig. 3.4, taken from Bennema (1977) who inter-

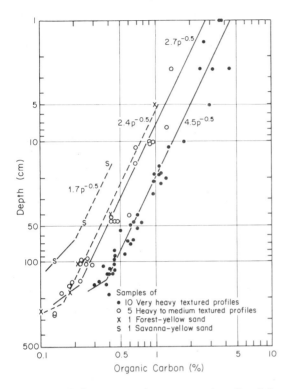

Figure 3.4 Soil organic carbon contents in soils of the Amazon region. [*From Bennema (1977). Academic Press, reprinted by permission.*]

preted carbon data collected in soils of the Amazon basin. Clayey iron oxide rich soils usually contain more organic carbon than light-textured soils, although their color may suggest the opposite. There are other examples: Verdade (1969) reported that clayey soils in São Paulo usually have double the N content of sandy soils. Allophane is known to be the most effective soil constituent to retard the decay of organic matter and cause the accumulation of high amounts of humus in volcanic ash soils.

Decomposition constants. The decomposition rates of soil organic matter are important parameters in models that are intended to predict the sustainability of cropping systems—i.e., in estimating the length of time that suitable physicochemical conditions can be maintained in topsoils or in calculating the amounts of nutrients that organic matter mineralization can supply to plants.

Decomposition constants vary with a number of climatic, biological, and edaphic factors such as the origin of the organic materials, soil temperature, activity of microorganisms, kind of soil fauna, soil moisture conditions, supply of oxygen, and properties of the mineral fraction of the soil. There is not sufficient information available to evaluate the impact of each of these factors.

The rates of mineralization of two soil organic pools, fresh organic residues (i.e., litter) and SOM, are discussed in the following sections.

Litter and crop residues. The decomposition rates of fresh organic matter are temperature dependent. In the lowland tropics the high temperatures accelerate the mineralization of plant residues beyond the rates that prevail in temperate regions.

Jenkinson and Ayanaba (1977) incubated fresh plant residues (*Lolium multiflorum*) for 2 years in polyethylene tubes placed in sandy topsoils in Nigeria (mean annual air temperature of 26.1°C) and found that the half-life of 70 percent of the plant material was 3 to 4 weeks. The remainder had a half-life of approximately 2 years. The data fit a two-compartment model $Y = 70.9\ e^{-11.32t} + 29.1\ e^{-0.35t}$, where Y is the percentage of the added plant residues that remain in the soil after t years. The decomposition constants in the sandy topsoils of Nigeria were *four times* greater than the ones obtained in England.

This model indicates that after 5 years only 5 percent of the original residue input would be left in the soil and could be considered to have *entered* the soil organic matter pool. The actual findings (Ayanaba and Jenkinson, 1990) showed that 8 percent of the original ryegrass carbon remained in the soil after 5 years, thus producing a close fit between the turnover model of organic C and the field observations.

Similar double exponential functions have been used to describe the mineralization of ^{14}C-labeled mature wheat straw in several soils of Costa Rica under different climatic conditions. The temperature regime had the strongest influence on the decomposition rates. The *seasonal distribution* of rainfall was also important. Farming practices and cropping history seemed to affect soil organic matter contents more than soil characteristics (Gonzalez and Sauerbeck, 1982).

Soil moisture contents also influence the decomposition rates; in the climatic conditions prevailing at Ibadan, Nigeria, a faster mineralization takes place in the shade than in the open because the shaded soil remains moist for longer periods, although the soil temperatures are cooler. The experiment also showed that in the 26 initial weeks of mineralization, more organic matter decomposes during the rainy season (77 percent) than during the dry season (59 percent). It should be noted that the incubation method that was used in the experiment does not allow the mesofauna to participate in the decomposition processes (Jenkinson and Ayanaba, 1977).

The sequence of transformations that residues undergo are strongly influenced by the mesofauna. For example, Ohiagu and Wood (1979) observed that the rate of litter disappearance (decomposition plus removal by invertebrates) in savannas is greatest during the dry season: 68.6 percent of the total amount of the grass litter that is lost in 1 year disappears in the 4 months when rainfall is negligible. The losses during that period are mainly due to consumption by fungus-growing termites.

Some soil characteristics influence the rates of chemical decomposition of organic residues, particularly in the early stages of mineralization. For example, soil acidity slows the decomposition of plant residues during the first year, but the differences become less marked later on (Ayanaba and Jenkinson, 1990). The effects of soil acidity, more specifically the toxicity of exchangeable aluminum, explain why litter layers, O-horizons, and root mats under mature rain forests are usually much thicker in acid soils than in profiles with moderate to high base saturations.

Table 3.2 gives leaf decomposition rates and half-lives of leaves at various locations in west Africa.

The decomposition rates of woody parts of litter in forests differ according to their sizes: At Kade, Ghana, twigs decompose at 83.3 percent per year, small wood at 77 percent per year, and medium-size wood at 11 percent per year (John, 1973).

Soil organic matter. Agronomists are most interested in the SOM because they are concerned about soil conservation and the sustainability of agriculture. The SOM confers permanent qualities to the soil and maintains its productive capacity over extended periods of time.

"Humified" soil organic matter resides in the soil for longer periods than fresh residues or litter. There are two reasons for this: The chemical composition of some components is such that they are more resistant to breakdown or some linkages with soil minerals make them more recalcitrant to mineralization.

Temperature is the most important soil property that controls the decomposition constant k of soil organic matter. The rates of chemical

TABLE 3.2   Leaf Decomposition Data from Various West African Locations

Location	Decomposition rate, k	Half life, months
Kade, Ghana	2.5	3.3
Ibadan, Nigeria	2.5	3.3
Olokomeji, Nigeria	5.0	1.7
Omo, Nigeria	2.5	3.3
Banco, Côte d'Ivoire	3.3	2.5
Mokwa, Nigeria	2.6	3.1

SOURCE: *Adapted from Collins (1981).*

decomposition processes increase with increasing temperature, and biological processes follow the same trend, at least in a large part of the soil temperature range.

To a large extent, the evidence for control of SOM decomposition rates by temperature comes from the study of the geographical distribution of soils with high organic matter contents in cool tropical high lands. Jenny et al. (1948) and Jenny (1950, 1980) described this temperature dependency for soils in Colombia and the United States (see Fig. 3.5). The general trends of the temperature influence on soil organic matter contents have since been verified in many other tropical regions.

The effects of soil temperature on decomposition rates and soil organic matter contents are more pronounced in the tropics than in the temperate regions. For example, in the Sierra Nevada of California the doubling of the soil nitrogen content requires a mean annual temperature drop of 14.6°C but only a 7°C decrease in India and only a 5.0°C difference in the Colombian Andes (Jenny, 1980).

Other conclusions that are often overlooked can be drawn from the SOM-temperature relationship. Figure 3.5 shows that at the same mean annual temperature, tropical soils contain *more* organic matter than temperate soils. The difference between the SOM contents illus-

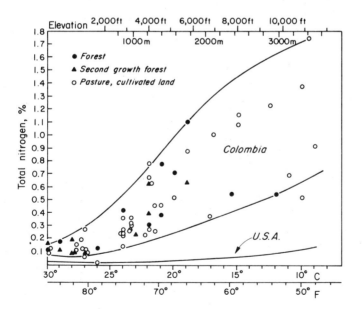

Figure 3.5   Relationship of soil nitrogen, elevation, and annual temperature. [*From Jenny (1948). Williams & Wilkins, reprinted by permission.*]

trated in Fig. 3.5 may be somewhat exaggerated because of the influence of allophane on the soil organic matter contents in the volcanic ash soils in Colombia. Jenny's findings, however, have been confirmed by Alexander and Pichott (1979) in a soil sequence that is little influenced by volcanic ash; the organic carbon content (C) to 1 m depth in kg/m^{-2} can be described by $C = 5.705\ e^{0.584\ A}$, where A is the altitude in m.

It is commonly accepted that short term (daily) fluctuations in soil temperature and soil moisture accelerate the decomposition of soil organic matter; the shrinking of soil materials and cracks would give microorganisms better access to physically protected organic substances. The same effects may be obtained by cultural practices. For example, tillage enhances mineralization. Meyer (1959) observed increased CO_2 and N release from plowed topsoils during approximately 20 days after the first tillage of recently cleared land. On the other hand, turning the organic matter layer under a subsoil material reduced the mineralization rate. The slower decomposition rate also lowered the leaching losses of N.

Soil constituents also influence the decomposition rates. Normally k decreases when the clay, iron oxide, or allophane content in the soil increases. Their effects on the SOM contents have been discussed in previous sections.

Mineralization rates vary among the chemical components of soil organic matter. Mueller-Harvey et al. (1985) computed the decomposition constants of organic C, N, P, and S contained in the 10-cm topsoil of a clayey-skeletal kaolinitic Alfisol under a no-till rotation of soybeans and maize during a period of 22 months after clearing a forest and obtained values of 0.179, 0.193, 0.136, and 0.262, respectively, giving half-lives of 3.5, 3.3, 4.7, and 2.3 years, respectively. The release of N in the 10-cm topsoils exceeded by more than 50 percent the amounts that are exported by the above-ground dry matter of the crops, and in the beginning months of the cropping cycle, substantial amounts of N are lost by leaching. The time that the N supply from these topsoils is sufficient to maintain maize crops that yield 1.5 Mg/ha of grain per season would not exceed 4 years.

3.1.3 Changes in soil organic matter contents

Human interventions that change the environmental conditions may have drastic effects on soil organic matter contents. In many soils, organic matter is one of the major soil attributes that control the sustainability of agricultural systems. Agronomists need to know the outcome of their recommendations on the conservation of organic matter. The following sections describe some predictors of changes in organic matter contents that occur when agroecological conditions are modified by humans. The predictors only take mineralization effects into account but not the erosion processes that may degrade the land so badly that the SOM decomposition models only play a minor role in the agronomic predictions.

When environmental conditions change, for example, because of land clearing or as a result of innovative cultural practices, the equilibrium conditions that maintain the steady-state soil organic matter level are broken and movement starts toward a new equilibrium, which is controlled by the modified ecological conditions. The basic equations have been used to estimate the dynamics of soil organic matter and to calculate decomposition constants. The same approach can be followed to predict the feasibility of soil organic matter restoration techniques in soils that have been misused.

Decreases of soil organic matter contents. Most decreases in soil organic matter take place when land under natural vegetation is cleared for cropping and when the soil is left unprotected against higher temperatures.

Some of the consequences of clearing land on soil organic matter dynamics can be estimated using the $[N_0 - N_E]$ term, which expresses the *N surplus over the equilibrium level* N_E to which the system is moving. It is a measure of the N-supplying power of the soil during the transition period from time t_0 to time t_1. $[N_t - N_E]$ estimates the amounts of N released between time t, and the time when no further decreases in soil organic matter take place.

Two conclusions can be drawn from the model depicted in Fig. 3.2: First, the largest amounts of nitrogen released by soil organic matter are to be expected in the beginning of the new cycle following the disruption of the previous one and the more abrupt the change, the more intensive the mineralization. Second, once the equilibrium level is reached, soil organic matter does not supply anything to the plants *without taking it back the same year*, i.e., at steady state, soils are said not to supply any nitrogen to the crops. Ecosystems at equilibrium or under climax vegetations are not productive in the sense that they operate as closed systems.

The last conclusion needs some qualifications. If at equilibrium the amount of nitrogen that is rotated or turned over the same year is adequate for crops to grow satisfactorily, then the system can be sustained, provided that the exports by harvests are replaced in due time. However, if one of the links in the cycle is broken, for example, by excessive erosion or by leaching of one of the major nutrients, there will be a decline in plant development, less biomass production, insufficient additions of soil organic matter, and loss of sustainability.

The quantity of nutrients (nitrogen and others) that rotates in one year between crops and soils is a critical component of the fertility of a cropping system. Their amounts and their availability depend on the total organic matter content of the soil and the decomposition rate k. This total organic matter content is a long-term function of k and the additions A that depend on the biomass production of the crop that is grown. The biomass production is usually the most critical factor in the maintenance of an adequate soil organic matter level in a cropping system. Biomass production is often limited by soil constraints and cannot exceed the amount set by the most severe limitation.

In the lowland tropics where either phosphorus, calcium, or other cations limit crop growth, the rotating amount of nitrogen plus the amount added by rain and by biological nitrogen fixation are usually sufficient to maintain a production of approximately 600 kg/ha wheat or 1000 kg/ha corn which corresponds roughly to a nitrogen uptake of 30 kg/ha.

However, if no other limiting factors restrict plant growth or reduce biological nitrogen fixation and if no erosion takes place, yields of approximately 3500 kg/ha maize can be obtained for several years (Grove et al., 1980). The *soil* nitrogen supply in this case is about 70 kg N/ha. This indicates that the *potential* nitrogen-supplying power of the soil organic matter is well above the level that is currently accepted in tropical agriculture. However, in order to achieve this potential, all other limitations for crop development have to be removed, as is often the case in nitrogen trials conducted at experiment stations. Thus, experience shows that tropical soils can supply nitrogen for a reasonable but not a maximum yield. It is important to note that the latter results (Grove et al., 1980) were obtained under experimental station conditions in the somewhat cooler tropics on the high plateaus of Brazil close to Brasilia.

Increases in soil organic matter contents.   Many soils have suffered from misuse and have deteriorated as a result of the depletion of nutrients, removal of the topsoil by erosion, or degradation of soil structure. Organic matter is often the critical soil constituent that is needed to restore adequate conditions for root growth.

Soil organic matter contents can be increased by cultural practices that create conditions that either maximize the additions A and/or reduce the decomposition constant k. For example, in shifting cultivation systems, when a secondary forest replaces a clean-weeded crop, the production of biomass by the vegetation and the cooler soil temperatures under the forest canopy act favorably on both A and k. The time needed for the restoration of the soil organic matter depends on the rate at which the forest fallow establishes itself; this rate in turn depends on, e.g., rainfall, soil physical and chemical conditions, topography, and erosion. In low-input systems that do not include amendments, the critical point is that the regrowth of the fallow vegetation is not delayed by unfavorable soil conditions because of the extended use of the land for crop production.

An example of extremely fast restoration of soil organic matter has been given by Smith et al. (1951). A desurfaced soil in Puerto Rico was cropped for 4½ years with tropical kudzu (*Pueraria phaseolides, Benth.*) and molasses grass (*Melinis minutiflora L.*). During that time the total nitrogen content in the surface layers (0–30 cm) increased from 2044 to 3388 kg/ha. The yearly decomposition constant calculated for soils under sugar cane in the same area was 0.07, and on the basis of the available data, the calculated annual additions of nitrogen to the soil organic matter amounted to 500 kg/ha.

Not all soils recover as quickly as the Puerto Rican soil mentioned in the previous paragraph. Some actually never return to their original status. The most frequent reasons for this are the incapacity of the pioneer plants to cover the soil because of low soil fertility, lack of water, uncontrolled burning, or erosion. The examples of nonrecovery are unfortunately more frequent than the examples where appropriate soil management practices maintain or increase soil organic matter contents. In rotations that include fallows the critical factor is the condition of the soil at the end of the cropping cycle. Soil conditions should not be allowed to drop below the level where the natural vegetation has little chance of reestablishing itself quickly. Appropriate crop-to-fallow ratios are given in following sections.

3.1.4 Maintenance of soil organic matter in cropping systems

Soil organic matter contributes substantially to the productivity of land as it is a source of plant nutrients and because it improves the physical conditions of the soil. Low organic matter contents may lead to severe limitations in plant growth and to the deterioration of cropland.

However, many functions of soil organic matter in plant production can be replaced by soil management practices, i.e., fertilizers may correct nutrient deficiencies. Although all amendments increase inputs either in capital or in labor, some systems are economically feasible either in low-input agriculture or capital-intensive enterprises.

Many farming and cropping systems that keep inputs to a minimum have been devised or experimented with. Soil organic matter maintenance is the key issue in these low-input agricultural systems. *It is not necessarily so in high-input technological packages* that almost completely replace the nutrient supplying and storage function of soil organic matter.

Shifting cultivation and natural fallow systems

Forest fallows. Experience has shown that at the end of a cropping cycle in a shifting cultivation rotation, forest fallows are best able to bring soil productivity close to its original level. The theory is that trees operate by deep rooting and pick up nutrients such as Ca, Mg, and K from the subsoil. Unlike grasses, trees contribute to the soil organic matter buildup without severe losses of N by burning. They also offer better protection against erosion.

There is no hard evidence that tree roots actually grow deep enough into subsoils to pump up nutrients. Burnham (1989) in his review mentioned that only few fine tree roots explore deep layers below 3 m depth and that the roots probably only play a minor role in nutrient supply.

The efficiency of a forest fallow depends essentially on its ability to create in a short time vegetation that protects the soil against high temperatures and erosion. Some soils do better than others, and cultivation practices differ in their ability to restore soil organic matter levels. Greenland and Nye (1959) estimated that in traditional shifting cultivation with minimum soil disturbance, a crop-to-fallow time ratio of 1:3 maintains the humus level in Alfisols of forested areas in west Africa at approximately 75 percent of the rain forest equilibrium level. The same authors estimate that the average length of the fallow period in the humid parts of tropical Africa and Asia is 10 years, and for the cropping period it is 2 years. On Oxisols that receive sufficient rain to allow forest regrowth in central Africa, Laudelout (1961) considers that 10 to 15 years fallow following a 3 to 4 year cropping period is adequate for reaching nearly the original organic matter level of the rain forest. Van Wambeke (1974) reported that in Oxisols of high-base status, crop-to-fallow time ratios of 6:15, 5:12, and 5:15 are the most common.

Not all fallow systems succeed in producing the highest soil organic matter increments during the first years as the basic equations would predict. Greenland and Nye (1959) pointed out that some time will usually be required for the natural vegetation to reestablish itself. Aweto's data (1981) rather indicate a decline in soil organic matter during the first 3 years of the fallow. During that time the tree vegetation was not able to occupy the fields on the very sandy topsoils derived from sandstones in Nigeria (Fig. 3.6).

Management practices that favor the regrowth of the forest favor soil organic matter buildup. It is therefore recommended not to end the cropping cycle with a clean-weeded crop. Cassava or bananas, which tolerate mixing with pioneer forest species, are more suitable transitions from crop to fallow. Plant residues that are rich in starch enhance nitrogen fixation and accelerate the restoration process.

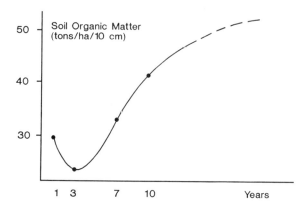

Figure 3.6 Soil organic matter buildup during fallow in evergreen forest in the 10 cm sandy topsoils in Nigeria. [*From Aweto (1981). Blackwell Scientific Publications Ltd., redrawn by permission.*]

Jurion and Henry (1969) reported that isolated trees, and forest strips that border fields, act as seed bearers and perches for birds that propagate trees. Tree stumps left in crops give a quick start to the forest regrowth by sending up shoots.

Grassland and woodland fallows. The natural savanna vegetation in the lowland tropics cannot create conditions that are similar to the forest environment. They consequently do not rebuild soil organic matter to the levels that are characteristic of rain forests. There are many reasons for this. For example, Ahn (1970) estimated that west African savannas probably do not supply more than 2.5 Mg/ha/year plant residues to the soil. This compares very unfavorably with the 5 to 15 Mg/ha/year litter produced by an established rain forest. Grassland usually does not cover the soil as efficiently as forests. Periodic burning also reduces the nitrogen supply to the soil. Natural grasslands and woodlands are therefore much slower than rain forests to rebuild organic matter in cropland that is left under fallow. Charreau (1974) mentioned that the traditional shifting cultivation practice in west Africa included a 10-year cropping cycle followed by about 50-year fallow of wooded savanna or dry forest. He believed that this sequence represented a relatively good balance between crop production and soil fertility.

Some light-textured topsoils in savanna regions suffer from poor aeration and mechanical impedance, which is detrimental to root growth. The role of grassland fallows on the improvement of soil productivity by increasing soil organic matter contents may relate more to the alleviation of physical constraints than to the correction of nutrient deficiencies. The capacity of grass roots to divide soil and improve soil structure is linked to the soil's tendency to crack. Stephens (1960) found that grassland fallows did not significantly improve the structure of soils in Ghana, and if they did, that their effect is not lasting. How much soil organic matter is needed to avoid detrimental hardening[1] and crusting of light-textured topsoils in the tropics is not known. Soils low in clay and iron oxides contents but with kaolinitic mineralogy do not shrink and swell significantly and are therefore the most subject to massive hardening.

Capital-intensive farming systems. The purpose of high-input agriculture is not to restore or maintain soil organic matter to the levels found in natural ecosystems. The objective is rather to obtain high yields or profits without deteriorating the soil physical conditions by reducing the organic matter in the plow layer. Very often, however, the higher inputs just aim at increased productivity without concern for the conservation of the soil resource.

In any cropping system the decomposition constant k and losses due to accelerated erosion put a "ceiling" on the organic matter levels that can be maintained in the topsoil. As long as the amounts of residues returned to the soil are inadequate for producing additions A equal to kN_E, maintenance of soil organic matter cannot be expected, although temporarily higher yields can be produced. For example, if k is 0.1 and the organic matter content to be maintained is 1.5 percent, then the annual amount to enter the organic matter pool of a 20-cm-deep topsoil with a bulk density of 1.2 g/cm^{-3} is 3.6 Mg/ha. If 50 percent of the organic matter would be lost during the transformation of plant residues to soil organic matter, then 7.2 Mg/ha/year dry matter are needed.

Jones (1971) found that annual applications of 7 to 8 Mg/ha farmyard manure[2] are needed to maintain a 1 percent soil organic matter level in sandy topsoils at Samaru, Nigeria. Another way to obtain similar results was to include a 3-year fertilized grass fallow in a 6-year rotation. The apparent retention by the soil of the applied organic carbon varied between 12 and 15 percent during the approximately 20-year duration of the experiment. Orders of magnitude of the amounts of dry matter and nitrogen in green manures are given in Table 3.3.

The effects of green manures on soil organic matter vary considerably with the composition of the organic substances in the soil and in the plant residues. The responses to organic amendments also depend

---

[1]"Hardening" in this text is a temporary increase in consistence of the soil upon drying that disappears when the soil is remoistened. The term should not be understood as irreversible induration into ironstone.

[2]70 percent dry matter of which 55 percent was C.

TABLE 3.3   Examples of Dry Matter and Nitrogen Contents of Green Manures

Type of green manure	Dry matter, Mg/ha	Nitrogen, kg/ha	Reference
Andropogon gayanus	10–16		Singh (1961)
Pennisetum purpureum	32		Singh (1961)
Mucuna	6.6	140	Vino (1063)
Mucuna (tops+roots)	9.3	277	Bowen (1987)
Soybean residue+roots	8.5	112	Bowen (1987)
Corn stover	5–7		Jones (1971)
Pennisetum purpureum (forest region)	27–60		Stephens (1960)
Pennisetum purpureum (savanna region)	6–16		Stephens (1960)

on soil properties such as nutrient deficiencies and toxicities. The incorporation of manures with high energy content may actually increase the activity of the microflora and thus accelerate the decomposition of soil organic matter, or make resistant soil organic components more vulnerable to decay. Agboola and Fayemi (1972) reported on such effects in soils of the rain forest zone of Nigeria; they pointed out the "falsity of the belief that green manures necessarily increase the humus content of the soil."

There is ample evidence that soil management practices that increase the amount of crop residues will also increase the soil organic matter content, *provided* that these practices do not expose the land to higher temperatures that lead to faster decomposition rates and increase the risks of accelerated erosion. The observations by Gokhale (1959) are an example of better conservation of soil organic matter in fertilized tea plantations; the fields that were fertilized maintained the highest soil organic matter levels. There are other examples. Alegre and Cassel (1986) found that they could preserve the original soil organic matter levels of 20-year-old secondary forests in Peru by appropriate fertilization and liming during a 23-month experiment with annual crops. In the unfertilized fields the soil organic carbon content dropped from 1.04 to 0.80 percent during the same period.

3.2   Chemical Composition of Soil Organic Matter

There are only a few studies that report on the chemical composition of organic matter in soils of the tropics that at the same time give due references to the environment and to the soils from which the samples were taken for analysis. It is therefore difficult to make interpretations with regard to soil development and genesis; indeed, the composition of organic matter is strongly dependent on the type of vegetation, kinds of soils, depth of sampling, and cultural practices.

3.2.1 Organic matter fractionation

Some terms need to be defined. In this section soil organic matter does not include the litter that lies on the soil surface; neither are living roots a part of the SOM. The SOM can be subdivided into *living* and *nonliving*. The living comprises the fauna and the microorganisms.

The dead SOM contains a *macroorganic* or "light" fraction that is obtained by sieving (250 μm) or flotation, and *humus*. The latter consists of *humic* and *nonhumic substances*. The nonhumic components are recognizable organic compounds, i.e., carbohydrates, lipids, acids, and proteins.

Humic substances can be fractionated into *humic acids, fulvic acids,* and *humin*. The acids are obtained by alkaline extractions; the solubilized acids in these extracts are separated into humic acids that precipitate by subsequent acid treatments and fulvic acids that remain in solution. The humic substances that are not extractable by the alkaline treatment are called humins. The fractionating methods that are most frequently used are the methods of Dabin (1971) or Schnitzer (1978).

The fulvic acids have in general lower molecular weights than humic acids and are considered the youngest and most reactive. Humic acids are the more condensed form, often linked to the mineral part of the soil and more resistant to microbial decomposition.

Volkoff (1977) studied the distribution of the soil organic matter fractions in an acid yellow Oxisol with low base saturation under a dense rain forest in the state of Bahia in Brazil. In this environment the organic residues at the surface mineralize rapidly and release large amounts of soluble precursors of humic substances. Some of them migrate downwards and react with each other and with the mineral fractions of the horizons. The resulting distribution of organic matter fractions with depth shows that *free* fulvic acids (solubilized by H_3PO_4) become the dominant form of soil organic matter in the subsoil, and contain from 40 to 60 percent of the total organic carbon. The amounts of these fulvic acids correlate well with the clay percentages and with the exchangeable aluminum, suggesting adsorption on the clay and the formation of aluminum organic complexes.

In well-drained soils the humic acid percentages, as well as the amounts of humin, in the soil organic matter tend to decrease with depth. The latter follows the total amount of organic carbon and is, therefore, more abundant in the surface horizons. This is not always the case in poorly drained soils in which there is less migration of the

fulvic acid precursors into the subsoil and, consequently, higher proportions of free fulvic acids in the topsoil. On the other hand, according to Volkoff and Cerri (1981), seasonal fluctuations of the water table promote the synthesis of highly polymerized forms of humic acids
in the subsoil.

Surface samples (0–50 cm) from a yellow latosol (Oxisol) under
dense rain forest in the Amazon region of Brazil contain primarily
humin and free fulvic acids (Volkoff et al., 1982). The organic matter
in spodic horizons that occur within the fluctuation of the water table
in Guyana (Rapaire and Turenne, 1977) are approximately 70 percent
fulvic acids and 6 percent humic acids.

Griffith et al. (1984) compared the composition of organic matter of
Vertisols and Andisols (Andepts) in the Caribbean. In the Vertisols
the fulvic-to-humic acid ratio (FA/HA) was always greater than 0.75;
in the Andepts, on the contrary, FA/HA was consistently less than 0.5.
The variations were mainly due to differences in the humic acid contents. In some calcium-saturated Vertisols only traces of humic acids
could be detected. This was interpreted as an indication of the presence of stable organomineral complexes in the Vertisols. It was also
shown that the elementary composition of purified FA and HA was
similar for all soils; however, the humic acids contained less oxygen
than the fulvic acids. The total acidity in the fulvic acids was greater
than in the humic acids, with the carboxylate groups being more important than the phenolic hydroxyl groups.

3.2.2 Exchange properties

The deprotonation of the carboxylate group (COOH) and of phenolic
and alcoholic hydroxyls (OH) results in negative charges and produces
cation exchange capacity. The charges are pH dependent: Abruña and
Vicente-Chandler (1955) measured an exchange capacity of 260
cmol(+)/kg C at pH 7, which dropped to 65 cmol(+)/kg C at pH 4.5.

The cation exchange capacity of soil organic matter also varies according to age. The younger and most decomposable materials carry
the highest charges per unit carbon. For example, fractions extracted
by flotation in soils of Puerto Rico may have charges as high as 700
cmol(−)/kg C at pH 7 (Abruña and Vicente-Chandler, 1955). The parts
of the soil organic matter that are more resistant to decay on the contrary have very little base exchange capacity.

References

Abruña, F., and J. Vicente-Chandler, "Organic matter activity of some tropical soils of
    Puerto Rico," *J. Agric. Univ. Puerto Rico* 39(2):65–76 (1955).

Ahn, P. M., *West African Soils*. Oxford University Press, London, 1970.

Alegre, J. C., and K. K. Cassel, "Effect of land-clearing methods and postclearing management on aggregate stability and organic carbon content of a soil in the humid tropics," *Soil Sci.* 142(5):289–295 (1986).

Alexander, E. B., and J. Pichott, "Soil organic matter in relation to altitude in equatorial Colombia," *Turrialba* 29(3):183–188 (1979).

Agboola, A. A., and A. A. Fayemi, "Effect of soil management on corn yield and soil nutrients in the rain forest zone of western Nigeria," *Agronomy J.* 64:641–644 (1972).

Aweto, A. O., "Organic matter buildup in fallow soil in a part of S-W Nigeria and its effect on soil properties," *J. Biogeography* 8:67–74 (1981).

Ayanaba, A., and D. S. Jenkinson, "Decomposition of carbon-14 labeled ryegrass and maize under tropical conditions," *Soil Sci. Soc. Amer. J.* 54:112–115 (1990).

Bennema, J., "Soils," in P. de T. Alvim and T. T. Kozlowski (eds.), *Ecophysiology of Tropical Crops*, Academic Press, New York, 1977, pp. 29–56.

Bowen, W. T., *Estimating the Nitrogen Contribution of Legumes to Succeeding Maize on an Oxisol in Brazil*. Thesis (Ph. D.) Cornell University, New York, 1987.

Brown, S., and A. E. Lugo, "The storage and production of organic matter in tropical forests and their role in the global carbon cycle," *Biotropica* 14(3):161–187 (1982).

Burnham, C. P., "Pedological processes and nutrient supply from parent material in tropical soils," in J. Proctor (ed.), *Mineral Nutrients in Tropical Forest and Savanna Ecosystems*, Special Publication #9 of the British Ecological Society, Blackwell Scientific Publications, Oxford, UK, 1989, pp. 27–41.

Charreau, C., *Soils of tropical dry and dry-wet climatic areas of West Africa and their use and management*, Agronomy mimeo 74-26, Department of Agronomy, Cornell University, Ithaca, NY, 1974.

Collins, N. M., "The role of termites in the decomposition of wood and leaf litter in the southern Guinea savanna of Nigeria," *Oecologia* (Berl) 51:309–389 (1981).

Dabin, B., "Etude d'une méthode de fractionnement des matières humiques du sol," *Science du Sol* 1:47–63 (1971).

Enright, N. J., "Litter production and nutrient partitioning in rainforests near Bulolo, Papua New Guinea," *The Malaysian Forester* 42(3):202–207 (1979).

Fournier, F., and A. Sasson, *Ecosystèmes Forestiers Tropicaux d'Afrique. Recherches sur les Ressources Naturelles*, No. 19, Orstom-Unesco, Paris, 1983.

Gokhale, N. G., "Soil nitrogen status under continuous cropping and with manuring in the case of unshaded tea," *Soil Sci.* 87:331–333 (1959).

Golley, F. B., J. T. McGinnis, and R. G. Clements, *Ciclagen de minerais en un Ecosistema de Floresta Tropical*, Pedagogia e Universitaria de São Paolo, São Paolo, Brazil, 1978.

Gonzalez, A., and M. A. Sauerbeck, "Decomposition of ^{14}C-labeled plant residues in different soils and climates of Costa Rica," *Regional Colloquium on Soil Organic Matter Studies*, Promocet, São Paulo, Brazil, 1982, pp. 141–146.

Greenland, D., and J. M. L. Kowal, "Nutrient content of the moist tropical forests of Ghana," *Plant Soil* 12:154–174 (1960).

Greenland, D. J., and P. H. Nye, "Increases in the carbon and nitrogen contents of tropical soils under natural fallows," *J. Soil Sci.* 10:284–299 (1959).

Griffith, S. M., M. B. Holder, and S. Munro, "Chemistry of organic matter colloids in Andepts and Vertisols of the Caribbean," *Trop. Agric.* (*Trinidad*) 61(3):213–220 (1984).

Grove, T. L., K. D. Ritchey, and G. C. Naderman, Jr., "Nitrogen fertilization of maize on an Oxisol of the cerrado of Brazil," *Agronomy J.* 72:261–265 (1980).

Jenkinson, D. S., and A. Ayanaba, "Decomposition of carbon-14 labeled plant material under tropical conditions," *Soil Sci. Soc. Amer. J.* 41:912–915 (1977).

Jenny, H., F. Bingham, and B. Padilla-Saravia, "Nitrogen and organic matter contents of equatorial soils of Colombia, South America," *Soil Sci.* 66:173–186 (1948).

Jenny, H., "Causes of the high nitrogen and organic matter content of certain tropical forest soils," *Soil Sci.* 69:63–69 (1950).

Jenny, H., *The Soil Resource. Origin and Behavior*, Springer-Verlag, New York, 1980.

John, D. M., "Accumulation and decay of litter and net production of forest in tropical West Africa," *Oikos* 24:430–435 (1973).

Jones, M. J., "The maintenance of soil organic matter under continuous cultivation at Samaru, Nigeria," *J. Agric. Sci.* 77:473–482 (1971).

Jordan, C. F., *Nutrient Cycling in Tropical Forest Ecosystems*, John Wiley & Sons, Chichester, 1985.

Jurion, F., and J. Henry, "Can primitive farming be modernised?" *INEAC Hors Serie* (Brussels, Belgium) (1969).

Klinge, H., W. A. Rodriguez, E. Brunig, and E. J. Fittkau, "Biomass and structure in a central amazonian rain forest," in F. B. Golley and E. Medina (eds.), *Tropical Ecological Systems*, Springer-Verlag, New York, 1975, pp. 115–122.

Laudelout, H., *Dynamics of Tropical Soils in Relation to Their Fallowing Techniques*, FAO publication 11266/E, Rome, Italy, 1961.

Medina, E., and H. Klinge, "Productivity of tropical forests and tropical woodlands," in O. L. Lange (ed.), *Physiological Plant Ecology. IV. Ecosystems Processes: Mineral Cycling, Productivity and Man's Influence*, Springer-Verlag, New York, 1983, pp. 318–465.

Menaut, J. C., and J. Cesar, "Savanna structure and primary productivity of Lamto savannas, Ivory Coast," *Ecology* 60(6):1197–1210 (1979).

Meyer, J. A., "Fluctuations de l'N minéral dans les sols sous cultures vivrières," *Comptes Rendus III⁰ Conférence Interafricaine des Sols*, Dalaba, CCTA Publications Bureau, Watergate House, London, 1959, 1:517–524.

Modenesi, M. C., E. Matsui, and B. Volkoff, "Relaçao $^{13}C/^{12}C$ nos horizontes humíferos superficiais e nos horizontes escuros profundos dos solos de campo e mata da região de Campos do Jordão, São Paulo, Brasil," *Regional Colloquium on Soil Organic Matter Studies*, Promocet, São Paulo, Brazil, 1982, pp. 155–160.

Mueller-Harvey, I., A. S. R. Juo, and A. Wild, "Soil organic C, N, S and P after forest clearance in Nigeria: mineralization rates and spatial variability," *J. Soil Sci.* 36: 585–591 (1985).

Nye, P. J., and D. J. Greenland, *The Soil under Shifting Cultivation*, Technical Communication #51, Commonwealth Bureau of Soils, Harpenden, England, 1960.

Ohiagu, C. E., and T. G. Wood, "Grass production and decomposition in southern Guinea Savanna, Nigeria," *Oecologia* (Berl) 40:155–165 (1979).

Repaire, J. L., and J. F. Turenne, "Mesure d'activité spécifique de fractions de matière organique appliquées à l'étude de l'évolution des sols de Guyane," in *Soil Organic Matter Studies*, vol. 2, International Atomic Energy Agency, Vienna, 1977, pp. 179–186.

Rodriguez, J. L. V. A., *Consideraciones Sobre la Biomasa, Composición Química y Dinámica del Bosque Pluvial Tropical de Colonias Bajas. Bajo Calima, Buenaventura, Colombia*, CONIF, Serie documentación #16, Bogotá, D.E. Colombia, 1989.

Saldarriaga, J. G., D. C. West, M. L. Tharp, and C. Uhl, "Long-term chronosequence of forest succession in the upper Rio Negro of Colombia and Venezuela," *J. Ecology* 76: 938–958 (1988).

Sanchez, P., *Properties and Management of Soils in the Tropics*, John Wiley & Sons, New York, 1976.

Sanchez, P., M. P. Gichuru, and L. B. Katz, "Organic matter in major soils of the tropical and temperate regions," *Transactions Xth International Congress of Soil Science* (New-Delhi, India), vol. 1–3:99–113 (1982).

Scharpenseel, H. W., "Organic matter characteristics," in M. Latham (ed.), *Land Development—Management of Acid Soils*, IBSRAM Proceedings #4, Bangkok, Thailand, 1987, pp. 83–100.

Schnitzer, M., "Humic substances: Chemistry and reactions," in M. Schnitzer and S. U. Kahn (eds.), *Soil Organic Matter*, Elsevier, 1978, pp. 1–64.

Singh, K., "Value of bush, grass or legume fallow in Ghana, 1961," *J. Sci. Food Agric.* 12:160–168 (1961).

Smith, R. M., G. Samuels, and C. F. Cernuda, "Organic matter and nitrogen buildups in some Puerto Rican soil profiles," *Soil Sci.* 72:409–427 (1951).

Stephens, D., "Three rotation experiments with grass fallows and fertilizers," *Empire J. Exper. Agric.* 28:165–178 (1960).

Verdade, da Costa, F., "Importance of nonsymbiotic organisms in the nitrogen economy of tropical soils," in *Biology and Ecology of Nitrogen*, Proceedings of a conference, National Academy of Sciences, Washington, D.C., 1969, pp. 129–152.

Vine, H., "Experiments on the maintenance of soil fertility at Ibadan, Nigeria, 1922–51," *Empire J. Exper. Agric.* 21(82):65–85 (1953).

Van Wambeke, A., *Management Properties of Ferralsols*, FAO Soil Bulletin #23, Rome, Italy, 1974.

Volkoff, B., "La matière organique des sols ferrallitiques du nordeste du Brésil," *Cah. ORSTOM, série Pédologie*, 15(3):275–290 (1977).

Volkoff, B., and C. C. Cerri, "Humus em solos da floresta amazónica na região do rio Madeira," *Revista Brasileira de Ciencia do Solo.* 5:15–21 (1981).

Volkoff, B., E. Matsui, and C. C. Cerri, "Discriminação isotópica do carbono nos humus de latossolo e podzol da região Amazónica do Brasil," *Regional Colloquium on Soil Organic Matter Studies*, Promocet, São Paulo, Brazil, 1982, pp. 147–153.

Westlake, D. F., "Comparisons of plant productivity," *Biological Review* 38:385–425 (1963).

4

Soil Horizon
Formation

Horizon formation results from the migration of soil constituents from one part of the profile to another without displacement of the parent materials as a whole and also includes the addition of substances such as organic matter or the accumulation of solutes contained in groundwater. Horizons are also recognized because of differences in structure or consistency between horizontal sections of a soil profile. The formation of horizons is often called *pedogenesis* in contrast with *geogenesis,* which was discussed in Chap. 2.

It is not always possible to make a clear distinction between geogenesis and pedogenesis. Some horizons develop by differential weathering as is the case in deep-weathering profiles. The differentiation of a profile into horizontal soil sections may also be the result of the sorting of particles by insects or by sediment transport. It is often difficult to distinguish between the two processes. Geogenesis and pedogenesis are not mutually exclusive, and these concepts are only used here to subdivide and organize the subject matter.

Not all horizons are discussed in this chapter. The diagnostic horizons of the major soil orders are described in the chapters of Part 2 that deal with these taxa. This is the case for the cambic, oxic, and kandic horizons. The intent here is to focus on soil features that are not diagnostic at the order level but are nevertheless important for land use and for understanding the genesis of the soil horizons that frequently occur in tropical environments.

The chapter is divided into two sections. The first section discusses the surface horizons or *epipedons* that form by the accumulation of organic substances on top of or in the upper part of the mineral soil. The second section deals with subsurface horizons.

4.1   Epipedons or Surface Horizons

Epipedons are horizons that form or that have been formed at the surface. All except two of them (some ochric and albic horizons) are darkened appreciably by organic matter.

The criteria that distinguish the epipedons are primarily their organic carbon contents, thicknesses, and base saturations. *Soil Taxonomy* (Soil Survey Staff, 1975) recognizes several classes, ranking them from the most to the least strongly developed as follows: the *histic, mollic,* and *umbric* epipedons (the latter two only differ by their base saturations), and finally the *ochric* epipedon.

4.1.1   The histic epipedon

The horizons that meet the definition of histic epipedon (Soil Survey Staff, 1975) should have *organic soil materials* to a depth of at least 20 cm. In addition they should be saturated with water for at least 1 month in a year. Organic soil materials are materials that contain between 12 and 18 percent organic carbon depending on the clay content of the mineral part of the horizon. When plowed, the materials contain between 8 and 16 percent organic carbon over a thickness of 25 cm. Organic soil materials include peats and mucks of earlier classification systems (see Appendix A).

Histic epipedons are only diagnostic for mineral soils. As soon as the organic soil materials exceed 40-cm thickness, *Soil Taxonomy* considers the soil to be an organic soil. For this reason, organic soil layers thicker than 40 cm are not included in the definition of the histic epipedon.

In the lowland tropics, high accumulations of organic carbon, as required by the definition of the histic epipedon, only form in places that are almost permanently saturated with water. They are then generally found as topsoils in rain forest swamps. In the isohyperthermic environments the organic matter decomposition rates in aerobic conditions are too fast to allow thick organic surface horizons to form in better drained soils. The acidity of the soil water may modify this general rule, and aluminum toxicities may be instrumental in the occurrence of histic epipedons in soils that are only periodically saturated with water.

In the cooler tropics that have soil temperature regimes that are isothermic or colder, the histic epipedons are generally found as transitions between upland soils and swamps that are characterized by soils that are entirely organic, named *Histosols*. Drainage class, soil temperature regime, and the length of the rainy season interact to control the distribution of soils with a histic epipedon.

4.1.2 The mollic epipedon

Mollic epipedons are surface horizons that typically form by the decomposition of the roots of gramineae that grow in parent materials well saturated with bases. The loess sediments of temperate regions under steppe vegetation and with cold winters are the classic examples of environments where mollic epipedons predominate.

There are only a very few locations in the tropics where conditions are favorable for the formation of mollic epipedons. Either the temperatures over the year are too high to conserve adequate amounts of humus in the soil or the rainfall in the cooler tropics is too high to avoid the leaching of bases.

Some valleys at high altitude such as the Altiplanos of the Andes Cordillera in South America may contain young sediments that are rich in weatherable minerals and have alternating dry and rainy seasons. Alluvial flats in these regions are often characterized by soils with mollic epipedons. This is particularly the case in regions that have been influenced by mafic volcanic ashes. The mollic epipedon is used to distinguish mollic subgroups in Ustands and Ustivitrands.

Other occurrences of mollic epipedons include the surface horizons of profiles that are developed on soft limestones or marls. In the semiarid tropics some concave or linear slopes at the foot of limestone hills or basaltic or other mafic rock outcrops may also contain soils with mollic epipedons. In many instances the finely structured self-mulching *Black cracking clay soils* or *Black cotton soils* that are grouped with the Vertisols order in *Soil Taxonomy* have mollic epipedons.

The definition of the mollic epipedon and of the umbric epipedon are similar. The only difference between the two diagnostic horizons is that the umbric epipedon should have a base saturation of less than 50 percent (by ammonium acetate, NH_4OAc). The base saturation of the mollic is always at least 50 percent. An abbreviated definition of an umbric horizon is given in the following section.

4.1.3 The umbric epipedon

Definition. The umbric epipedon is a surface horizon that does not dry into a hard or very hard massive soil material.[1] In the tropics umbric epipedons have the following properties:

1. Moist and dry, broken and crushed soil fragments have darker colors than underlying C horizons. *Soil Taxonomy* (Soil Survey Staff,

---

[1]In the case of the umbric epipedon, this restriction is only valid for epipedons that become dry under field conditions.

1975) gives precise color chart notations to evaluate the color criteria.

2. A base saturation of less than 50 percent by the NH_4OAc method.

3. An organic carbon content of at least 0.6 percent.

4. A thickness of more than 25 cm in soils that have a solum that is thicker than 75 cm; in other soils with loamy or clayey epipedons, umbric epipedons should be thicker than one-third of the solum and at least 18 cm deep. In coarse-textured soils the required depth is 25 cm.

Occurrence. Umbric epipedons are characteristic for the cool tropics at high elevations. They occur under forests as well as in grasslands. Most of the soils are characterized by isothermic or colder temperature regimes. Soils with umbric epipedons are distributed over broad geographical regions, and the distribution is not spotty as is the case for mollic epipedon. The mineral part of the soil may influence the thickness of the umbric epipedon. For example, allophane and other amorphous clays usually produce darker and deeper surface horizons than phyllosilicates. The slope shapes, more specifically their concavity or convexity, also influence the thickness of the epipedon.

The aeration of the surface layers is another factor that controls the formation of umbric epipedons. Many isohyperthermic soils that remain saturated with water during most of the year have umbric epipedons.

4.1.4   The ochric epipedon

The ochric epipedon has a default definition. All horizons that do not meet the requirements of the histic, mollic, or umbric epipedon, and present soil structure rather than rock structure or thin bedding, are named ochric.

The ochric epipedon is the typical epipedon of the lowland, well-drained soils of the tropics. Either the rapid decomposition of soil organic matter, the deficient production of biomass because of drought conditions, the truncation by erosion, or the indirect effects of bush fires prevent stronger expressions of surface horizons than the ochric epipedon. Exceptions to this general rule are caused by clay minerals such as allophane that retard organic matter decomposition or by temporary anaerobic conditions in surface horizons.

4.2   Subsurface Horizons

4.2.1   Plinthite

Definition. Plinthite is defined in *Soil Taxonomy* (Soil Survey Staff, 1975) as "an iron-rich, humus-poor mixture of clay with quartz and other dilutents. It commonly occurs as dark red mottles, which usually are in platy, polygonal, or reticulate patterns. Plinthite changes irreversibly to an ironstone hardpan or to irregular aggregates on exposure to repeated wetting and drying, especially if it is exposed also to heat from the sun." This definition was written to cover the original concept of laterite (from the Latin *later*, meaning brick) that was first described by Buchanan (1807).

In a moist soil, plinthite is soft enough that it can be cut with a spade. After irreversible induration it is no longer considered plinthite but is called *ironstone* or *petroplinthite* (Sys, 1968).

Genesis of plinthite. The formation of plinthite is a result of an accumulation of iron hydroxides, which cause the cementation of soil constituents into a material that indurates into ironstone when exposed to repeated drying. The enrichment in Fe may in theory be the result of an in situ *relative* or residual accumulation during desilication or it may be the product of *absolute* accumulation by direct import of Fe from outside (D'hoore, 1954). It is commonly accepted, however, that most plinthites form by absolute accumulation.

The formation of plinthite requires a large supply of Fe in solution. The main source of this Fe is groundwater that either receives it from the weathering of primary minerals containing bivalent Fe^{2+}, or, less likely, that obtains it from the reduction of ferric oxyhydrates at low enough oxidation-reduction potentials with organic matter acting as an electron donor.

Groundwater alone is not sufficient to explain the origin of plinthite. Figure 4.1 illustrates a model that assumes a mechanism of iron accumulation that includes the lateral movement of groundwater constantly feeding the zone of enrichment with Fe^{2+} (part 1 of Fig. 4.1). The most favorable conditions for the accumulation of iron hydroxides occur in lowlands that border highlands where weathering saprolites release Fe^{2+}.

In the depressions Fe^{2+} is trapped at the contact of the groundwater with soil air by oxidation and precipitation as Fe^{3+}. The profile section where the water table fluctuates between the dry and the rainy seasons offers the most favorable environment for Fe accumulation. Fe is brought into the soil layers when the groundwater rises. Fe is precipitated as the water table drops and air enters into the layers that were supplied with Fe during the preceding wet season (part 2 of Fig. 4.1). When this mechanism is repeated year after year, it results in large concentrations of iron oxides that correspond to plinthite. Most of them are underlain by permanent groundwater from which frequently wells spring just under the lower limit of the plinthic layers. According to Daniels et al. (1971) plinthite will begin to form if the horizon is saturated with water between 50 and 75 percent of the time.

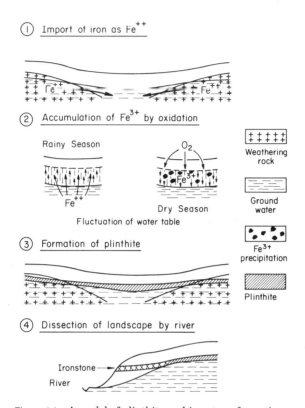

Figure 4.1   A model of plinthite and ironstone formation.

Plinthite may form in any kind of material but is seldom found in heavy clays. Plinthite may develop in saprolites or alluvium. The smaller the specific surface of the impregnated material, the more susceptible it is for cementation and induration upon drying. Plinthite formation is also favored by high Fe contents in the parent materials, and mafic rocks are more likely to provide the necessary amounts of Fe needed to produce plinthite than felsic rocks. Most clay minerals in plinthites are kaolinitic.

Kinds of plinthite.   There are many forms of plinthite. Some of them relate to the kinds of parent materials, more specifically the nature of the embedded gravel or sand size particles. In certain cases they may be ironstone gravel layers derived from the disintegration of plinthites that were formed during earlier geomorphic cycles.

Many plinthites display vesicular patterns that show similarities with the networks of galleries that connect termite nests with water-saturated layers where these insects find water and clay (De Barros

Machado, 1983). The mesofauna may, thus, contribute to the shaping of certain morphological features in plinthites. For example, walls of insect galleries in contact with surface air offer favorable conditions for the precipitation of Fe^{2+} that is contained in the groundwater. Other kinds of plinthite are defined by their degree of development, such as the thickness of the horizon and its continuity. Further subdivisions can be made on the basis of whether the potential for induration affects the whole horizon (*carapace* in the French literature) and would form a continuous crust or whether it is restricted to certain parts of the horizon.

Products derived from plinthite: Ironstone. The most common induration product of plinthite is ironstone, which may form continuous hardpans that are very resistant to erosion and disintegration. The induration assumes the permanent fall of the water table as a consequence of the dissection of the landscape by rivers (part 4 of Fig. 4.1). The geomorphology is usually inverted by this process, the ironstone crusts occupying the highest parts of the topography above the level of the dissecting rivers. Many surfaces in the tropics are marked by ironstone levels, variously called laterites (Sivarajasingham et al., 1962), cuirasses (Maignien, 1958), or ferricretes. McFarlane (1976) described many forms of ironstones and discussed in detail their position in the landscape.

Ironstone nodules contain more iron oxides than soft nodules from the same site (Comerma et al., 1977). For each situation there is a critical iron content that has to be reached to achieve induration upon repeated wetting and drying. D'hoore (1954) considered that no hardening takes place as long as the free Fe_2O_3 content is less than 12 percent of the clay content. Micromorphological examinations show that ironstones contain quartz as skeletal grains embedded in goethite cutans. These goethans are formed by a closely packed parallel arrangement of tubular crystals perpendicular to the walls of the external surfaces (Comerma et al., 1977). The crystallinity of goethite apparently is the property that best correlates with the hardness of ironstones. Slightly hardened concentrations of goethite show less crystallinity; the soft materials have no goethans but isotropic sesquioxide coatings. Goethite is not the only iron oxide mineral that participates in plinthite or ironstone formation. There are ironstones that contain mostly hematite.

Ironstone hardpans may disintegrate into gravel that accumulates in layers on pediment surfaces or is incorporated in stone lines. Rounding of the gravel during sediment transport followed by recementation and induration results in the formation of pisolithic ironstones. In such compound structures, goethite is found to be the major cementing agent; an example from peninsular Malaysia is reported by Debaveye and De Dapper (1987). Not all pisoliths however originate from ironstone gravel but may form directly as concretions in soils.

4.2.2 Clay increase horizons

Textural differentiations in profiles have received several names in soil classification systems and in horizon designations. The most common names are B_t *horizons* for materials that have accumulated silicate clays. *Argillic, kandic* and *natric horizons* (Soil Survey Staff, 1987), and *argic horizons* (FAO, 1988) are all terms that indicate increases in clay contents with depth. They may or may not be associated with decreases in clay in the lower part of the profile. When the clay migration produces a clay maximum or *clay bulge,* it is called *lessivage.* When no decrease in clay can be observed in the lower part of the profile, the process is known as *appauvrissement* (impoverishment) by French pedologists.

Several processes, none of which is mutually exclusive, may be at the origin of clay increase subsurface horizons:

1. The vertical downward translocation of clay within profiles with *or without* its accumulation in an illuvial horizon

2. The selective removal of the finest fraction from the topsoil either by erosion or the mesofauna

3. The destruction of clays in the topsoil as a result of ferrolysis, a process by which seasonal flooding creates strongly acid conditions in the surface layers that affect the stability of the clay minerals (Brinkman, 1970)

Lithological discontinuities may also be responsible for textural changes.

There is much evidence of clay migration in soils of the tropics. For example, clay cutans that result from the deposition of clay particles on ped surfaces or in voids have been reported in many instances. The present section mainly discusses clay migration in soils dominated by low-activity clays that are not affected by high exchangeable sodium percentages.

Clays migrate in soils as colloidal suspensions in water. A prerequisite for the translocation is *dispersion,* which depends either on the thickness of the double layer around the suspended particles or their protection against flocculation by organic substances. The double layer is a cloud of ions attracted by oppositely charged particles; the thicker it is, the more it prevents particles from coming close enough to each other for van der Waals attraction to cause flocculation.

The low-activity clays have a small permanent negative charge that in most cases is complemented by a highly variable, pH-dependent charge. The clays are therefore not likely to attract double layers that are thick enough to form stable suspensions, especially in very dilute solutions; hence, this provides the theoretical basis of the hypothesis that low-activity clay soils are less affected by illuviation than soils that are dominated by 2:1 clays. This does not mean, however, that clay migration is absent in materials dominated by low-activity clays. Micromorphological evidence shows the contrary.

The processes that led to the formation of clay increase horizons in soils of the tropics may have taken place under environmental conditions that are different from the present ones. Dijkerman and Miedema (1988), for example, suggested that in Sierra Leone, illuviation was more active during a dry period of Pleistocene times (20,000 to 12,500 years before present) than under the actual, more humid conditions. Their interpretation is based on micromorphological examination of thin sections.

The *intensity* of clay migration (amount of moving clay per unit time and unit distance) depend on several factors: the amount of clay that is present in the original material, the charge characteristics of the clay, the available energy to detach and disperse the clay, a vector (water) to carry the clay suspension, and, as stated before, a mechanism to protect the clay against flocculation. The amount of clay that actually *accumulates* in an illuvial horizon depends on the intensity of the clay migration as well as on the rate at which this intensity decreases with depth as conditions for migration become less favorable. Finally, the difference in clay content between the eluvial and the illuvial horizon varies according to the original texture and thickness of the layers that receive the moving clay. The mathematical modeling of these processes has been developed by Van Wambeke (1972, 1976). Medium-textured materials are more likely to more strongly show the effects of clay migration than extremely sandy or clayey sediments.

Appauvrissement. Roose (1981) measured the amounts of solids in suspension that percolate through soils of west Africa, where annual rainfall varies between 800 and 2150 mm and the length of the dry season ranges from 3 to 8.5 months. The soil clays in the region under study are dominantly kaolinitic. The amounts of solids in the percolating water varied between 20 and 300 mg/L, the lowest concentrations being found at the highest water fluxes. The solids in suspension were mainly quartz and kaolinite particles coated by organic matter; the source of these materials was thought to be the surface layers where the colloidal fractions were detached by raindrop impact. Tessens (1984) considered that the open structure of fulvic acids with voids of 0.2 μm (Schnitzer and Kodama, 1975) are able to trap very fine clay and protect it against flocculation.

The detachment of clay from soil aggregates in surface horizons by raindrop impact seems to be the key process that initiates the formation of light-textured topsoils in low-activity clay soils. The detached clay in suspension is in part carried downslope by runoff water leaving behind coarser sand and silt particles, which form a thin protective mulch on the surface. This mulch would prevent further detachment of clay if the mesofauna would not constantly mix topsoil materials and bring up fresh soil aggregates. The mesofauna, thus, contributes considerably to the deepening of the lighter textured topsoil.

The mechanism that is described above does not necessarily lead to an enrichment of translocated clay in the subsoil. It only explains the formation of a lighter textured surface or, in French terminology, to *appauvrissement* or impoverishment. Some suspended clay, however, does penetrate into the lower horizons with percolating water. Roose (1981) distinguished between migration of clay through macropores and the illuviation (enrichment) in micropores.

Translocation through macropores. According to Roose's theory (1981), clay in macropores is continuously pushed down by infiltrating water during each heavy rainstorm. Macropores act like wide drainage channels: i.e., as long as they are present, water passes through them without "unloading" the clay they carry. If they do unload the clay, it does not stick to the walls of the pores and is mobilized again by the following rain. When no lower limit of the macropores exists within the depth of the profile and if there is a continuing downward movement of percolating water, no clay accumulates in the soil unless physicochemical conditions change drastically. At large migration intensities, the soils lose clay, and the losses are the highest in the surface horizons. According to this theory, and starting from a uniform parent material, the clay distribution with depth shows a gradual increase in clay from the lowest clay percentages in the topsoil to the highest clay content in the subsoil. Many low-activity clay soils under tropical rain forests have such gradual textural transitions.

Actual clay enrichment by translocation through macropores is only possible at the depth where the macropores become narrower. In soils with perudic moisture regimes, tree roots may maintain macropores to a great depth for considerable time, and clay illuviation only shows at lithological discontinuities or at abrupt changes in structure or texture. Many profiles contain the thickest, not necessarily the most frequent, clay cutans just under the zone of major root penetration and biological activity or in the upper part of the underlying saprolite.

Illuviation in drying micropores. Micropores adsorb clay suspensions slowly and trap clay when the soil dries out as a result of evapotranspiration. The clay that is deposited in micropores is difficult to remove mechanically and for this reason accumulates in the profile at the depth where the downward movement of water is retarded or stops completely. Seasonal drying of moist soils also favors cracking and facilitates the penetration of water into the subsoil. The shallower the moistening of the soil, the shallower the illuvial horizon and the stronger the textural differentiation. The occurrence of pronounced textural changes in soils of the tropics increases as the dry seasons become longer or more severe. The drier the climate, the sharper the textural contrasts and the shallower the upper boundary of the textural B horizon. The physicochemical properties of the horizon that accumulates the clay may be different from the ones that prevail in the surface horizons, and this may enhance the enrichment in clay.

Physicochemical flocculation. Clays that are transported as dispersed particles may flocculate because of changes in the physicochemical conditions that maintain the suspension. For example, a simple increase in the soil-to-water ratio of a suspension may cause its flocculation.

The mobility of clays in soil horizons varies significantly. Topsoils usually contain more water-dispersible clay than subsoils. Gombeer (1977) found that the mobility of clay suspensions extracted from Alfisols of Brazil drops from very highly mobile in the A horizons to almost nil in the illuvial B horizons; the analytical procedures that measured the amounts of mobile clays in soils at moisture contents that occur under field conditions did not detect dispersed clay in the B horizons. In the same instance, horizons that have the lowest clay mobility show under the microscope the most developed and the best oriented clay cutans.

The accumulation of illuvial clay in many cases implies a reduction in the mobility of the clay during migration and, consequently, variations in the physicochemical conditions of the soil solution. However, it is not always possible to identify the primary cause of the illuviation and to decide whether mechanisms other than changes in physicochemical conditions are involved.

Tessens (1984) argued that only the physical deposition of dispersed phyllosilicate particles (this section) produces optically oriented clay skins and that the absence of the argillans in many soils of the tropics does not necessarily mean that clay accumulation by physicochemical flocculation has to be excluded. Theoretically, soil clays with high proportions of pH-dependent charges may flocculate edge-to-face in subsurface horizons as the ionic strength of the soil solution decreases to values below 0.01 M (Tessens, 1984) or where the organic substances that protect the mobile clay decompose.

The additions by flocculation may take different forms. Some of them are described as birefringent circular walls around micropeds (Embrechts and Stoops, 1987). These coatings may either remain indefinitely in the illuvial horizon or be translocated again when favorable conditions for dispersion are restored. Most of the clays around the micropeds are water dispersible and contain less dithionite-extractable Fe and Al than the core of the microaggregates (Embrechts and Stoops, 1987).

The routine analyses of what is called water-dispersible clays are difficult to interpret. There are, for example, no data available on the organic carbon content of the water-dispersible clays extracted by end-over-end shaking of untreated samples.

4.2.3 Sombric horizons

Sombric horizons were originally defined as horizons that have less than 50 percent base saturation by NH_4OAc and that contain illuvial humus not associated with aluminum. According to the definition, sombric horizons do not underlie albic horizons. They may have formed in argillic, cambic, or oxic horizons. They are restricted to the soils of the cool high plateaus and mountains of the tropics that usually have isothermic or colder temperature regimes (Soil Survey Staff,

Figure 4.2 Exposure showing subsurface organic matter accumulations in soils with isothermic temperature regime in Rwanda.

1975). The actual definition of the sombric horizon has not changed (Soil Survey Staff, 1990), although the absence of any association of illuvial humus with aluminum has not always been verified.

The most frequent occurrences of sombric horizons are at altitudes between 1400 and 3000 m above sea level. Their moisture regime is udic, seldom ustic. The presence of dark subsurface horizons at high altitudes in central Africa was first mentioned by Kellogg and Davol (1949). Ruhe and Cady (1954) recognized Dark-Horizon Latosols that were characterized by the concentration of organic matter in a B_2 horizon. Frankart (1983) described several kinds of sombric horizons in Rwanda and Burundi. The presence of dark subsurface horizons has been reported in other high-altitude regions of the tropics.

There are many, often conflicting theories about the genesis of dark horizons. This is not surprising because the environment in which they occur is very diverse with regard to topography, age, sediment transport mechanisms, volcanic activity, and human intervention. On the other hand, dark horizons have different modes of occurrences (Figs. 4.2 and 4.3) and have either not been strictly defined or the definitions are not rigorously applied by pedologists. The controversy is whether the sombric horizon is an illuvial horizon or a buried surface layer.

The unifying attribute of sombric horizons is their exclusive geographic location at high altitudes in the tropics. There are no major

Figure 4.3   Toposequence of soils with sombric horizons at high elevation in Rwanda.

extensions of dark horizon soils in the lowland tropics. These facts explain the hypothesis that climate more than sediment transport is responsible for the presence of dark horizons and the theory that they form by migration and accumulation of organic matter in the subsoil. Low soil temperatures, however, are also favorable for the conservation of buried surface horizons, and the zonal distribution of sombric horizons does not completely exclude valid explanations of their genesis by colluviation of sediments on top of older surface horizons. In addition, many areas where sombric horizons occur have been covered by volcanic ashes that may spread uniform blankets over the landscapes.

There are examples of dark horizon profiles where there is no evidence of lithological discontinuities. The data of Table 4.1 summarize the most important chemical characteristics of a Sombrihumult sampled in Rwanda (NSSL, 1985) at 1900 m altitude. The soil developed in materials derived from granite. The dark subsurface horizon in this profile meets the chemical criteria for spodic horizons because more than one-half of the Al plus Fe that is extractable by dithionite-citrate is also extractable by Na pyrophosphates [see (5) in Table 4.1]. However, not all sombric horizons have this characteristic.

There is micromorphological evidence that at least some of the organic matter in sombric horizons is illuvial because organic matter illuviation cutans often alternate with clay cutans in the same coatings on the surfaces of peds in sombric horizons. The sequence in these multilayer cutans indicates that the organic matter accumulation took place in the past and is not an actually dominant soil-forming process (Mutwewingabo, 1989).

The organic carbon extracted from sombric horizons is approximately 12,000 years old, a time that corresponds to the Late Pleistocene–Holocene transition. During that period the climatic conditions were

TABLE 4.1 Sombric Horizon (Bth) Data of a Sombrihumult in Rwanda

Horizon	Depth, cm	Organic C,%	Clay, %	Fe, % (1)*	(2)†	Al, % (3)*	(4)†	(5)‡
A	15	1.66	33.2	1.2	0.5	0.3	0.3	0.5
Bo1	40	1.26	34.4	1.4	0.5	0.4	0.3	0.4
Bo2	66	1.19	38.5	1.8	0.7	0.5	0.3	0.4
Bth1	91	1.37	48.8	2.2	1.0	0.6	0.4	0.5
Bth2	121	1.60	49.7	2.2	1.1	0.6	0.5	0.6
Bth3	150	1.14	50.0	1.8	1.0	0.5	0.4	0.6
Bt4	190	0.74	52.0	2.1	1.0	0.5	0.3	0.5

*(1) and (3) are extracted by dithionite-citrate.
†(2) and (4) are extracted by Na pyrophosphate.
‡(5) = [(2) + (4)]/[(1) + (3)]

colder than the present ones. For example, in Ecuador, during the last glaciation the snowline was 1600 m lower than its present level of about 4700 m elevation (Sauer, 1965); in the Peruvian Andes radiocarbon data indicate that deglaciation began approximately 12,000 years ago (Birkeland et al., 1989). At that time conditions may have been comparable to the temperature regimes that now prevail at altitudes above 3000 m where spodic horizons in soils with isotemperature regimes actually form. Spodosols may have been the precursors of some sombric profiles that, in that case, could be considered paleosols of past climates. Ruhe (1956) suggested this explanation of the genesis of sombric horizons.

Other explanations are possible. For example, we may assume that during the cooler climatic period, thick dark surface A_1 horizons formed that placed organic substances at considerable depths in the profiles. The humus in the subsurface layers would be either stable enough to resist decomposition more strongly or less exposed to oxygen than in the topsoil. In this way, as the climate became warmer, a lighter colored, more recent surface horizon would develop above the remnants of the old A_1 horizon. This theory also implies a cooler soil temperature regime than the present one during the time of the "sombric" organic matter formation.

4.2.4 Pans

A number of pans have been recognized in soils of the tropics. Pans differ from horizons in that they form barriers for root penetration or they drastically retard or completely stop the infiltration of water into the subsoil. However, the terms "pan" and "horizon" are not mutually exclusive and the definitions overlap. The following formations need to be mentioned: duripans, densipans, and fragipans. The placic horizon and ortstein could also be included in this section. A detailed description of these pedological features is given in *Soil Taxonomy* (Soil Survey Staff, 1975).

Duripans. *Soil Taxonomy* (Soil Survey Staff, 1975) defines duripans as "a subsurface horizon that is cemented by silica to a degree that fragments from air-dry horizons do not slake during prolonged soaking in water or in HCl. Coatings of silica that are soluble in hot concentrated KOH or in alternating acid and alkali are present in some pores and on some structural faces."

In the tropics duripans occur most frequently in soils derived from volcanic ash in which weathering releases large amounts of silica that accumulates in the subsoil where water percolation is retarded. The soils usually have an ustic moisture regime. They have been given several local names such as *cangahua* in Ecuador (Frei, 1964) and *talpetate* or *talpuja* in Central America. However, not all formations that carry these names are duripans, and other origins are possible.

Densipans. Densipans were first recognized by Smith et al. (1975) in soils of Trinidad. Densipans are defined as massive, silty, firm or extremely firm albic[2] horizons with a very high bulk density (more than 1.8 g/cm^3). Wet or dry fragments of densipans slake in water. They have a very low saturated hydraulic conductivity and are almost free of roots. They are impenetrable to an auger or spade at any moisture content. They occur close to the soil surface and overlie argillic horizons.

The silt and sand fractions of the typifying specimen are almost all quartz; its organic carbon content is 0.1 percent, and the pore space is 27 percent. The densipan and fragipan have comparable physical effects on water and root penetration.

The origin of densipans is obscure. Densipan is considered a very old albic horizon that for long periods has not been penetrated by roots. The densipans contain few plant nutrients and are devoid of oxygen for many months. Their low clay content explains the absence of shrinking and swelling and their massive structure. A single drying of broken materials resettles the horizon at bulk densities greater than 1.7 g/cm^3. Under field conditions, it is assumed that the densipan would reform within a year (Smith et al., 1975).

Fragipans. Fragipans are defined (Soil Survey Staff, 1975) as "a loamy or uncommonly a sandy subsurface horizon that has a very low content of organic matter and has a high bulk density relative to the horizons above it. It is seemingly cemented when dry, having then a hard or very hard consistence. Fragipans normally have bleached vertical streaks that form a roughly polygonal pattern in a horizontal plane. If the pan is near the wilting point the matrix between the streaks is very firm. When moist a fragipan has moderate to weak brittleness. A dry fragment slakes or fractures when placed in water. A fragipan is slowly or very slowly permeable to water. Fine feeder roots are virtually absent in the brittle parts of a fragipan."

The origin of fragipans in temperate regions has been explained by glacial or periglacial phenomena. The properties of fragipans are considered the result of the drying of water-saturated materials; the water is lost because of its extraction by overlying or underlying ice

---

[2]Albic horizons are subsurface horizons the color of which is determined by the color of the uncoated mineral grains.

masses. Similar processes may be responsible for the existence of fragipans at high elevations in the tropics. These fragipans probably formed when the snow line was at a lower altitude than the present one. The presence of fragipans has also been reported in the lowlands tropics—for example, in the northeast of Brazil. No satisfactory explanation based on periglacial phenomena is possible, and the mode of formation of these lowland tropical fragipans needs further research.

References

Birkeland, P. W., D. T. Rodbell, and S. K. Short, "Radiocarbon dates on deglaciation, Cordillera Central, northern Peruvian Andes," *Quaternary Research* 32:111–113 (1989).

Brinkman, R., "Ferrolysis, a hydromorphic soil forming process," *Geoderma* 3:199–206 (1970).

Buchanan, F., *A Journey from Madras through the Countries of Mysore, Kanara and Malabar*, East India Company, London, 1807.

Comerma, J. A., H. Eswaran, and U. Schwertmann, "A study of plinthite and ironstone from Venezuela," *Proceedings Clamatrops Conference on Classification and Management of Tropical Soils*, Malaysian Society of Soil Science, Kuala Lumpur, Malaysia, 1977, pp. 27–36.

Daniels, R. B., E. E. Gamble, and J. G. Cady, "The relation between geomorphology and soil morphology and genesis," *Adv. Agron.* 23:51–88 (1971).

Daniels, R. B., H. F. Perkins, B. F. Hajek, and E. E. Gamble, "Morphology of discontinuous phase plinthite and criteria for its identification in the southeastern United States," *Soil Sci. Soc. Amer. J.* 42:944–949 (1978).

De Barros Machado, A., "The contribution of termites to the formation of laterites," in A. J. Melfi and A. Carvalho (eds.), *Laterisation Processes*, Proceedings of the II International Seminar on Laterisation Processes, University of São Paulo, São Paulo, Brazil, 1983, pp. 261–270.

Debaveye, J. and M. De Dapper, "Laterite, soil and landform development in Kedah, Peninsular Malaysia," *Z. Geomorph. N.F.* (Suppl.-Bd.) 64:145–161 (1987).

D'hoore, J., *L'accumulation des Sesquioxides Libres dans les Sols Tropicaux*, Publ. Inst. Nat. Etudes Agronomiques Congo Belge. Série Sc. 62, Brussels, Belgium, 1954.

Dijkerman, J.C. and R. Miedema, "An Ustult-Aquult-Tropept catena in Sierra Leone, West Africa, I. Characteristics, genesis and classification," *Geoderma* 42:1–27 (1988).

Embrechts, J. and G. Stoops, "Microscopic identification and quantitative determination of microstructure and potentially mobile clay in a soil catena in a humid tropical environment," *Soil Micromorphology* 1:157–162 (1987).

FAO, Fao-Unesco Soil Map of the World, Revised Legend, World Soil Resources Report 60, FAO, Rome, 1988.

Frankart, R., "The soils with sombric horizons in Rwanda and Burundi," *Proceedings Fourth International Soil Classification Workshop*, Agricultural Editions 4, ABOS-AGCD, Brussels, Belgium, 1983, 1:48–61.

Frei, E., "Micromorphology of some tropical mountain soils," in *Soil Micromorphology*, Proceedings of the Second International Working Meeting on Soil Micromorphology, Elsevier, 1964, pp. 307–311.

Gombeer, R., "Potential and effective clay mobility in tropical soils," *Proceedings Clamatrops Conference on Classification and Management of Tropical Soils*, Malaysian Society of Soil Science, Kuala Lumpur, Malaysia, 1977, pp. 9–19.

Kellogg, C. and F. D. Davol, *An Exploratory Study of Soil Groups in the Belgian Congo*, Publications INEAC, Série Scientifique No. 46, Brussels, Belgium, 1949.

Maignien, R., "Le cuirassement des sols en Guinée, Afrique Occidentale," *Mém. Serv. Carte Géol. Alsace et Lorraine 16* (1958).

McFarlane, M. J., *Laterite and Landscape,* Academic Press, New York, 1976.

Mutwewingabo, B., "Génèse, caractéristiques et contraintes d'aménagement des sols acides à horizon sombre de profondeur de la région de haute altitude du Rwanda," in *Soltrop 89,* Editions de l'Orstom, Collection Colloques et Séminaires, Paris, France, 1989, pp. 353–385.

NSSL (National Soil Survey Laboratory), Analysis of pedon No 84P 330, Murambi Series, Rwanda, Soil Conservation Service, USDA, Lincoln, NE, 1985.

Roose, E., *Dynamique Actuelle des Sols Ferrallitiques et Ferrugineux Tropicaux d'Afrique Occidentale,* Travaux et documents de l'ORSTOM #130, Paris, France, 1981.

Ruhe, R. V., "Landscape evolution in the high Ituri Belgian Congo," *Publications INEAC, Série Scientifique No. 66,* Brussels, Belgium, 1956.

Ruhe, R. V. and J. G. Cady, "Latosolic soils of central African interior high plateaus," *Transactions Fifth International Congress of Soil Science,* Kinshasa, Zaire, 1954, IV: 401–407.

Sauer, W., *Geologia del Ecuador,* Editorial del Ministerio de Educación, Quito, Ecuador, 1965.

Schnitzer, M. and R. Kodama, "An electron microscope examination of fulvic acids," *Geoderma* 13:279–287 (1975).

Sivarajasingham, S., L. T. Alexander, J. G. Cady, and M. G. Cline, "Laterite," *Adv. Agron.* 14:1–60 (1962).

Smith, G. D., L. M. Arya, and J. Stark, "The densipan, a diagnostic horizon of Densiaquults for Soil Taxonomy," *Soil Sci. Soc. Amer. Proc.* 39:369–370 (1975).

Soil Survey Staff, *Soil Taxonomy: A Basic System of Soil Classification for Making and Interpreting Soil Surveys,* Agriculture Handbook No. 436, Soil Conservation Service, U.S. Government Printing Office, Washington, D.C., 1975.

Soil Survey Staff, *Keys to Soil Taxonomy,* SMSS Technical Monograph #6, International Soils, Department of Agronomy, Cornell University, Ithaca, N.Y., 1987.

Soil Survey Staff, *Keys to Soil Taxonomy,* SMSS Technical Monograph #19, 4th ed., Virginia Polytechnic Institute and State University, Blacksburg, Va., 1990.

Sys, C., "Suggestions for the classification of tropical soils with laterite materials in the American classification," *Pédologie* 18:189–198 (1968).

Tessens, E., "Clay migration in upland soils of Malaysia," *J. Soil Sci.* 35:615–624 (1984).

Van Wambeke, A., "Mathematical expression of eluvial-illuvial processes and the computation of the effects of clay migration in homogeneous parent materials," *J. Soil Sci.* 23(3):325–332 (1972).

Van Wambeke, A., "A mathematical model for the differential movement of two constituents into illuvial horizons," *J. Soil Sci.* 27(1):111–120 (1976).

Termites and Mesofauna

Termites, other insects, and the mesofauna in general interfere with many soil-forming processes that take place in tropical regions. They participate in the accumulation of parent materials by nest building, either obliterating or enhancing soil horizon formation by mixing or sorting soil constituents, and lead the decomposition of organic residues through distinct metabolic pathways.

Most termites live in the tropical and subtropical areas, but they also extend to about 45° N and S latitudes (Lee and Wood, 1971). Arid regions have few termites. Many large termite mounds on the African continent are fossil. In the central part of Zaïre the abandoned large termite mounds were probably built during a drier period of the early quaternary (Meyer, 1960).

Lee and Wood (1971) reviewed the literature on termite activities in soils and pointed to the specific effects termites have on the distribution of organic matter and the modifications they bring about in profile morphology. Lee and Wood's study on the influence of termites in soil formation is often cited.

There are other organisms that are active in soil formation. Leaf-cutter ants are an example; they participate in the pedoturbation of surface and subsurface horizons by building nests and digging tunnels. Earthworms transform topsoils and contribute to the mixing of soil materials. Their activities will be briefly described in this chapter.

5.1 Biology of Termites

5.1.1 Food

Termites are social insects that live in nests made of earthy materials. The termite population in Miombo woodland may reach 16 million individuals per hectare (Aloni et al., 1989). Their food is collected by workers and consists of either living or dead plant materials. Living wood is only used by a few species. Some feed on sound dead wood but most use rotten wood or decaying wood as a source of food. Several termites are dependent on wood-rotting fungi.

Many species are harvesters or scavengers and collect leaves, twigs, or other plant debris lying on the ground. Some species feed on well-decomposed materials such as dung. The size of the material that is collected seldom exceeds 1 cm, with the average length being approximately 0.5 cm. The area explored by termites around the mounds may be as large as 0.25 hectares and may reach as far as 50 m from the mound (Aloni et al., 1989).

In deciduous woodland savannas in Nigeria termites remove almost 25 percent of the total wood and leaf fall, which amounts to approximately 3800 kg/ha/year; bush fires, on the other hand, burn 31 percent of it. The diet of the insects consists essentially (92 percent) of wood (Collins, 1981).

There is evidence that soil rich in humus is a part of the diet of certain termite species. Some termites grow fungi for nutritional purposes.

5.1.2 Nests, mounds, and galleries

Termite mounds generally contain more clay than the surrounding soils. The density of the mounds is often a function of the clay content of the soils on which they are established. For example, Sys (1955) recorded approximately 5 mounds per hectare on soils with 60 percent clay, and only 2.5 mounds per hectare when the clay content was only 30 percent. The area they covered dropped from 7.8 to 4.3 percent. The average volume of one termite mound built by *Macrotermes falciger* in the Miombo forest of southern Africa is about 250 m^3 (Aloni et al., 1989).

The diameter of the particles that termites transport is limited by the maximum size they can carry in their mandibles. Commonly no gravel size grains can be transported and used in termite constructions. The finer particles that are too small to fit the width between the mandibles may be ingested and regurgitated or excreted. Cementing materials are either saliva or excrements. The amount of soil that termites displace per hectare and per year varies between 300 and 1000 kg, the highest values being observed in the wettest climates (Aloni et al., 1989).

Termites produce *carton*, which consists largely of excreta. They use it to build labyrinths of laminar or alveolar structures or to line galleries. Termites also construct covered runways. The usual radius of the runways from the nests is 30 m. Subterranean galleries are mainly concentrated in the surface horizons, although to reach water or moist clay, termites have built deeper galleries of 23 m (Watson, 1972), 30 m (Marais, 1937), and 50 m (Lepage, 1974). (Fig. 5.1)

The size and shape of the mounds depend on the species and the environmental conditions. They vary from small domed (mushroom) or conical structures of a few centimeter to large mounds 10 m high and 30 m diameter. The maximum weight reported is 2.4 million kg/ha (Meyer, 1960), equivalent on 1 ha to a soil layer that is 20 cm thick.

The distribution and abundance of the mounds mainly depend on the soil moisture conditions and the texture of the soils. Some species prefer shallow water tables and may be restricted to specific drainage classes. Texture, more specifically the clay content, may influence the size and the density of the mounds. Termite mounds are not found on

Figure 5.1   Large termite mound in road cut in Zimbabwe.

pure sands; neither do termites occupy swelling clays (Vertisols, defined later). Termites seem to avoid shallow soils and gravel layers that are difficult to penetrate.

Soil temperature also has an effect on the distribution of termites. Generally termite mounds are restricted to isohyperthermic soil temperature regimes and their number decreases with increasing elevation. Termites are absent in montane soils.

5.2   Effect on Soil Formation and Soil Properties

5.2.1   Particle size sorting

Termite mounds that are built on soils low in clay usually contain more clay than the surrounding soil. Termites also sort sand particles according to size. Stoops (1964) found that 100- to 500- μm particles are not as common in termite constructions as they are in the original soil. His interpretation is that the fraction smaller than 100 μm can be swallowed and transported in the crop of the insects; on the other hand, sand particles of less than 500 μm are too small to be carried between the mandibles. Nye (1955) found that the maximum size of particles in termite mounds near Ibadan is approximately 4 mm.

The casts of earthworms that live in light-textured surface horizons contain more fine silt and clay than the original topsoil (De Vleeschauwer and Lal, 1981).

5.2.2   Surface mantles

Most termite species dig into subsoil horizons and constantly bring new materials to the surface. The depth from which soil is extracted may differ according to the depth of moistening, the position of the water table, and the presence of impenetrable layers. The availability of plant residues may also be a determining factor.

Nye (1955) calculated that the mounds of *Macrotermes bellicosus* on soils derived from gneiss near Ibadan, Nigeria, could account for the accumulation of a 30-cm surface mantle above the stone line. The time required for the accumulation was estimated at 12,000 years. There are many examples of almost gravel-free sediments on top of stone lines and saprolites in the tropics. Although the stone line indicates a lithological discontinuity, there is usually a close relationship between the mineralogy of the sand fractions above and below the stone line: When the underlying parent rock changes, the characteristics of the surface mantles change accordingly. When this is the case the hypothesis of transportation of surface mantle particles over long horizontal distances is not valid. Termite activity provides a reasonable explanation for parent material formation by upward transport. Other mechanisms would generally not account for the very abrupt boundary between the gravelly stone line and the gravel-free surface mantle.

Lee and Wood (1971) mentioned rates of surface soil accumulation reported in the literature that vary between 0.02 and 0.1 mm/year. Some exceptionally rapid accumulations that reach 1 mm/day were reported by Lepage (1984) in shallow soils under savannas and dry woodlands.

It is difficult to determine when the nests that contributed to surface soil accumulation were built. Most investigators date the constructions to dry periods, 4000 to 8000 years ago.

Leaf-cutter ants build one or two nests per hectare in forest land. One nest covers approximately an area of 50 m^2. In cultivated fields there are usually more, but smaller, nests (Alvarado et al., 1981). Weber (1972) reports that *Atta sexdens* is capable of bringing 40 tons of subsoil to the surface in 77 months. The area that is disturbed by the ant is much larger than the nests because of tunnel building around them. The ants also carry large amounts of plant tissues below the surface.

5.2.3 Soil horizons

Mesofauna indirectly contribute to the formation of light-textured topsoils. They bring soil materials to the surface where the impact of raindrops disrupts the aggregates and detaches clay from them. The latter is usually removed as a suspension by runoff water, leaving behind the coarse particles that form the light-textured topsoil.

Roose (1981) estimates that *Trinervitermes geminatus Wassman* in west Africa bring approximately 1200 kg/ha/year soil materials to the surface. Their major activity takes place in the upper 30 cm. The amount transported is 3 to 10 times greater than the normal rate of erosion. In another region the weight of soil that earthworms deposit on the surface is estimated at 50 t/ha/year; the materials also originate from the top 30 cm. The continuous mixing of soil and its exposure to raindrop impact and selective erosion contribute to the formation of empoverished surface horizons. Folster (1964) believed that termites constantly replenish the supply of soil materials at the surface where it is separated by raindrop impact into various particle size fractions.

5.2.4 Chemical properties

Nearly all termite mounds reported in the literature contain more calcium than the surrounding soils. Eventually calcium carbonate concretions may be found, sometimes in quantities that are adequate for their use as soil amendment. In many cases the calcium is brought up from deep soil horizons under the stone line.

Termite mounds may change the rate of leaching of soluble salts by the "umbrella effect" (Nye, 1955 as quoted by Lee and Wood, 1971). A large proportion of the precipitation is removed by runoff from the mound and less water percolates through the soil under the mound.

Termite mounds modify their immediately surrounding environments. In many cases the erosion of the mounds distributes soil materials over the adjacent land area. Arshad (1982) observed marked differences in soil fertility over a 10-m-wide ring around the mounds of *Macrotermes michaelseni* established in semiarid savanna ecosystems in Kenya. Calcium and phosphorus amounts were approximately 30 percent higher in the ring than in the nonaffected soils, i.e., 15 versus 10 cmol(+)/kg soil for calcium and 190 versus 150 mg/kg soil for phosphorus. The biomass production close to the mound was three times higher than the average amount of 300 g/m/year. The botanical composition was also different, with a predominance of *Cynodon dactylon* that was completely absent in the surrounding savanna. These environments attract animals that add nutrients to the sites and sometimes contribute to the improvement of soil fertility (Arshad, 1982).

The higher calcium, magnesium, and potassium contents in termite mounds usually correlate with higher organic matter contents. The same is true for earthworm casts that may contain two to six times more organic matter, two to four times more nitrogen, and two to eight times more Bray phosphorus than the surrounding materials (De Vleeschauwer and Lal, 1981).

5.2.5 Physical properties

The soil materials in large-dome termite mounds have higher bulk densities than the surrounding soils. Arshad (1982) reported 1.76 g/cm^3. Mound surfaces are usually very hard. Water infiltration from above is impeded by surface compactness. Termite mounds are more macroporous because of the presence of chambers, galleries, and tunnels. In deserted mounds root penetration often follows the course of filled chambers and galleries. Outside tunneling of surrounding soils increases the microvariability of the hydraulic conductivity, which makes measurements of water infiltration rates by routine methods often impractical and unrealistic.

Large termite mounds at high densities create irregular surfaces in fields that may decrease their suitability for mechanized agriculture.

Leveling of these fields without preserving the favorably structured topsoil and keeping it at the surface results in very irregular physical conditions that are characterized by patches of impenetrable soil. These occur at the locations from which the mounds have been removed or where the materials from the nests have been piled on top of the original surface layers around the mounds. Leveling techniques that include the conservation of the topsoils have been described by Meyer (1960).

Termites are attracted by the dead organic matter contained in mulches. The insects create stable macropores in crusted soils covered with plant residues. The infiltration rate of the soil is increased in this way and less runoff occurs (Chase and Boudouresque, 1989).

References

Aloni, K., F. Malaisse, and Mbenza Muaka, "Comportement et activité de récolte de *Macrotermes falciger* dans une forêt-claire zambézienne du Shaba (Zaïre)," *Bull. Séanc. Acad. Sci. Outre-Mer.* 35(3):301-322 (1989).

Alvarado, A., C. W. Berish, and F. Peralta, "Leaf-cutter ant (*Atta cephalotes*) influence on the morphology of Andepts in Costa Rica," *Soil Sci. Soc. Amer. J.* 45(4):790–794 (1981).

Arshad, M. A., "Influence of the termite *Macrotermes michaelseni* on soil fertility and vegetation in a semi-arid savannah ecosystem," *Agro Ecosystems* 8:47–58 (1982).

Chase, R. G. and E. Boudouresque, "A study of methods for the revegetation of barren crusted Sahelian forest soils," in *ICRISAT: Soil, Crop and Water Management Systems for Rainfed Agriculture in the Sudano-Sahelian Zone*, Patancheru, India, 1989, pp. 125–135.

Collins, N. M., "The role of termites in decomposition of wood and leaf litter in the Southern Guinea Savanna of Nigeria," *Oecologia* 51:389–399 (1981).

De Vleeschauwer, D. and R. Lal, "Properties of worm casts under secondary tropical forest regrowth," *Soil Sci.* 132(2):175–181 (1981).

Folster, H., "Die pedi-sedimente der Sudsudanesischen pediplane. Herkunft und bodenbildung," *Pédologie* 14:68–84 (1964).

Lee, K. E. and T. G. Wood, *Termites and Soils*, Academic Press, London, 1971.

Lepage, M., ""Recherches écologiques sur une savane sahélienne du Ferlo septentrional, Sénégal: influence de la sécheresse sur le peuplement en termites," *La Terre et la Vie* 28:76–94 (1974).

Lepage, M, "Distribution, density and evolution of *Macrotermes bellicosus* nests in the north-east of Ivory Coast," *J. Animal Ecology* 53:107–117 (1984).

Marais, E. N., *The Soul of the White Ant*, Methuen, London, 1937.

Meyer, J. A., "Résultats agronomiques d'un essai de nivellement des termitières dans la cuvette centrale congolaise," *Bull. Agric. Congo Belge* 51:1047–1059 (1960).

Nye, P. H., "Some soil-forming processes in the humid tropics. IV. The action of the soil fauna," *J. Soil Sci.* 6:73–83 (1955).

Roose, E., *Dynamique Actuelle des Sols Ferrallitiques et Ferrugineux Tropicaux d'Afrique Occidentale*, Travaux et documents de l'ORSTOM #130, Paris, France, 1981.

Stoops, G., "Application of some pedological methods to the analysis of termite mounds," in A. Bouillon (ed.), *Etudes sur les Termites Africains*, Université Lovanium, Kinshasa, Zaïre, 1964, pp. 379–398.

Sys, C., "The importance of termites in the formation of latosols," *Sols Africains* 3:392–395 (1955).

Watson, J. P., "A termite mound in an iron age burial ground in Rhodesia," *J. Ecology* 55:663–669 (1967).

Watson, J. P., "The distribution of gold in termite mounds at a gold anomaly in Kalahari sands," *Soil Sci.* 113:317–321 (1972).

Weber, N. A., *Gardening Ants, the Attines,* American Philosophical Society, XVII, Memoirs 92, Philadelphia, PA, 1972.

Yakushev, V. M., "Influence of termite activity on the development of laterite soil," *Soviet Soil Sci.* (1):109–111 (1968).

The Soil Orders

The purpose of Part 2 is to describe the major kinds of soils of the tropics. Several aspects will be highlighted in each description: first, the criteria that are needed to identify the soils; second, their general geographic distribution; third, their relationships with soil-forming factors; and fourth, their management qualities or limitations, including a discussion of the most appropriate land use types.

An accurate soil classification system is needed to achieve the first objective. Chapter 6 explains the system that will be used to classify and name the soils.

Part 2 does not discuss all the soil orders that have been recognized by the classification system. Only those that are characteristic of the humid and semiarid tropics are included—e.g., the Aridisols, Histosols, and Mollisols are not described in detail, either because they seldom occur in the tropics or they do not present specific management problems.

6

Soil Classification

Soil Taxonomy (Soil Survey Staff, 1975, 1990), prepared by the United States Department of Agriculture, is used in this text as a reference system. It is a multicategorical classification that subdivides the soil universe into classes, called taxa. *Soil Taxonomy* provides names that allow scientists to designate soils in a reproducible manner and, by doing so, it facilitates the transfer of technical information. The class of the highest category of *Soil Taxonomy,* the *order* level, has been taken as the basis for subdividing Part 2 into chapters.

Most soil classifications have been developed by pedologists who are primarily interested in soil geography. For that reason "genetic" properties have been used more frequently than others to differentiate units. These properties arc not always those that are the most important for management. Special technical interpretations are additional steps that are often needed to arrive at practical recommendations.

Soil Taxonomy has its own vocabulary. A brief discussion of the relationships between classes and soil-forming factors is necessary to understand the classification. Certain terms that relate to soil climatic regimes also need to be introduced and defined before explaining the rationale of the system.

6.1 Soil Temperature Regimes

As noted in Chapter 1, tropical climates are characterized by the small difference that exists between their summer and winter temperatures. *Soil Taxonomy* uses this attribute to define *isotemperature* regimes, a term that covers a concept comparable to "tropical." The definition reads as follows: "when the mean soil temperature at a depth of 50 cm calculated for the periods June, July, and August and for December, January, and February differ by less than 5 degrees C absolute, the soils are considered to have an isotemperature regime."

The prefix "iso" is used to designate temperature regimes that characterize soils of the tropics; they include regimes that are recognized on the basis of their *mean annual soil temperature* (MAST) as follows: *isomesic* (8°C ≤ MAST < 15°C), *isothermic* (15°C ≤ MAST < 22°C), and *isohyperthermic* (MAST ≥ 22°C). The equivalent temperature regimes in temperate regions would be called mesic, thermic, and hyperthermic. When the mean annual soil temperature in isoregimes is less than 8°C but higher than 0°C, the temperature regime is called *cryic*.

The subdivisions of the isotemperature regimes parallel the altitudinal vegetation belts that characterize tropical mountains (section 1.1.2). The isohyperthermic regime roughly corresponds to the hot lowland tropics; the isothermic regime is dominant in the cool mountainous areas, e.g., the *tierra templada* in the South American Andes. In the humid tropics of Africa the boundary between the isohyperthermic and isothermic regimes is located at approximately 1600 m altitude. The isomesic regime characterizes the *tierra fria* or páramos of South America; in Cameroon they occur at elevations above 3100 m. The lower geographic limit of the cryic soil temperature regime would be at approximately 4600 m in Cameroon (Embrechts and Tavernier, 1986). It has been pointed out that temperatures have changed dramatically during the Pleistocene and Holocene and that in the past the mountain climates in the tropics were much colder than they are at present (Fig. 1.4).

The temperature regimes as defined by *Soil Taxonomy* are very broad and are not sufficient to determine the suitability of land for crop production in detailed studies. Additional soil temperature classes would be needed for that purpose. For example, in the isohyperthermic range, there is an area at ±5° N and S of the equator where cloudiness causes the rainy season to be cooler than the dry season; at higher latitudes, bordering the nonisoregimes, on the contrary, the dry season may be cooler than the rainy season because cool air masses from temperate regions penetrate into the region during the winter time. The occurrence of the cooler season may offer possibilities of growing C_3 plants such as wheat. There are other examples: Some researchers proposed an *isomegathermic* regime that is characterized by mean annual soil temperatures higher than 28°C. These hot regimes are usually preferred for C_4 plants, e.g., sugar cane. It would be useful to recognize these subdivisions (Comerma and Sanchez, 1980).

6.2   Soil Moisture Regimes

Seasonal variations in soil moisture are used in *Soil Taxonomy* to classify soils on the basis of soil climate. The classification has its own nomenclature for designating broad moisture regimes classes that are essentially based on the time during which a *moisture control section* (MCS) in the profile is completely moist, partly dry, or completely dry. The role of moisture regime classes in *Soil Taxonomy* is to evaluate seasonal trends in moisture conditions and to estimate what major soil-forming processes actually take place in the MCS. The soil moisture control section in medium-textured soils is located approximately between 30 and 60 cm depth; these limits vary with texture but always include a soil section that can store 50 mm of water.

Soil moisture regimes do not provide sufficient information for determining the suitability of land for crop production because their definitions do not consider the moisture conditions in the surface layers from which most plants take their water. Moisture regime classes only allow rough estimates regarding the land use potentials of broad geographical areas.

The definitions of soil moisture regimes are given in a simplified key that is only valid for well-drained soils. The key assumes that the required conditions are met in more than 5 out of 10 years.

6.2.1 Key to soil moisture regimes (except aquic)

1. The moisture control section is completely dry more than one-half of the time (cumulative) that the soil temperature is over 5°C.
   If true, go to 2. If false, go to 3.

2. When the soil temperature is over 8°C, the moisture control section is partly or completely moist for 90 consecutive days or more.
   If true, go to 3. If false: *aridic*.

3. The mean annual soil temperature is less than 22°C.
   If true, go to 4. If false, go to 7.

4. The difference between winter and summer[1] soil temperatures at 50 cm depth is equal to or greater than 5°C.
   If true, go to 5. If false, go to 7.

5. Within the 4 months that follow the summer solstice,[2] the moisture control section is completely dry for at least 45 consecutive days.
   If true, go to 6. If false, go to 7.

6. Within the 4 months that follow the winter solstice, the moisture control section is completely moist for at least 45 consecutive days.
   If true: *xeric*. If false, go to 7.

---

[1]Winter or summer temperatures refer to averages of December, January, and February or June, July, and August.

[2]The 4-month period begins the first of the month after the occurrence of the solstice.

7. The moisture control section is completely dry or partly dry for 90 cumulative days or more.

    If true, go to 8. If false, go to 9.

8. The soil temperature regime is warmer than cryic.

    If true: *ustic*. If false: *undefined*.

9. The precipitation exceeds evapotranspiration in all months.

    If true: *perudic*. If false: *udic*.

6.2.2   The aquic moisture regime

Soil Taxonomy (Soil Survey Staff, 1975) defines the aquic moisture regime as follows: "The aquic moisture regime implies a reducing regime that is virtually free of dissolved oxygen because the soil is saturated by groundwater or by water of the capillary fringe." The aquic moisture regime is a reducing one, by the presence of electron donors provided by organic matter. A soil has an aquic moisture regime when it is completely saturated with water during sufficient time to create reducing conditions; they need not have reducing conditions to last the whole year.

6.2.3   Moisture regimes, vegetation, and climate

The soil moisture regime classes have been defined in such a way that they broadly coincide with agroecological zones. In the tropics, the following correlations are intended.

Udic regimes are the regimes of soils that have enough moisture for trees to dominate in the pioneer vegetations that invade fields in cropland cleared from rain forests. Ustic moisture regimes, on the other hand, are characteristic of soils that support savannas, woodlands, and shrubs and do not have the potential to restore a rain forest after clearance. Land with soils having ustic moisture regimes generally reverts to grasslands after cultivation.

Soils with ustic moisture regimes have sufficient water to support at least one crop per year without supplemental irrigation. The aridic regimes, on the contrary, are regimes that do not allow annual crops to be grown because of lack of water.

There are no soils with xeric moisture regimes at low latitudes because of the summer concentration of rain in the tropics. Xeric regimes have dry hot summers and rainy cold winters.

Aquic moisture regimes usually correspond to permanent or temporary swamps. They would be considered very poorly to poorly drained soils. However, in monsoon climates, profiles with an aquic moisture regime may completely dry out during the dry season.

A small-scale map of the dominant soil moisture regimes of Africa is shown in Fig. 6.1. Figure 6.2 is a similar map for South America. The intent is to depict the distribution of the soil moisture regime classes on these two continents and to exemplify the very general character that the moisture regime classes have in *Soil Taxonomy.*

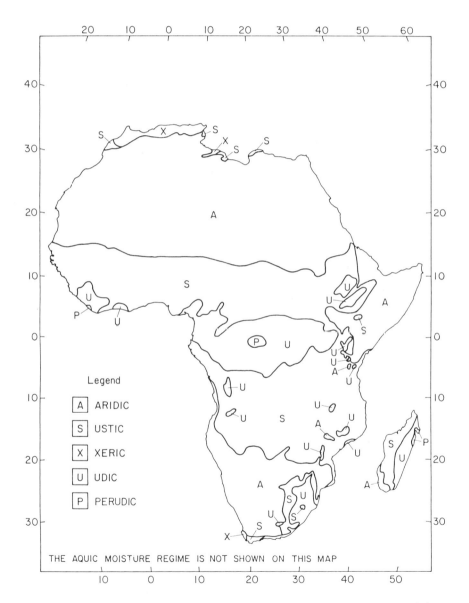

Figure 6.1   Distribution of soil moisture regime classes in Africa. [*From Van Wambeke (1982).*]

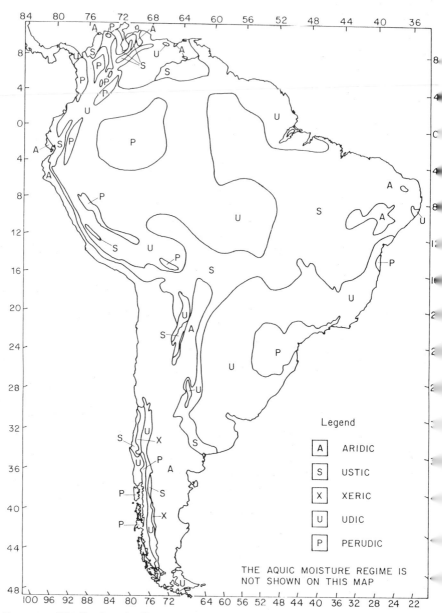

Figure 6.2 Distribution of soil moisture regime classes in South America. [*From Van Wambeke* (*1982*).]

6.3   The Soil Orders

The relationships between soil-forming factors (parent materials, soil temperature and moisture regimes), soil-forming processes, diagnostic features, and soil orders is explained by using a cross section of a hypothetical continent that is shown in Fig. 6.3. The graph has six levels. The top level refers to temperature and includes the division between iso- and nonisoregimes. The left-hand side thus represents the temperate regions. The extreme right corresponds to the cool tropics of high altitudes.

The second level relates to the soil moisture regimes: The complete cross section can be thought of as a sequence starting in humid equatorial central Africa, continuing over the wet and dry savannas of west Africa, crossing the aridic Sahara desert before reaching the mediterranean area (xeric moisture regime with winter rains). The extreme left of the graph would correspond with the high latitudes of Europe where humid temperate climates prevail. The loop that connects the udic and ustic regimes is a representation of the changes in climate that took place during the Pleistocene. They included alternations of dry interpluvial periods separated by pluvial eras.

The parent materials are shown in the third level called landscapes. The cross section includes a continental platform that is subdivided into a shield area, a pediment, and a clayey depression. This part is placed under the sequence of udic, ustic, and aridic moisture regimes. The landscape section also contains an orogenic belt that is shown on the right-hand side; the extreme left represents calcareous loess plains under a dry steppe vegetation that grades into humid temperate forest land.

The soil-forming processes that gave soils their most striking characteristics are shown in level four of Fig. 6.3. They are at the origin of the soil properties that have been selected by pedologists as differentiating characteristics or diagnostic features (level five) to identify the soil orders (level six).

6.3.1   Diagnostic features

The developers of *Soil Taxonomy* have selected *weathering* as the most active soil-forming process on old continental shields in the tropics. Strong weathering and desilication lead to the residual concentration of kaolinite and sesquioxides; once crystallized these minerals are practically inert, and, unless they are subjected to strong acid or reducing conditions, they tend to persist indefinitely in soils.

Soil Taxonomy has defined the *oxic* horizon to reflect the results of the desilication process. The definition (Chap. 7) has been written in such a way that it characterizes materials that are completely weath-

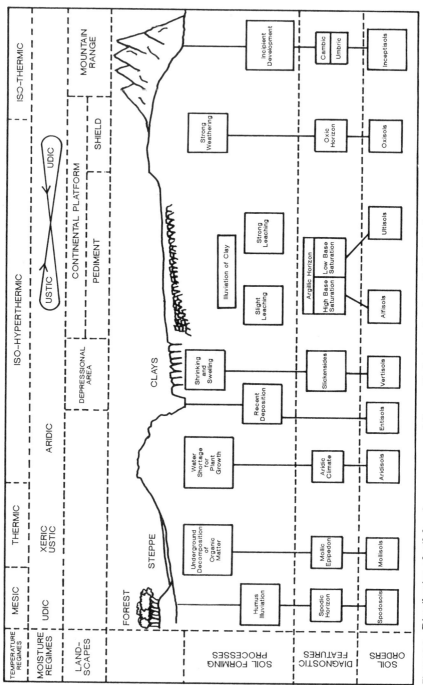

Figure 6.3 Distribution of soil-forming processes and diagnostic features in a hypothetical landscape.

ered. The oxic horizon is diagnostic for most of the *Oxisols* that cover the old landscapes in the tropics. In some cases *Soil Taxonomy* also uses the *kandic* horizon (Sec. 8.1.3) to classify soils in the Oxisols. Soil materials that are not completely weathered usually contain clays other than kaolinite that migrate more easily from surface layers into the subsoil. This illuviation process is most pronounced in soils that are subject to periodic wetting and drying. It results in the formation of the *argillic* horizon. Its strongest development is reached in soils with ustic moisture regimes. A special kind of argillic horizon that results from its intense desilication is called the *kandic* horizon. Two soil orders have been recognized in *Soil Taxonomy* on the basis of the argillic or the kandic horizons. The first, *Ultisols,* groups soils from which bases such as Ca, Mg, K, or Na have been almost completely leached. The second order of soils with clay increase horizons that have not suffered such severe leaching are the *Alfisols*. They occur in the drier parts of the tropics, and Ca, Mg, and K dominate their exchange complex. Many Ultisols and Alfisols in the tropics essentially contain low-activity clays and are then classified in one of the *kandi-* great groups; there are many Alfisols, however, that have appreciable amounts of 2:1 clay minerals.

Further subdivisions into orders recognize recent sediments of montmorillonitic clays that were deposited in semiarid climates. Under these conditions the clays undergo strong shrinking and swelling. As a result of these processes, they develop peculiar soil structures, showing oblique stress surfaces known as *slickensides* that separate wedge-shaped aggregates. They are indicative of a soil order with *vertic* properties, called *Vertisols*.

Soil Taxonomy not only uses properties that are the result of genesis, it also uses properties that are causes of genesis. One of these is the *aridic* moisture regime that is characteristic of extremely dry climatic conditions prevailing in deserts. The aridic moisture regime has been used as a criterion in the classification system to recognize the *Aridisols*.

The *mollic* epipedon seldom occurs in the tropics. It forms under grass vegetation on materials rich in Ca where climate is cool enough to retard organic matter decomposition. The mollic epipedon is used to recognize the *Mollisols*, which in the tropics are mainly found at high elevation.

The last horizon that is diagnostic of mature soils is the *spodic* horizon that forms by illuviation of organic matter in sandy deposits in a perhumid climate. Spodic horizon are diagnostic for *Spodosols*.

Some diagnostic features are linked to the kind of parent material in which the soils develop. One of them is the dominance of amorphous clays that characterizes soils that form in volcanic ashes. They are called *andic properties*. The order that is established to group them is called *Andisols*. They are most frequently found in the orogenic belts shown on the right side of Fig. 6.3.

The diagnostic horizons discussed until now identify "mature" soils. Some profiles, however, do not show enough development to justify their classification in any mature soils. Two orders have been set apart to distinguish those that only have an incipient development, the first being the *Inceptisols*. The major diagnostic properties are the *cambic* horizon and the *umbric* epipedon. These young soils are mostly found in the orogenic belts, in subsiding zones of geosynclines, or on the presently forming erosion surfaces. The soils that do not show any of the diagnostic features described above are grouped into the *Entisols*.

6.3.2 Key to the soil orders

The genetic concepts alone are inadequate to identify soils in the classification system. Measurable soil properties that reflect these genetic processes are necessary. To secure consistent classifications among groups of people, the diagnostic properties are organized into structured definitions and keys that allow the unambiguous naming of the soils. A simplified key extracted from *Keys to Soil Taxonomy* (Soil Survey Staff, 1990) is given below. At the order level it should correctly classify 85 percent of the soils of the tropics.

Simplified key to orders of mineral soils

B. Soils that have andic soil properties in at least 35 cm of the top sixty centimeters of the mineral soil.

...ANDISOLS

C. Other soils that have a spodic horizon whose upper boundary is within two meters of the surface, *or*
that have a placic horizon that rests on a spodic horizon

...SPODOSOLS

D. Other soils that within 150 cm of the soil surface *either*
1. have an oxic horizon, but *do not have* a kandic clay increase[3]; *or*
2. have 40% or more clay in the upper 18 cm, *and, either* an oxic *or* a kandic horizon that meets the weatherable mineral requirements of an oxic horizon.

...OXISOLS

---

[3]See on page 168 the definition of kandic horizon requiring a given clay increase within a distance of 15 cm.

E. Other soils, deeper than 50 cm that have, after mixing the top 18 cm, 30% clay or more in all sub-horizons to at least 50 cm, *and*
   1. have, unless irrigated or cultivated, open cracks (> 1 cm) at 50 cm depth at some time, *and*
   2. have one or more of the following:
      a. gilgai microrelief; *or*
      b. slickensides close enough to intersect at some depth between 25–100 cm, *or*
      c. wedge-shaped tilted aggregates at some depths between 25 and 100 cm.

   ...VERTISOLS

F. Other soils that have an ochric or anthropic epipedon and either
   1. Do not have an argillic or a natric horizon but
      a. are saturated with water within one meter of the surface and have a salic horizon with its upper boundary within 75 cm of the surface, *or*
      b. have within one meter depth either a petrocalcic, calcic, petrogypsic, or cambic horizon or a duripan; *and* have an aridic moisture regime, *or*
   2. have an argillic or natric horizon and have
      a. an aridic moisture regime, *and*
      b. an epipedon that is not both massive and hard to very hard when dry.

   ...ARIDISOLS

G. Other soils that have an isomesic or warmer temperature regime and have an argillic or kandic horizon with <35% base saturation (by sum of cations) at 125 cm below the upper boundary of the argillic or kandic horizon, or immediately above a lithic or paralithic contact, or 1.8 meter below the surface of the soil, whichever is shallower.

   ...ULTISOLS

H. Other soils that have
   1. A mollic epipedon, *and*
   2. A base saturation >50% (by NH_4OAc) in all subsurface horizons to a depth of 1.25 m below the top of the argillic or kandic horizon, or to a depth of 1.8 m below the surface if no argillic or kandic horizon is present, or to a shallower lithic or paralithic contact.

   ...MOLLISOLS

I. Other soils that have an argillic or a kandic horizon

   ...ALFISOLS

J. Other soils that have no sulfidic material within 50 cm depth and have *one or more* of the following:
   1. an umbric, mollic or histic epipedon;
   2. a cambic horizon;
   3. a calcic, petrocalcic, gypsic, petrogypsic, placic horizon, or a duripan within 1 meter of the surface;
   4. an oxic horizon with its upper boundary between 150 and 200 cm depth;
   5. a sulfuric horizon whose upper boundary is at less than 50 cm depth;

6. sodium saturation of 15% or more in the major part of the first 50 cm, that decreases with depth below 50 cm, and groundwater within a depth of 1 m at some period during the year.

...INCEPTISOLS

K. Other soils

...ENTISOLS

6.4 Surface Areas of Major Soil Taxa

The geographical importance of mapping units that are described in terms used by *Soil Taxonomy* is given in Table 6.1. It refers to the tropics only. The areas are rough estimates because they are based on measurements made on small-scale maps and some of them have been mapped according to older classification systems that have been reinterpreted. The units have been ranked in decreasing order of importance. The most uncertain figure concerns the high percentage given to Oxisols that is based on data collected before 1987 when ma-

TABLE 6.1   Areas of Major Map Units in the Tropics

Soil map unit dominated by	Area, million ha	Percent of tropics
Oxisols	1100	22.5
Ustox	350	7.5
Udox	750	15.0
Aridisols	900	18.4
Alfisols	800	16.2
Ustalfs	760	15.4
Udalfs	40	0.8
Ultisols	520	10.6
Aquults	10	0.2
Ustults	100	2.2
Udults	410	8.2
Entisols	490	10,2
Aquents	10	0.2
Psamments	480	10.0
Inceptisols	243	5.0
Aquepts	128	2.7
Tropepts	115	2.3
Vertisols	200	4.0
Andisols	57	1.1
Mollisols	50	1.0
Histosols	31	0.6
Spodosols	6	0.1

jor changes in the definitions of the oxic and kandic horizons were introduced. The precision of Table 6.1 can be estimated by verifying that the sum of all the percentages only reaches 85 percent.

References

Comerma, J. and J. Sanchez, C., "Consideraciones sobre el régimen de temperatura del suelo en Venezuela," VI Congreso Latinoamericano de la Ciencia del Suelo, San José, Costa Rica, 1980.

Embrechts, J. and R. Tavernier, "Soil temperature regimes in Cameroon as defined in Soil Taxonomy," *Geoderma* 37:149–155 (1986).

Soil Survey Staff, *Soil Taxonomy: A Basic System of Soil Classification for Making and Interpreting Soil Surveys,* Agricultural Handbook No. 436, Soil Conservation Service, U.S.D.A., Washington, D.C., 1975.

Soil Survey Staff, *Keys to Soil Taxonomy,* SMSS Technical Monograph #6, International Soils, Agronomy Department, Cornell University, Ithaca, NY, 1987.

Soil Survey Staff, *Keys to Soil Taxonomy,* SMSS Technical monograph #19, 4th ed., Virginia Polytechnic Institute and State University, Blacksburg, VA, 1990.

Van Wambeke, A., *Calculated Soil Moisture and Temperature Regimes of South America,* SMSS Technical Monograph #2, Soil Conservation Service, U.S.D.A., Washington, D.C., 1981.

Van Wambeke, A., *Calculated Soil Moisture and Temperature Regimes of Africa,* SMSS Technical Monograph #3, Soil Conservation Service, U.S.D.A., Washington, D.C., 1982.

Oxisols

Oxisols are soils that carry the marks of strong weathering and desilication in humid, free-draining environments. They generally cover old geomorphic surfaces or characterize soils in recently deposited preweathered sediments that originate from ancient regoliths. They also occur in younger materials that weather rapidly, such as mafic rocks.

Oxisols are found under almost any kind of tropical vegetation; their moisture regimes include all types but very seldom include the xeric regime. The Oxisols that have an aridic moisture regime are considered to be relics. The temperature regimes of most Oxisols range from isomesic to isohyperthermic. In the southern part of South America and Africa they may have hyperthermic or thermic temperature regimes.

In the Brazilian classification the well drained Oxisols are called *Latosols*. The FAO-Unesco soil map legends (FAO 1974, 1988) identify them as *Ferralsols*. Precursors of the order are *Red Earths,* as used in the British literature. In the French system they are referred to as *Sols ferrallitiques typiques.*

The Oxisol order is not the only order that contains soils that are completely weathered. For example, *Soil Taxonomy* (Soil Survey Staff, 1987) does not accept as Oxisols strongly weathered profiles that have a sharp[1] increase in clay under a light-textured[2] surface horizon. In this case the soils are classified as kandic great groups of Ultisols and Alfisols.

---

[1]A sharp clay increase is defined as the within 15 cm clay increase needed for the *kandic* horizon described in Chap. 8.

[2]"Light-textured" in this sentence means less than 40 percent clay.

7.1 Definition

According to *Keys to Soil Taxonomy* (Soil Survey Staff, 1987), Oxisols are mineral soils that have no spodic horizon and *either*

1. Have an oxic horizon with its upper boundary within 150 cm of the soil surface *and* do not have a clay[3] content increase necessary to define the upper boundary of a kandic[4] horizon within the same depth; *or*

2. Have 40 percent or more clay in the surface 18 cm after mixing, *and,* with its upper boundary within 150 cm of the soil surface, *either* an *oxic* or a *kandic* horizon that meets the weatherable mineral requirements of an oxic horizon.

The definition that is given above dates from 1987 and has not been changed since. It modified considerably the older concept of Oxisols that was described in the 1975 version of the classification system (Soil Survey Staff, 1975). This should be kept in mind when consulting literature published before 1987.

The major change in the 1987 classification is that many strongly weathered soils with more than 40 percent clay in the surface horizons and not classified as Oxisols on the basis of the presence of an argillic horizon are now accepted as Oxisols. The reason for the change was that the identification of the argillic horizon in these materials was extremely difficult, was not reproducible, and often led to conflicting classifications of the profiles.

The present text follows the 1990 version of the *Keys to Soil Taxonomy* (Soil Survey Staff, 1990).

7.2 Profile Descriptions and Analyses

The descriptions and analyses of three Oxisols are included in App. B. They are all in Brazil, between 15° and 23° S latitude and at least 600 m above sea level. Two of them are at more than 1000 m elevation. They have isohyperthermic temperature regimes that border the isothermic limit. The mean annual air temperature in the region is 20.4, 20.5, and 21.9°C at the sampled profile sites with some monthly air temperatures as low as 16.6°C. These temperatures are probably

---

[3]Measured clay by standard U.S.D.A. methods (1984) or three times the 1500 kPa water content, whichever is greater but <100.

[4]The kandic horizon is defined in Chap. 8.

responsible for the somewhat higher organic carbon contents than is usually accepted as being representative of isohyperthermic Oxisols.

The profiles have been selected for the properties of their mineral constituents. The oxic horizons in these profiles all have some subhorizons without dispersible clay, low fine silt contents, and water retention of less than 10 percent by volume.

Some of the structures are described as "pseudostructureless" by the authors, a term that is meant to designate a strong, extremely fine granular structure that looks massive to the naked eye. It was also described as "without discernible structure" in *Soil Taxonomy* (Soil Survey Staff, 1975). This structure is discussed in some detail in Sec. 7.3.3.

The volumetric nutrient contents per hectare and at 1 m depth of the three profiles are given in Table 7.1. The low values for cations in the Anionic Acrustox are also representative for many Typic Haplustox (see Fig. 7.6). The second profile, the Humic Rhodic Kandiudox has been limed and fertilized for sugarcane.

7.3   The Oxic Horizon

The oxic horizon is the major differentiating characteristic of the Oxisol order. In certain cases the oxic horizon can be replaced as a classification criterion by the kandic horizon. The two horizons have many properties in common, primarily the low cation exchange capacity of the dominantly kaolinitic clay. Only the oxic horizon is discussed in this chapter; the kandic horizon is defined in Chap. 8 on Alfisols and Ultisols.

7.3.1   Definition

The oxic horizon is a subsurface horizon, at least 30 cm thick, that has less than 5 percent of its volume that shows rock structure. The oxic horizon has the following characteristics:

TABLE 7.1   Volumetric Nutrient Content per Hectare and Meter Depth of the Three Oxisols Included In App. B

	C*	N†	Ca†	Mg†	K†	Na†
Anionic Acrustox	141	7,356	101	57	116	23
Humic Rhodic Kandiudox	164	11,910	8,689	1,088	992	7
Humic Rhodic Eutrustox	213	13,126	7,462	366	219	55

*Units of Mg/ha/m.
†Units of kg/ha/m.

1. The fine-earth fraction has an apparent[5] ECEC (*effective cation exchange capacity*) equal to or less than 12 cmol(+)/kg clay (bases extractable with NH_4OAc plus aluminum extractable with 1 M KCl), and an apparent CEC (*cation exchange capacity* using NH_4OAc at pH 7) that does not exceed 16 cmol(+)/kg clay (see footnote 3).

2. The horizon has a diffuse upper particle size boundary, i.e., <1.2 times the clay content within a vertical distance of 15 cm if the surface horizon contains 20 to 40 percent clay; less than a 4 percent absolute clay increase if the surface contains ≤20 percent clay; or <8 percent absolute clay increase if the surface contains ≥40 percent clay.

3. The horizon has <10 percent weatherable minerals in the 50- to 200-μm fraction.

4. The fine earth fraction has a texture of sandy loam or finer.

5. The horizon does not have andic properties (defined later in Chap. 10 on Andisols).

The weathering of the mineral fractions of oxic horizons practically does not release nutrients or exchangeable aluminum. No neoformation of clay is expected to occur in oxic horizons.

In summary, oxic horizons are chemically poor but in most cases physically well structured. The clays that are present in oxic horizons do not disperse easily in water. There is little osmotic swelling because the cation concentration at the surface of the kaolinitic clay is low. Marks of pronounced shrinking and swelling are not characteristics of oxic horizons and cracks seldom occur in them.

7.3.2 Mineralogy of the clay fraction

Kaolinite crystals in oxic horizons are usually small and are coated by iron oxides or iron oxyhydroxides. Fripiat (1958) described the clay as poorly crystallized "fire clay." The structural defects in the lattices are related to substitutions by Fe or inclusions of interlayer K^+ (Herbillon, 1980). Clays separated from oxic horizons by standard U.S.D.A. methods (1984) usually contain 40 to 80 percent fine clay (smaller than 0.2 μm).

Deferrified kaolinites from horizons with oxic properties have a specific surface measured by BET-N ranging from 32 to 46 m^2/g; when the oxides are not removed the measurements result in values rang-

---

[5]Value obtained by dividing ECEC of total soil by the clay percentage without correcting for the contribution of organic matter to the ECEC.

ing between 34 and 79 m²/g (Gallez et al., 1976). (For details on the
measurement of the specific surface see Brunauer et al., 1938.)

The most common free-iron constituents are goethite and hematite.
The latter crystallizes from precipitates (ferrihydrite or amorphous
iron hydroxides) or iron hydroxides that are released by the rapid
weathering of iron-rich rocks. For example, well-drained soils derived
from basic rocks, such as basalts, have dusky red colors because of the
presence of hematite.

Much of the hematite and goethite in oxic horizons is aluminum
substituted. Bigham et al. (1978a) reported that goethites from 0.2-
μm clay contain 30 to 40 mol % aluminum.

Many oxide crystals in oxic horizons are so small that they have often been considered amorphous (±5 nm). Their specific surface, measured by N_2 adsorption, may reach 324 m²/g. This area varies according to the size of the particles, existence of discrete concentrations of
oxides, or linkages with other surfaces (Gallez et al., 1976). Some iron
oxides in Oxisols of Brazil (Bigham et al., 1978b) have a smaller specific surface, ranging between 60 and 100 m²/g. It is suggested that in
this case the oxides precipitate on themselves or on silicate surfaces
and therefore their specific surface is smaller than expected.

Gibbsite is the most common crystallized aluminum hydroxide in
oxic horizons. It has an extremely low solubility product ($10^{-36.3}$).
Gibbsite forms preferentially outside the oxic horizon at sites where
the silica concentration is low and the pH is high enough to promote
the precipitation of aluminum hydroxides. It is often observed close to
weathering primary minerals in saprolites. It may or may not
resilicate into kaolinite. Some of the gibbsite found in oxic horizons
was probably formed in underlying saprolites; in certain cases it may
have its origin in the decomposition of kaolinite crystals in the oxic
horizon itself.

7.3.3 Physical properties

Structure. The primary structure of oxic horizons consists of
microaggregates that have the size of silts and sands. They form a typical very fine crumb structure. The microaggregates are near spherical in the size range of 0.03 to 0.15 mm (Trapnell and Webster, 1986).
They have a high stability since they resist 4 h of end-over-end shaking in water; however, they disperse in hexametaphosphate (Ahn,
1972). The microaggregates may cluster into materials that show
weakly developed secondary subangular blocky structures. Typical
oxic horizons fail to do so and form macroscopically massive soil ma-

terials. The horizons with blocky structures are usually indicative of younger parent materials.

Clays in *typical* oxic horizons do not disperse, and this property gives the microaggregates a high stability that increases the more the soil solutions are diluted. This is especially true for oxic horizons that have a high proportion of clays whose charge density diminishes with decreasing ionic strength of the soil solution.

The most favorable structures for root penetration are usually found in Oxisols with the highest clay contents. They contain less sand and silt size particles that can accommodate each other to form barriers better than light-textured soils. Clayey surface horizons show less tendency to harden upon drying than light-textured topsoils.

Root penetration in Oxisols cannot take place through a mechanism of soil shrinking at the proximity of the growing root tip. Roots mostly use macropores larger than 50 μm that are the result of biological activity. Although plowing of virgin Oxisols is usually not necessary for root development, 2 or 3 years of cropping or trampling by animals may make deep cultivation desirable, especially for increasing the water acceptance. Experience has shown that when the bulk density of the soil is higher than 1.35 g/cm^3 mechanical impedance may restrict root development. In oxic horizons rich in clay, lower bulk density values may be critical. For example, Arruda Mendes (1989) found that bulk densities above 1.10 g/cm^3 restrict root penetration in oxic horizons containing 64 percent clay.

The micropores of less than 50 μm are very stable in soils containing more than 40 percent clay. The micropores are only slightly affected by cultural practices.

Water-holding capacity. The weight percentage of water held by oxic horizons at permanent wilting point is approximately 0.24 times that of the clay percentage. This standard relationship is reasonably accurate in undisturbed as well as disturbed samples of oxic horizons. The water content retained at 1500 kPa can be computed using the following regression equation: weight percent water = 8.2 + 0.26 × percent clay. On a volume basis the regression equation is: volume percent water = 11.2 + 0.275 × percent clay (Fig. 7.1).

The water content at field capacity depends on the texture as well as on the structure of the soil. The water held at 10 kPa is highest in soils with the largest percentages of fine sand and silt. Since most oxic horizons contain little silt (2 to 20 μm), their water content at field capacity is usually low. The water content at field capacity is related to the bulk density by the equation: weight percent water = 84 − 46 × bulk density (g/cm^3).

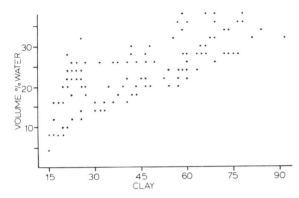

Figure 7.1 Volume percentage water held at 1500 kPa by oxic horizons with variable clay contents.

The capacity of oxic horizons to hold water between the field capacity and permanent wilting point is estimated at less than 1 mm water/cm soil or, in other words, less than 10 percent water by volume. This amount, however, increases with the percentage fine silt in the soil. The regression equation reads as follows: volume percent water = 4.5 + 0.7 × percent fine silt (Fig. 7.2).

Figure 7.2 illustrates the low water contents held by oxic horizons between 33 and 1500 kPa. It has been argued that the 10 kPa tension is a better limit than 33 kPa for estimating the available water content in Oxisols. A reasonable regression equation for water held between 10 and 1500 kPa would be: volume percent water = 7.5 + 0.7 × percent fine silt.

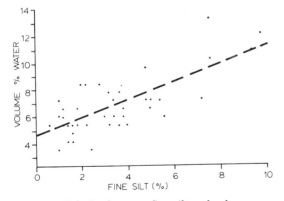

Figure 7.2 Relation between fine silt and volume percent water held by oxic horizons between tensions of 33 and 1500 kPa.

Because of the low water-holding capacities of oxic horizons, *annual* crops that grow in deep Oxisols are more exposed to drought than in other soils of comparable clay content. This situation is particularly critical at the beginning of the growing season when crops have not fully developed their root systems and have not penetrated into deeper layers. Some oxic horizons overlie less weathered subsoils that contain more silt. These layers can usually store more available water than the oxic horizon and are able to increase considerably the water-holding capacity of the Oxisols.

Perennial crops are less exposed to drought stress in Oxisols than annual crops because they do not have to establish a new root system at the beginning of each rainy season. On the other hand, tree crops with deeper root systems may take advantage of subsoil water reserves.

In summary, the water-holding characteristics of oxic horizons are comparable to those of sands, but oxic horizons are physiologically dry at a considerably higher water content. Part of the unavailable water is present in the microaggregates. It may only serve as a source of water vapor that may move by distillation from warm sites to cooler sites. Mulching may have such an effect on water movement in Oxisols.

Hydraulic conductivity.   Water movement in fully saturated oxic horizons is fast: approximately 8 to 15 cm/h. The hydraulic conductivity, however, decreases abruptly upon drying and is strongly dependent on the water content itself. Not only is the water-holding capacity of oxic horizons small, the ability of water to move to the roots is rapidly reduced as the soil dries out. Therefore, the management strategy should try to have roots growing as early as possible to the water, rather than expect the water to flow to the roots.

Water flow through Oxisols under forests is often called *biphasic* with a first component of rapid percolating water through the macropores and a second component of slow water transfer in the microporosity. This division of soil water explains why in certain cases the solute concentrations of groundwaters under rain forests that grow on Oxisols do not reflect the chemical composition of the water retained in the micropores (Nortcliff and Thornes, 1989).

7.3.4 Chemical properties

Ion exchange reactions

Theoretical background.   The clays in oxic horizons, except those that are exclusively formed by sesquioxides, generate their charges from a mixture of permanent and pH-dependent charges. The *permanent* charges are the consequence of isomorphous substitutions in the lattices of the clay minerals. Whatever the kinds of electrolytes or the pH of the soil solution, these charges remain unchanged. The definition of the oxic horizons requires that the permanent *negative* charges do not exceed 12 cmol(+)/kg clay (Sec. 7.3.1). This criterion was selected to exclude soils from Oxisols whose clay fraction is not primarily dominated by kaolinite and sesquioxides.

It may be useful at this stage to point out that the concept of charge per amount of clay should not be equated with charge density: although the negative charges per unit weight of kaolinite are low, the charge density on the clay particles may be high because of the low specific surface of kaolinite. The charge density is important for an understanding of specific adsorptions of ions on the clay mineral surfaces.

The *pH-dependent* negative charges develop by deprotonation of hydroxide groups on clays and on organic matter. These charges vary according to the pH of the soil and the concentration of potential-determining ions in the soil solution. They are also called *variable* charges. Pratt and Alvahydo (1966) observed that the ratio between the pH-dependent and permanent negative charges in surface horizons of some Oxisols in Brazil varies between 1.6 and 5.3. Given the high proportion of variable charges in oxic horizons, the determination of the cation exchange properties with 1 M NH$_4$OAc at pH 7 strongly overestimates the CEC that exists under field conditions. Conversely, it grossly underestimates the base saturation. Gillman and Bell (1977) mentioned values five times in excess of the field CEC and base saturations that drop from 100 percent in the field to 9 percent in the laboratory as a result of the analytical procedures. The maximum cation exchange capacity of 16 cmol(+)/kg clay for an oxic horizon (see Sec. 7.3.1) is a criterion that is intended to estimate the dominance of kaolinite in the clay fraction, not the field cation exchange properties.

The clays in oxic horizons also develop *positive* charges. A few of these are assumed to be the result of isomorphous substitution by Ti^{4+} (Tessens and Zauyah, 1982) in iron oxides and would consequently be permanent. Most of them, however, are variable and are generated by iron and aluminum oxides and hydroxides. The pH at which no net positive or negative charges develop on oxide surfaces is called the isoelectric point and is approximately 7.5 for goethite and 8.5 for hematite. Therefore, these minerals produce positive charges in oxic horizons that have pH values lower than their isoelectric points.

An example of the distribution of positive and negative charges in an oxic horizon of a Typic Acrustox is given in Fig. 7.3 (Marcano-Martinez and McBride, 1989). The permanent negative charges are

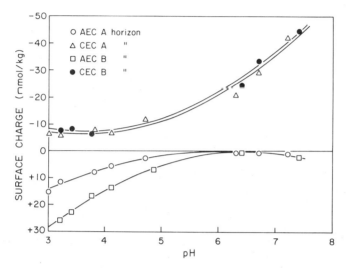

Figure 7.3 Example of ion exchange properties of an oxic horizon. [*From Marcano-Martinez and McBride* (1989). *Soil Science Society of America, reprinted with permission.*]

estimated at 0.73 cmol(−)/kg soil. The soil sample contains 65 percent clay and 0.86 percent organic carbon. It should be noted that oxic horizons by definition may have up to 12 cmol(−)/kg clay and that the amounts shown in Fig. 7.3 are at the low end of the range observed in oxic horizons. The example also illustrates how the positive charges vary as a function of pH. The pH at which the positive and the negative charges balance, called the PZNC (*point of zero net charge*), is 4.45 for the analyzed sample.

The position of the PZNC on the pH scale is a function of the amount of permanent negative charges generated by isomorphous substitutions in the clay minerals, the oxide content of the clay fraction, and the organic matter content. Soils with 2:1 clay minerals have low PZNCs, usually between 2 and 3. When kaolinite is the dominant clay mineral, the PZNC varies between 4 and 5. For high oxide contents, values of 6 and higher are possible. Organic matter that has the potential to develop negative charges with increasing pH may shift the PZNC to lower values. Gillman (1985) estimated that the PZNC is reduced by 1 pH unit for each 1 percent increase in organic carbon.

The ion exchange properties of oxic horizons primarily depend on the position of the soil pH with respect to the PZNC. In most cases the pH of the soil is higher than the PZNC and the soil has a net negative charge or, in other words, some cation exchange capacity. When the pH is lower than the PZNC, there is a net positive charge, and the

AEC (*anion exchange capacity*) is greater than the CEC. In *Soil Taxonomy* the latter condition is often referred to as *acric, acrudoxic,* or *acrustoxic*; soils at this stage have practically no cation exchange capacity.

When placed in a soil-formation perspective, desilication drives the clay fractions to higher PZNCs, close to the isoelectric point of sesquioxides. At the same time, leaching tends to displace the pH of the soil to the PZNC. The end point of this evolution results in conditions where the pH of the soil equals the PZNC and the soil has practically no CEC and only retains very small amounts of cations, even at neutral pHs. For example, oxic soil materials rich in free iron oxides (>15 percent Fe_2O_3) may reach a pH-H_2O of 6 and have practically no capacity to adsorb cations. They may, however, show a strong affinity for phosphates.

Soil management aspects. Figure 7.4 illustrates the agronomic properties that result from the ion exchange characteristics of oxic horizons. The diagram shows on the positive side of the y-axis the cation exchange capacity that corresponds to the negative charges. The x-axis contains the pH values of the soil water. The anion exchange capacity that corresponds to the positive charges is shown on the negative side of the y-axis.

The horizontal line ab in Fig. 7.4 marks the permanent negative charge that is developed by the clay and indicates the cation exchange capacity that does not vary with pH. Line cd is a simplified representation of the increase in negative charges by deprotonation of hydroxide groups on the clay mineral's broken edges and the organic matter.

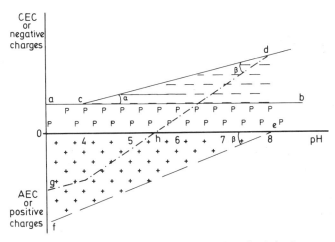

Figure 7.4 Changes in surface charge properties of oxic horizons.

When the pH increases, more negative charges develop, and the size of the increments is mainly a function of the amount of organic matter that is present in the soil. The angle α between lines cd and cb is proportional to the size of these increments. Line ef represents the reduction of positive charges as the pH increases. Point e could be the isoelectric point of one of the iron oxides. The angle β of fe with the x-axis varies with the amount of oxides present. The broken line dg symbolizes the *net* charge obtained by subtraction of positive and negative charges; where it crosses the x-axis at point h it represents the PZNC.

Figure 7.4 is a useful illustration since it shows that the PZNC would shift to higher pH values if there are fewer permanent negative charges than in the example shown (i.e., a decrease of the distance of line ab to the x-axis), when the amount of oxides is greater (and angle β is wider), or when the organic matter content is decreased (and angle α is narrower). In Oxisols, lowest PZNCs are to be preferred, because they often correspond to higher CECs, more organic matter, and less active phosphorus adsorption.

Soil management can only act on a few of the components of ion exchange. The most important techniques are briefly discussed below:

1. The low nutrient retention capacity of oxic horizons is one of the major weaknesses of Oxisols. There are some cultural practices that can alleviate this deficiency. Negative charges (i.e., cation exchange capacity) can be developed by organic matter additions. Freshly decomposed organic matter has a higher CEC per unit carbon than older forms of humus. The contribution of organic matter to the CEC is relatively more important in Oxisols than in other soils, because of the extremely low CEC of the mineral fraction of the oxic horizons. The protection of the topsoil against erosion is therefore the major critical aspect of the utilization of Oxisols. Their management has often been called the management of soil organic matter because the organic matter is a source of nitrogen and because it not only releases phosphorus by mineralization, but it also blocks anion adsorption sites on the clay that may retain phosphorus too strongly for plant uptake.

2. Addition of base causes more hydroxide groups to deprotonate and produces additional negative charges. Liming is one of the techniques of achieving this. It is beneficial as long as it does not at the same time unduly accelerate the decomposition of organic matter and create minor element, such as zinc, deficiencies. The main objective of liming Oxisols is to supply calcium as a nutrient. More details on application rates are given in Sec. 7.3.4.

3. The positive charges are responsible for anion, i.e., phosphate, adsorption. They are most noticeable in soils with iron oxide contents that exceed 15 percent of the clay fraction. In capital-intensive systems it is often advised to block these charges by large initial broadcast applications of phosphorus-fertilizers and to continue with banded maintenance treatments for subsequent crops. Additional information on phosphorus rates is given in Sec. 7.3.4. Silica may also be used to block the positive charges. However, no conclusive results on the use of silica in Oxisols that are not derived from volcanic ash are available. Organic matter may also counteract the effects of the positive charges as shown in Fig. 7.5.

Cation exchange. Oxic horizons are poor cation retainers, and for that reason the soil solutions in Oxisols are very dilute. Under these conditions kaolinite may show a preferential affinity for elements that depart from the normal lyotrophic series. For example, kaolinitic soils in certain cases seem to hold potassium more strongly than calcium.

Levy et al. (1988) assume that the high charge density of kaolinite and the low energy of adsorption of water on K are responsible for the preference for K, which can approach the adsorption surface more closely than other cations. The high charge density is a result of the smallness of the specific surface of the clay particles. Other hypotheses suggest that impurities of vermiculite or micas may cause the preferential adsorption of K at low-exchangeable-K activities. Earlier Ca and K exchange studies on kaolinitic soils indicate that high Al saturations increase the selectivity for K with respect to Ca; Al would reduce the amount of Ca-specific sites available for Ca and K exchange (Nye et al., 1961; Delvaux et al., 1989).

The selectivity for K could in part explain why extremely leached oxic horizons become Ca deficient before they show evidences of K depletion. Other explanations are given by Bennema (1977) who assumed that the high mobility of K in the organic matter turnover between plants and soils keeps K in the system more effectively than the other cations. In this hypothesis small amounts of K may be adequate for plant growth in most natural ecosystems. They may not be sufficient in more intensive cropping systems that utilize large amounts of fertilizers.

Ca depletion most frequently occurs in oxic horizons that have reached the ultimate stage in the weathering sequence, labeled acric, acrudoxic, or acrustoxic in Sec. 7.3.4. Light applications of approximately 1/2 to 2 t/ha lime are usually sufficient to correct the deficiency. In humic subgroups of Oxisols, previously known as Humox, larger amounts, from 4 to 6 Mg/ha, may be more appropriate. Soil scientists in Brazil recommend 1.5 ton Ca/cmol(+) Al/kg soil. In all cases dolomitic lime that contains Mg is preferable because it avoids Ca and Mg imbalances that may cause difficulties in the translocation of P in oil-producing crops. The incorporation of lime in the subsoil or the use of more soluble Ca salts (i.e., gypsum) is recommended in Ca-depleted soils.

Al toxicities are usually not the major problem in oxic horizons. There are no primary minerals that weather and that would thus produce exchangeable Al. On the other hand, the kaolinitic clays are relatively stable, and do not release toxic amounts of Al either. However, small amounts of exchangeable Al may be displaced into the soil solution when oxic horizons are heavily fertilized without neutralizing the acidifying effect of the chemical amendments by liming. It is not always clear whether plants growing in Oxisols suffer more from nutrient deficiencies than from toxicities; it is known that toxic levels of Mn frequently occur in Oxisols when their pH decreases after extended cultivation, especially in soils that are poor in organic matter.

Anion exchange. Most oxic horizons react with phosphate ions in such a way that they either adsorb them or chemically bind them on sesquioxidic surfaces of the clay. Phosphates may also form independent aluminum ($AlPO_4$) or iron ($FePO_4$) precipitates. These reactions have a two-fold effect on soil behavior. First, practically no added phosphate is lost from the soil by leaching, and the additions can be considered investments in the creation of an agricultural soil. Second, some soils retain phosphate so tightly that they make little of it available to plants.

Measurements of phosphate-sorption capacities are useful estimates of the amounts of sorbed phosphates that are needed to achieve a certain concentration of phosphate in the soil solution (intensity factor); for example, a concentration of 0.2 mg P/L soil solution seems adequate for a number of crops (Juo and Fox, 1977; Juo, 1981). Delta-P is defined as the amount of phosphate, expressed as P, that needs to be added to the soil to raise the solution concentration to 0.2 mg P/L (Moshi, Wild, and Greenland, 1974). This amount depends on the strength of adsorption of P on the colloid surfaces, the dimension of this surface, and the amounts of P already attached to it. Figure 7.5 illustrates the relationship between positive charges, phosphate adsorption, and organic carbon content in a topsoil in Kenya.

Most experimental work indicates that the phosphate-sorption capacity of oxic horizons is a function of their specific surface; exchangeable aluminum that would precipitate as $AlPO_4$ seems to play a lesser role than it does in Ultisols. Juo (1981) found a close correlation between BET-N surface area measurements and phosphate adsorption;

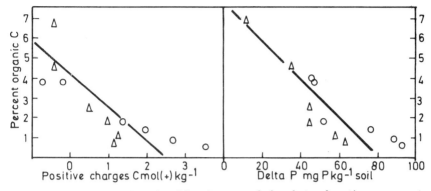

Figure 7.5 Linear regression of positive charges and phosphate adsorption on percent carbon. [*From Moshi et al. (1974). Elsevier Science Publishers, reprinted by permission.*]

furthermore, the BET-N surface is considered to be essentially related to the surface area produced by the sesquioxides.

Specific surface area measurements of tropical soils are seldom available. For practical purposes the clay content is a useful substitute to approximate the P-adsorption characteristics in Oxisols. Lopes and Cox (1979) found a highly significant correlation between the P-adsorption maxima and clay contents in surface soils under cerrado vegetation in Brazil, i.e., P (mg/g soil) = 0.051 × percent clay − 0.0003 (clay percent)^2. Catani and Gloria (1964) obtained the following relationship: P (mg/kg soil) = 5 + 22.6 × percent clay.

Most phosphorus fertilizer programs for Oxisols that are cleared from savannas recommend an initial broadcast application of seven times the clay percentage, i.e., P_2O_5 requirement (kg/ha) = 7 × clay percent, followed by annual banded applications of one-quarter of this amount for 3 or 4 years; further applications should be based on chemical soil test results. The factor seven in this formula is intended to produce a P concentration of 0.2 mg/L in the soil solution. Smaller factors such as three may be adequate for less demanding crops, such as cassava and pineapples. A detailed discussion of P fertilization of Oxisols is outside the scope of this text. The properties of the organic matter, additions of lime, and kinds of crops to be grown all influence the correct choice of a P-fertilization program. An excellent practical review of the possible alternatives for low-input agriculture is given in Sanchez and Salinas (1981).

The choice of an appropriate extracting agent to test the P requirements of Oxisols is difficult because of the diverse chemical forms under which it is present in the soils. The Benchmark Soils Project (Silva, 1985) recommended a Truog method modified by Ayres and Hagihara (1952) that uses a 0.02 M H_2SO_4 plus 0.03% $(NH_4)_2SO_4$ solution. Little crop response to P from corn on Oxisols could be expected above a critical value of 18 mg/kg modified Truog extractable P.

7.4   Agronomic Properties of Oxisols

7.4.1   Nutrient-supplying power

The most striking characteristic of Oxisols under humid climates is their very low volumetric nutrient content. Selected analytical data of a Typic Haplustox in Zambia are shown in Fig. 7.6. The profile is a good example of the widespread poverty of many Oxisols. It contains approximately 250 kg Ca, 120 kg Mg, and 199 kg K per hectare and per meter depth. For Ca this is approximately 175 times less than the content of the Typic Hapludoll from the United States that is also shown in the figure. Although the quantities in the Zambian example are larger than the needs of most annual crops to produce reasonable yields, the root systems would have to extract the cations from all the locations in the entire profile to achieve these yields.

Within the group of Oxisols that are poor in nutrients, agronomists often distinguish various subclasses on the basis of the cations that saturate the exchange complex. One criterion is the percent base saturation [100 × (Ca + Mg + K + Na)/CEC at pH 7] to separate *eutrophic* from *dystrophic* soils at the 50 percent level. Another diagnosis considers the percentage aluminum [100 × Al/(Ca + Mg + K + Na + Al)] that saturates the effective cation exchange capacity. It is used to identify *alic* classes in dystrophic profiles when the Al saturation exceeds 50 percent. Both criteria are the result of long experience by the Brazilian National Soil Survey. An example of the effects of

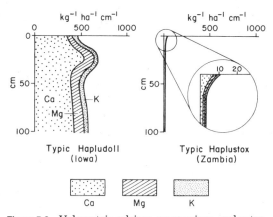

Figure 7.6   Volumetric calcium, magnesium, and potassium contents in an Oxisol compared to a Mollisol in Iowa (USA).

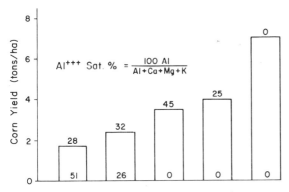

Figure 7.7 Corn yields ranked according to the percent aluminum saturation in A horizons (top of bars) and B horizons (bottom of bars). [*From Olmos et al. (1971).*]

these soil properties on "best fertilizer" corn yields is given in Fig. 7.7 (Olmos et al., 1971). All profiles, except the last one in the figure, are Oxisols of comparable textures. The left bar is of an alic soil; the second and third are of dystrophic soils; and the fourth is of an eutrophic soil. The extreme right bar is of an Alfisol (Rhodic Paleudalf).

7.4.2 Interpretations of soil fertility experiments

In spite of the severe nutrient poverty of Oxisols, there are numerous references in the soil science literature claiming that productivity of Oxisols can be raised to levels that are among the highest in the world. Several reports contend that small amounts of a given fertilizer, e.g., nitrogen, can raise yields to 90 percent of the *maximum yield*. The experiments on which the above statements are based are valid, but before transferring them to field conditions, they have to be placed in their context: In many cases the reports fail to specify the actual dimensions of these maxima, which in absolute terms may be extremely low. As a consequence, the investment in fertilizers is often out of proportion to the actual yields, so that no real benefits can be obtained from the application of fertilizers.

Other misleading statements omit that the controls in the fertilizer experiments received complete fertilization except for the elements that are being tested. Finally, some strategies recommend growing crops that are not demanding in essential nutrients and consequently produce high yields of extremely poor or very low quality products.

Although all reported research findings are scientifically sound and accurate, they seldom fit into a comprehensive development strategy that would meet the requirements of a sustained agricultural program. Some confusion also originates from the inconsistent use of concepts such as "potential, performance, costs, productivity, etc." In this text these terms are used in the following sense: potential = performance − costs. In other words, the potential of land is measured by the benefits that a farmer derives from his or her enterprise and it corresponds to the difference between the price obtained from the yield (i.e., performance) *minus* the costs that had to be paid to achieve this yield. For most individuals the potential is the yardstick that farmers use to make decisions, evaluate parcels of land, and compare their qualities with others. In this respect many Oxisols are at a disadvantage with other soils because of the high costs of fertilizers that are needed to correct the nutrient deficiencies and allow Oxisols to compete with other available soils in an open market economy.

The paragraphs above do not deny the fact that the major management issues for Oxisols are related to soil fertility and that fertilization is necessary for their sustained use. If Oxisols are to support prosperous farming communities on a continuing basis, an "agricultural" soil will have to be created. This can only be achieved by large investments in labor and/or capital; the effort will not necessarily be greater than in other soils that have been transformed by people, such as sandy soils in Europe or steeplands that have been terraced in Asia.

7.4.3 Management practices

Management recommendations on appropriate practices for Oxisols vary considerably with their environment. The type of vegetation, climate, soil parent materials, thickness of the oxic horizon, and degree of erosion all influence the choice of the proper technology. The management of Oxisols is therefore best discussed in a context of well-defined agroecological zones, as components of specific soil landscapes and under given sets of socioeconomic conditions. Only some general guidelines are given in the following sections. They are mainly applicable in soils that have isohyperthermic temperature regimes or, in other words, that occur in hot, lowland tropics.

Udox. The Udox are the Oxisols that have a udic moisture regime. They are located in regions with no, or with only a short, dry season. In each year the soil moisture control section is completely moist for at least 9 months. The Udox usually occur under rain forests. The soil moisture conditions are such that the tree vegetation spontaneously comes back after clearing the forest and cropping the land; exceptions to this rule would only occur when sloping land is left unprotected against erosion. Sandy topsoils or deep light-textured subsoils together with frequent exposure to fierce bush or savanna fires may also impede the return of the forest.

Most plant nutrients in isohyperthermic Udox under rain forests are concentrated in shallow topsoils where they are associated with organic matter. The topsoils are seldom more than 10 cm thick. They are very fragile and cannot persist when, after clearing, they are exposed to erosion and high temperatures that accelerate the decomposition of organic matter. Annual crops that leave the surface bare for· extended periods are therefore not the appropriate use of Oxisols. Instead, *tree crops* that parallel the soil cover provided by the rain forest are ecologically best suited for this agroecological zone. They protect the land against raindrop impact and direct insolation; their root systems are *permanently* exploring a large volume of soil to tap nutrients and to overcome short drought periods. There is no need to develop a new root system as annual crops have to do each season. Perennial crops that are adapted to the regional climate warrant sustainability and should be the major components of farming systems on Udox. In plantation agriculture, given the fragility of the topsoils, it is recommended that trees be planted immediately after the forest is cleared, intermediate food crops not be planted until the trees come into production, and cover crops be planted between the lines to protect the land.

The annual crops that are needed to feed the populations who live in these areas require opening land. Topsoils are exposed to rain and sunshine. If better soils are accessible, it may be desirable to locate the food crops on them. Otherwise special care must be taken to conserve the soil resource that has to support the people. The appropriate management practices depend on the economic and social conditions under which the farms are operating.

The nutrient-supplying power of Oxisols declines very rapidly within 2 or 3 years of clean-weeded cultivation without chemical fertilizers or other amendments, and no profitable crops can be grown after that time. If the population density is low enough, *shifting cultivation* can be practiced, whereby long periods of tree fallows alternate with short cropping cycles. People move to new locations when the yields that are produced become inadequate. Chapter 3 discussed the optimum crop-to-fallow ratios most commonly used in lowland rain forest areas. Shifting cultivation is an acceptable land use alternative as long as there is enough land and the costs and efforts to open new land are less than the costs and risks of applying fertilizers and pesticides on existing cropland.

Shifting cultivation becomes impractical when the population pressure is such that insufficient land is available to keep it under fallow during the time that is needed to replenish the soil nutrient storage. People are often forced to lengthen the cropping cycle and shorten the fallow period. To maintain a more intensive use of the land, additional inputs are needed: they may include fertilizers, lime, pesticides, green manures, and mulches. Another or additional approach would be to make the fallows more effective. To finance these inputs some cash crops have to be introduced in the rotations, and to sustain these systems, soil conservation measures are vital. The intensification of the traditional subsistence agriculture is not possible without moving into a market economy, in which cost-to-benefit ratios and the conservation of land resources determine the success of the operation.

The fertilizer needs of Oxisols under continuous cultivation of annual crops are comprehensive in the sense that at a certain time Oxisols require additions of *all* major and minor nutrients. The sequence of the deficiencies to be corrected may vary, but the general trend follows a series that starts with phosphorus and calcium, and continues with nitrogen, potassium, magnesium, and minor elements. Calcium is usually needed as a nutrient and small applications of approximately 2 t/ha of dolomitic lime may be adequate; as discussed in previous sections, rough estimates of the amounts of P-fertilizers needed can be made on the basis of the clay content of the soil.

Farmers who are established on Oxisols are at a comparative disadvantage when they have to compete economically in the production of commodities that can be grown elsewhere on other soils. This is because Oxisols are inherently poor in nutrients and require considerably larger inputs than many other soils.

Ustox.   The Ustox are Oxisols that have an ustic moisture regime; the time that the soil moisture control section is partly or completely dry is at least 3 months. However, the length of the dry season varies considerably, and the period of moisture stress may last up to 9 months. The typical vegetation on Ustox is savanna or woodlands in which grasses are dominant in the understory. Some Ustox, however, are covered by rain forests that may have developed during more humid climatic periods.

The general guidelines given for Udox in the previous section are valid for the Ustox under rain forest vegetation. However, special precautions for stimulating the forest regrowth after the cropping cycle have to be taken. For example, soils with light-textured surface horizons are to be avoided as well as soils on convex slopes. Forested land should border fields at all sides for protection against the invasion by savannas and against fire. If selection of land is possible, only the best land should be used for annual, clean-weeded crops.

Ustox under savannas or woodlands are often severely eroded. Their nutrient deficiencies, combined with frequent periods of water shortages, make them very poorly suited for crop production without amendments. Under these conditions they are mainly used for extensive grazing.

A traditional system that intends to locally improve the fertility of Ustox includes *kraaling*. In this system livestock is used to concentrate nutrients from a large area into one small location by bringing the animals periodically in a kraal. Another system is the *Chitemene* in which farmers cut the branches from the savanna trees, pile them up, usually around termite mounds, and burn them to produce ash to fertilize the soil.

There are many ways to intensify farming on savanna Ustox by more advanced technologies. Pasture management, including the introduction of legumes adapted to the low nutrient levels, periodic grazing, and light fertilization may improve the carrying capacity for livestock of the savannas. Extensive grazing by animals that can be moved periodically to better locations or can be fed by supplements to overcome specific nutrient deficiencies is a frequently applied land use alternative.

Savannah Ustox on level uplands have often attracted large farming enterprises because of the ease of clearing the land, the possibilities of mechanization, and the generally favorable soil structure for root development (Fig. 7.8). There are many examples of such enter-

Figure 7.8   Mechanized seed bed preparation of Oxisols in savanna areas of Brazil.

prises in South America. The key issues in these operations are the economics of input versus output and the conservation of the land resources that are dangerously exposed to erosion at the beginning of the rainy season. In many instances these large mechanized farms, unless subsidized, find it difficult to overcome unfavorable market conditions because of the high costs of fertilizer inputs

References

Ahn, P., "Contribution to the knowledge of Ghana forest soils," Doctoral thesis, University of Ghent, Belgium, 1972.

Arruda Mendes, R. C., "Restriçoes físicas ao crecimento radicular em latossolo muito argiloso," Ph.D. thesis, University of Brasilia, Brazil, 1989.

Ayres, A. A. and H. H. Hagihara, "Available phosphorus in Hawaiian soil profiles," *Hawaiian Planters' Record* 44:81–99 (1952).

Bennema, J., "Soils," in P. de T. Alvim and T. T. Kozlowski (eds.), *Ecophysiology of Tropical Crops*, Academic Press, 1977, pp. 29–56.

Bigham, J. M., D. C. Golden, L. H. Bowen, S. W. Buol, and S. B. Weed, "Iron oxide mineralogy of well-drained Ultisols and Oxisols: I. Characterization of iron oxides in soil clays by Moessbauer spectroscopy, X-ray diffractometry, and selected chemical techniques," *Soil Sci. Soc. Amer. J.* 42:816–825 (1978a).

Bigham, J. M., D. C. Golden, L. H. Bowen, S. W. Buol, and S. B. Weed, "Iron oxide mineralogy of well-drained Ultisols and Oxisols: II. Influence on color, surface area and phosphate retention," *Soil Sci. Soc. Amer. J.* 42:825–830 (1978b).

Brunauer, S., P. H. Emmett, and E. Teller, "Adsorption of gases in multimolecular layers," *J. Amer. Chem. Soc.* 60:309–319 (1938).

Catani, R. A. and N. A. Gloria, "Evaluation of the capacity of P fixation by the soil through the isotopic exchange using ^{32}P," *An. Esc. Sup. Agr. L. de Queiroz* 21:230–237 (1964).

Delvaux, B. J., E. Dufey, L. Vielvoye, and A. J. Herbillon, "Potassium exchange behavior in a weathering sequence on volcanic ash soils," *Soil Sci. Amer. J.* 53:1679–1684 (1989).

FAO, *FAO-Unesco Soil Map of the World. 1:5,000,000*, vol. I, Unesco, Paris, 1974.

FAO, *FAO-Unesco Soil Map of the World. Revised Legend*, World Soil Resources Report 60, FAO, Rome, 1988.

Fripiat, J. J., "Les argiles des sols tropicaux," *Silicates Ind.* 23:623–632 (1958).

Gallez, A., A. S. R. Juo, and A. J. Herbillon, "Surface and charge characteristics of selected soils of the tropics," *Soil Sci. Soc. Amer. J.* 40:601–608 (1976).

Gillman, G. P., "Influence of organic matter and P content on the PZC of variable charge components of oxidic soils," *Aust. J. Soil Res.* 23:643–646 (1985).

Gillman, G. P. and L. C. Bell, "Soil solution studies in relation to classification and fertility of soils," *Proceedings Conference on Classification and Management of Tropical Soils*, Malaysian Soil Science Society, Kuala Lumpur, Malaysia, 1977, pp. 279–286.

Herbillon, A., "Mineralogy of Oxisols and oxic materials," in B. K. G. Theng (ed.), *Soils with Variable Charge*, New Zealand Society of Soil Science, Lower Hut, New Zealand, 1980, pp. 109–126.

Juo, A. S. R., "Chemical characteristics," in D. J. Greenland (ed.), *Characterization of Soils*, Oxford University Press, New York, 1981, pp. 51–79.

Juo, A. S. R. and R. L. Fox, "Phosphate sorption capacity of some bench-mark soils in West Africa," *Soil Sci.* 124:370–376 (1977).

Levy, G. J., H. v. H. Van der Watt, I. Shainberg, and H. M. Du Plessis, "Potassium-calcium and sodium-calcium exchange on kaolinite and kaolinitic soils," *Soil Sci. Soc. Amer. J.* 52:1259–1264 (1988).

Lopes, A. S. and F. R. Cox, "Relaçao de características físicas, químicas e mineralógicas com fixação de fósforo em sólos sob cerrados," *Rev. Bras. Ci. Solo* 3:82–88 (1979).

Marcano-Martinez, E., "CaSO$_4$ adsorption and surface charge properties of Oxisols," Master of Science thesis, Cornell University, Ithaca, NY, 1987.

Marcano-Martinez, E. and M. B. McBride, "Comparison of titration and ion adsorption methods for surface charge measurement in Oxisols," *Soil Sci. Soc. Amer. J.* 53: 1040–1045 (1989).

Moshi, A. O., A. Wild, and D. J. Greenland, "Effect of organic matter on the charge and phosphate adsorption characteristics of Kikuyu red clay from Kenya," *Geoderma* 11: 275–285 (1974).

Nortcliff, S. and J. B. Thornes, "Variations in soil nutrients in relation to soil moisture status in a tropical forested ecosystem," in J. Proctor (ed.), *Mineral Nutrients in Tropical Forest and Savanna Ecosystems,* Special Publication #9 of the British Ecological Society, Blackwell Scientific Publications, Oxford, 1989, pp. 43–54.

Nye, P., D. Craig, N. T. Coleman, and J. L. Ragland, "Ion exchange equilibria involving aluminum," *Soil Sci. Soc. Amer. Proc.* 25:14–17 (1961).

Olmos, I. L., C. A. Scotti, O. Muzilli, and L. Turkiewicz, Estudo comparativo da produtividade de alguns sólos álicos e não álicos do estado do Paraná, *Bol. Univ. Fed. do Paraná, Agronomia #6,* Curitiba, Brazil, 1971.

Pratt, P. F. and R. Alvahydo, "Cation exchange characteristics of soils of São Paulo, Brazil," *IRI Bulletin #31,* IRI Research Institute, NY, 1966.

Sanchez, P. and J. G. Salinas, "Low-input technology for managing Oxisols and Ultisols in tropical America," *Advances Agronomy* 34:279–406 (1981).

Silva, J. A., *Soil-Based Agrotechnology Transfer,* Benchmark Soils Project, College of Tropical Agriculture and Human Resources, University of Hawaii, 1985.

Soil Survey Staff, *Soil Taxonomy, A Basic System of Soil Classification for Making and Interpreting Soil Surveys,* Agriculture Handbook No. 436, Soil Conservation Service, U.S.D.A., Washington, D.C.

Soil Survey Staff, *Keys to Soil Taxonomy,* SMSS Technical Monograph #6. International Soils, Agronomy Department, Cornell University, Ithaca, NY, 1987.

Soil Survey Staff, *Keys to Soil Taxonomy,* SMSS Technical Monograph #19, 4th ed., Virginia Polytechnic Institute and State University, Blacksburg, VA, 1990.

Tessens, E. and S. Zauyah, "Positive permanent charge in oxisols," *Soil Sci. Soc. Amer. J.* 46:1103–1106 (1982).

Trapnell, C. G. and R. Webster, "Microaggregates in red earths and related soils in east and central Africa, their classification and occurrence," *J. Soil Sci.* 37:109–123 (1986).

U.S.D.A., *Soil Survey Laboratory Methods and Procedures for Collecting Soil Samples,* Soil Conservation Service, U.S. Government Printing Office, Washington, D.C., 1984.

8

Alfisols and Ultisols

The descriptions of soils that belong to Alfisols and Ultisols have been grouped into one chapter. The main reason for doing this is that many tropical Alfisols and Ultisols share common properties. Another reason is that the differentiating characteristic between the two orders, base saturation, is not an efficient discriminator in soils that have low *cation exchange capacities* (CEC), which is the case for most tropical Alfisols and Ultisols.

Alfisols and Ultisols are soils with an argillic or kandic horizon. The Ultisols definition requires that they have a low base saturation in at least the lower part of the profile. On old geomorphic surfaces in the tropics most Alfisols and Ultisols contain clays that are dominated by kaolinite often mixed with minor amounts of other silicate clays. For this reason most Alfisols, Ultisols, and Oxisols belong to the *low-activity clay soils*. Some weatherable minerals may occur in the sand and silt fractions. Ultisols usually occupy younger geomorphic surfaces than the Oxisols with which they are often associated in landscapes.

Ultisols occur under almost any type of tropical vegetation. Their acidity links them to high-precipitation regions or to areas with highly concentrated seasonal rainfall that produces strong leaching. The prefix "Ulti" connotes *extreme leaching*. Many tropical Ultisols have an ustic moisture regime that enhances clay illuviation processes by opening cracks in the soils through which clays move more easily. The soil temperature regimes of tropical Ultisols vary from isomesic to isohyperthermic.

The Alfisols generally occupy regions under savanna vegetation, woodlands, or shrubs, although they may also occur under rain forests where present day weathering of mafic rocks or the dissolution of limestones supplies adequate amounts of cations to keep the base saturation of the exchange complex above the critical level of Alfisols. The most frequent moisture regime of tropical Alfisols is ustic, with many occurrences bordering soils that have aridic moisture regimes. Alfisols are often associated in landscapes with Vertisols.

The revised definitions of taxa in *Soil Taxonomy* (Soil Survey Staff, 1987) have changed the classification of many soils that were classified as Alfisols or Ultisols before 1987, especially those that contain more than 40 percent clay in the topsoil. Some of them that contain a kandic horizon and were previously called *Paleustults, Rhodustults, Paleudults,* and *Rhodudults* or *Paleustalfs, Rhodustalfs, Paleudalfs,* and *Rhodudalfs* are now grouped with the Oxisols. Some caution should be exercised in the interpretation of land units when dealing with soil resource inventories published before 1987.

Approximate equivalents of tropic Ultisols in other systems are *Acrisols* and *Nitisols* (FAO, 1974, 1988); in earlier classifications most Ultisols were named *Red-Yellow Podzolics,* a term which has been translated to *Podzolico Vermelho-Amarelo* to designate similar soils in the Brazilian classification system. The French system would classify most Ultisols of tropical areas as *Sols ferrallitiques lessivés* or *Sols ferrallitiques fortement désaturés-appauvris.*

Most Alfisols of tropical regions are classified as *Lixisols* or *Nitisols* in the 1988 FAO-Unesco classification, but were mapped as *Luvisols* in the 1974 FAO-Unesco soil map of the world legend. The French system would classify most Alfisols of tropical areas as *Sols ferrugineux tropicaux lessivés.*

8.1 Definitions

8.1.1 Ultisols

Ultisols are mineral soils that have a soil temperature regime warmer than cryic and that both:

1. Fail to meet the requirements of Spodosols, Oxisols, Vertisols, and Aridisols[1], *and*
2. Have a kandic or an argillic horizon with less than 35 percent base saturation (by sum of cations) at the shallowest end of one of the following depths:
   a. 125 cm below the upper boundary of the argillic or kandic horizon, *or*
   b. Immediately above a lithic or paralithic contact, *or*
   c. 1.8 m below the soil surface

---

[1]See identification key in Chap. 6.

A wide range of soils meet the definition of Ultisols. It is not possible to describe them as an homogeneous group in a way that would highlight common soil characteristics that are important for land use. For that purpose some additional subdivisions have to be introduced.

At the suborder level *Soil Taxonomy* (Soil Survey Staff, 1987) recognizes the *Humults* that are characteristic of the cool tropics and usually include soils of the highlands with isothermic temperature regimes. In the lowland tropics the Ultisols are represented by *Ustults* for regions having a dry season and *Udults* in the permanently humid parts. The poorly drained Ultisols usually belong to the *Aquults*.

The suborders are subdivided into several great groups. Some of them are of special interest in the tropics: The *kandic* great groups, variously labeled by the prefix "kan" or "kanhapl," are characterized by a typical horizon sequence that includes a light-textured topsoil overlying a strongly weathered heavier subsoil. The nonkandic great groups include the *rhodic* soils having dusky red colors, usually developed from mafic rocks or limestone, e.g., the *Rhodudults,* as well as the common or *haplic* great group, e.g., the *Hapludults.* Parallel great groups are recognized in the Ustults.

Several land qualities characterize the agronomic attributes of Ultisols: the light-textured surface layers that rest on heavier subsoils and the soil acidity that is often related to high amounts of exchangeable aluminum. Other agronomic properties of Ultisols are given in the sections on the diagnostic horizons.

8.1.2 Alfisols

The identification criteria of Ultisols apply to Alfisols except for the low base saturation requirement in line 2 of the definition. According to the *Soil Taxonomy* key, Alfisols *cannot* have a mollic epipedon.

The diversity of Ultisols profiles also exists in Alfisols, and some subdivisions are needed to arrive at meaningful soil management interpretations.

At the suborder level *Soil Taxonomy* (Soil Survey Staff, 1987) recognizes *Ustalfs* for regions with a marked dry season and *Udalfs* in the permanently humid parts of the lowland tropics. The poorly drained Alfisols usually belong to the *Aqualfs*. The soils that are rich in organic matter in the cool tropical highlands are either strongly leached and then belong to the Ultisols or they have a mollic epipedon and are classified as Mollisols. For this reason, no Humalfs have been retained in the classification.

The suborders are subdivided into several great groups. Some of them are of special interest in the tropics: The *kandic* great groups, variously labeled by the prefix "kan" or "kanhapl," are again characterized by a typical horizon sequence that includes a light-textured top-soil overlying a strongly weathered heavier subsoil. The nonkandic great groups include the *rhodic* soils (Rhodudalts) having dusky red colors, usually developed from basic rocks or limestone, e.g., the *Rhodudalfs*, and the common or *haplic* great group, e.g., the *Hapludalfs*.

Many agronomic attributes of tropical Ultisols are shared by Alfisols, although their expression may vary in intensity. The most important common property is the light-textured surface layer covering either the argillic or the kandic horizon. The most striking differences with Ultisols are the base saturation and the soil reaction that eliminates excessive amounts of aluminum in the soil solution.

8.1.3 Diagnostic horizons: Textural clay increases

The processes that lead to textural differentiations in soil profiles have been discussed in Chap. 4. Two clay increase horizons, the argillic and kandic horizons, are diagnostic for Ultisols and Alfisols (Soil Survey Staff, 1987). *Soil Taxonomy* defines the argillic horizon to be the result of clay illuviation. There are no restrictions in the definition on the degree of weathering reached by the argillic horizon.

The kandic horizon (Soil Survey Staff, 1987) is also a clay increase horizon. The mineralogy of the *clay fraction* in kandic horizons is *always* indicative of strong weathering, just as in the case of oxic horizons. The sands and silts in kandic horizons, however, may contain weatherable minerals.

The textural increase of the kandic horizon may be the result of several processes that include sedimentation, lithological discontinuities, and clay migration. Weathering and illuviation processes are not necessarily synchronous; i.e., the clay translocation is usually more intense during dry climatic periods than during permanently humid conditions under which weathering prevails. The clay increases observed in kandic horizons may have originated during dry interpluvials of pleistocene times involving 2:1 clays that subsequently weathered into kaolinite during a more humid phase.

The argillic horizon. A summary definition of the argillic horizon, based on the one in *Soil Taxonomy* (Soil Survey Staff, 1975), is as follows: The argillic horizon does not show rock structure in more than one-half of its volume and contains illuvial layer–lattice clays. The argillic horizon has the following additional properties:

1. If there is an eluvial horizon and there is no lithological discontinuity, the argillic horizon contains more total and fine clay than

the eluvial horizon. The increases are reached over a vertical distance of *30* cm or less. In addition, one of the following occurs:

   a. If any part of the eluvial horizon has less than 15 percent total clay, the argillic horizon must contain at least 3 percent more clay (e.g., 13 versus 10 percent).
   b. If the eluvial horizon has between 15 and 40 percent clay, the ratio of the clay percentages between the argillic and the eluvial horizons should be 1.2 or more.
   c. If the eluvial horizon has more than 40 percent total clay, the argillic horizon must contain at least 8 percent more clay (e.g., 41 versus 49 percent).

2. If peds are present, an argillic horizon should meet one of the following requirements:

   a. The horizon should have clay skins on some of both the vertical and horizontal ped surfaces and in the fine pores.
   b. If the horizon is clayey, the clay is kaolinitic, and the surface horizon has >40 percent clay, the horizon should have some clay skins on peds and in pores in the lower part of the horizon that has blocky or prismatic structure.

3. In structureless soils, the argillic horizon has oriented clay bridges between the sand grains and also in some pores.

4. If there is a lithological discontinuity or only a plow layer overlies the argillic horizon, the argillic horizon needs to have clay skins only in some parts.

The definition given above is incomplete but is adequate to discuss the agronomic properties of argillic horizons in landscapes that are domi nated by low-activity clay soils such as tropical Ultisols and Alfisols.

The clay skins are indications of clay migration, which suggests the presence of 2:1 clay minerals. The argillic horizon therefore implies that the soil may have a somewhat larger agricultural potential than the Oxisols it may be associated with. The reasons are a larger cation exchange capacity of the clay and the possible presence of weatherable minerals in the sand or silt fractions. The decomposition of these minerals may release plant nutrients, although aluminum may enter the soil solution through the same process. If leaching is intensive, which is true for Ultisols, aluminum toxicities may have to be eliminated by liming.

The clay increase at the upper boundary of the argillic horizon causes differences in hydraulic properties between the topsoil and the subsoil. The permeability of the argillic horizon is slower than the eluvial surface layer, which saturates rapidly under heavy rains, induces runoff, and increases erosion hazards. On the other hand, the less advanced weathering stage of the subsoil and the frequent presence of silt in the lower part of the horizon may improve the available water-holding capacity of Alfisols and Ultisols.

The kandic horizon. The definition given in *Keys to Soil Taxonomy* (Soil Survey Staff, 1987) is summarized here. The horizon is a subsurface horizon that has all of the following properties:

1. Has a cation exchange capacity of <16 cmol($+$)/kg clay (by 1 M NH_4OAc at pH 7) and an effective cation exchange capacity <12 cmol($+$)/kg clay (by sum of bases extracted with 1 M NH_4OAc at pH 7 plus 1 M KCl-extractable Al), *and*

2. Has a texture of loamy very fine sand or finer, *and*

3. Underlies a coarser textured surface horizon, *and*

4. Has more total clay than the overlying surface horizon, and the increase is reached within a vertical distance of *15* cm. The percentage increase requirements are the same as for the argillic horizon except for the first clause which requires 4 percent more clay absolute when the surface layer contains less than 20 percent total clay.

The kandic horizon is only diagnostic for Alfisols and Ultisols when the surface horizon contains less than 40 percent clay.

The kandic horizon does not have the advantages of high cation exchange capacities that some argillic horizons may have. The mineralogy of the clay fraction is the same as in the oxic horizon.

8.2   Profile Descriptions and Analyses

The three Ultisols that have been included in App. B have a kandic horizon, and many of their properties are derived from their advanced weathering stage. The generally low water-holding capacities, the small amounts of fine silt (2 to 20 μm), and the cation exchange capacities of the clays are indicative of this. The three profiles have light-textured surface horizons.

Two of the Alfisols (*Typic Kandiustalfs*) in App. B have a kandic horizon with the characteristic of extremely low water retentions, sometimes as low as 0.2 mm water/cm soil.

The cation exchange capacities of the clays of the two other Alfisols are too high to meet the definition of kandic horizons: i.e., the x-ray and the mineralogical analyses detected micas and feldspars in the *Kandic Paleustalf.* The *Rhodic Paleudalf* derived from mafic rocks contrasts sharply with the other examples given by its large nutrient content and would by most standards be considered one of the best soils of the tropics.

8.3 Aquic Suborders

There are large areas of Ultisols and Alfisols that suffer seasonally from impeded drainage. Water saturation and reducing conditions are unique attributes that impose special management approaches to land use. These soils have been recognized as *Aquults* and *Aqualfs* when the groundwater affects the whole profile or as *aquic* subgroups in other suborders when only the lower horizons are periodically saturated with water. The Aquults and Aqualfs often occur in association with Tropaquepts, a great soil group of Inceptisols.

8.3.1 Occurrences

The Aqualfs and Aquults usually cover low-lying areas such as alluvial plains, small depressions on old continental shields, and large subsiding zones that border mountain ranges consisting dominantly of felsic rocks. The most typical examples occur in climates with sharply contrasting wet and dry seasons such as monsoon climates where direct rainfall is the main cause of inundations rather than overflow by rivers. When the latter carry suspended solids that are deposited as sediments in depressions, they retard the horizonation processes that form the diagnostic criteria of Alfisols and Ultisols.

There are some typical landscapes sites in which Aquults frequently occur. One of them is the *dambos* of southern Africa. *Gray dambo soils* (Magai, 1985) are typical soils of old depressions in ancient erosion surfaces, most of which are covered or underlaid by thick ironstone crusts. These low-lying areas contain many different soils; Aquults are frequent components of the soil associations that characterize the *dambos* in north Zambia and the southern part of Zaïre.

8.3.2 Soil-forming processes

The reduction of Fe and Mn is the controlling process that gives soils of aquic taxa their most striking characteristics; it is usually known as *gleying*. It preferentially takes place in the presence of organic matter that acts as an electron donor, the absence of O_2, and environments that are favorable for the growth of microorganisms. The dissolution of Fe is enhanced by acidity, because the reduction of Fe and Mn requires milder reducing conditions (lower Eh) in acid soils than in alkaline soils.

Reduced iron as Fe^{2+} is soluble and migrates with groundwater that does not contain dissolved O_2. Upon contact with air in the capillary fringe, Fe^{2+} readily oxidizes and precipitates as compounds containing Fe^{3+}, creating colored patterns that usually consist of yellow and red mottlings on a grayish background. If the process is repeated over several years, the accumulation may be such that it results in the formation of plinthite. Some Aquults contain plinthite in the subsoil in amounts that are sufficient to include them into a special great soil group named *Plinthaquults*. Most soils that belong to that taxon were formerly known as Groundwater Laterites.

The second important consequence of aquic regimes is the retardation of organic matter decomposition. Anaerobic microorganisms are inefficient decomposers and are largely responsible for the longer residence times of organic substances in topsoils. When saturation by water is permanent, histic epipedons (see Sec. 4.1.1) develop that mark many very poorly drained soils in the lowland tropics. In isohyperthermic soils that only temporarily undergo aquic conditions and that are completely dry for several months during the dry season, the ochric epipedons are the most common.

Aqualfs and Aquults have clay increase horizons. Aquic moisture regimes influence the processes that form these horizons and may obliterate or exaggerate the marks they leave in soils. Wilding and Rehage (1985) considered that the amounts of clay that are translocated in soils with aquic moisture regimes have often been overestimated. Lithological discontinuities, ferrolysis, and neoformation or destruction of clay are mentioned as other possible mechanisms for clay increases with depth. These authors also pointed at frequent misinterpretations of bleached ped coatings, organic films, and pressure faces that are erroneously described as illuviation cutans.

Ferrolysis (Brinkman, 1970) is a mineral-destructive process that contributes to textural differentiations in profiles. It starts with the reduction of Fe^{3+} to Fe^{2+} that, being soluble, displaces bases from the exchange complex and subsequently removes the bases from the topsoils by leaching. When the soils dry, the oxidation to Fe^{3+} produces H^+ that intensifies the weathering of clays and primary minerals. Repeated cycles of flooding and drying thus accelerate the acidification processes that enhance clay destruction and cause stronger textural increases than observed in soils that are not exposed to water table fluctuations.

Ferrolysis is most effective in Aquults that are dominated by low-activity clays. In Aqualfs the solubility of the Fe^{2+} is reduced by the higher pHs (Somasiri, 1985), and processes other than ferrolysis probably take a major part in causing the textural differentiations in wet Alfisols.

8.3.3 Land use

The need for drainage is the most important limitation of Aquults and Aqualfs. It is an unavoidable requirement, at least during the time that most crops, except rice, are grown. For moisture-tolerant crops it can be achieved by building raised beds and furrows, and this is often practiced in narrow valleys for sweet potatoes, particularly in soils with isothermic regimes where higher organic matter contents reduce the tendency of the topsoil to harden.

The low hydraulic conductivity of the clay increase horizons may reduce the feasibility of drainage works. A shallow, abrupt textural change is usually a sign of unfavorable physical conditions. If tree crops are grown, the hydraulic conductivity should be such that water from intense rains can be drained from the fields in a couple of days, leaving enough topsoil aerated for the development of the root system.

Paddy rice is the normal crop on Aquults and Aqualfs in southeast Asia. It is grown during the rains in bunded fields. It is usually difficult to include an upland crop in the paddy rice rotation if the topsoil is light textured and hardens upon drying.

Aqualfs offer more land use alternatives than Aquults because of their better nutrient-supplying power. The prospects of draining Aquults are seldom encouraging, because acute soil fertility problems, including aluminum and manganese toxicities, are likely to occur. Large extensions of Aquults are used during the dry season for grazing. On the other hand, when Aqualfs contain excessive amounts of sodium, their physical properties may prohibit their use without costly reclamation works.

8.4   Management Properties of Tropical Alfisols and Ultisols

8.4.1   Light-textured surface horizons

The eluvial horizons that cover the argillic or the kandic horizons in Alfisols and Ultisols have a tendency to harden upon drying, often making tillage of dry soil difficult or impossible with the tools that are available to small farmers. The hardening is the most frequently reported physical problem in Ultisols and Alfisols that have light-textured surface layers (less than 40 percent clay). The phenomenon is also known as *hard-setting, compaction,* or *prise en masse.* According to French pedologists the most serious cases affect soils that have less than 18 percent clay. The compaction is a temporary condition that disappears when the soils are rewetted; however, their hardness when dry is a serious constraint because it reduces the time available for seed bed preparation, often retards sowing, and thus shortens the length of the growing period. The problems are the most severe in areas where the dryness is the most pronounced and in soils that contain large amounts of very fine sand and coarse silt. The compaction constraint is the most critical in Alfisols, but Ultisols are not immune to the problem, especially when cultural practices expose the topsoils to raindrop impact and direct sunshine.

There are other physical problems with surface horizons in Alfisols and Ultisols. *Sealing* or *crusting* is the formation of a thin (1 to 5 mm) layer that is very dense and hard when dry, without any porosity, sometimes water repellent, and located close to or at the surface. Valentin (1985) distinguished several mechanisms for crust formation that include detachment of fine particles from small aggregates under the impact of raindrops, followed by migration into the soil, or sorting and deposition by runoff water into microdepressions. Under natural conditions microhorizons form at less than 1 cm depth as shown in Fig. 8.1. The voids in the crusts are mainly vesicles that are formed by entrapped air. Crusts affect infiltration, runoff, and seed emergence. For annual crops the crusts require interventions such as harrowing.

As for compaction, sealing is most frequent in soils that contain dispersible fine particles and high percentages of very fine sand and silt. The particle size distribution of the sands usually shows little sorting (Hoogmoed, 1985). More precise estimates of the risks of crusting have been proposed by FAO (1979) using an index of crusting that equals:

$$\frac{1.5 \, Zf + 0.75 \, Zc}{C + (10 \times OM)}$$

where Zf, Zc, C, and OM are the percent fine silt, coarse silt, clay, and organic matter, respectively. However, the indices may need to be calibrated to suit local conditions.

Several soil management techniques are available to alleviate the constraints imposed by topsoil hardening or crusting on crop production. The practices act either on soil temperature and moisture, nutrient availability, or a combination of these. They use one or more of the following techniques: cover crops, mulches, organic matter incorporation, fertilizers, lime, minimum tillage, deep plowing, postharvest plowing, seed bed configurations, or grass fallows. They aim at improving the physical conditions of the rooting zone so that roots can explore a larger volume of soil. Other strategies work less on the soil's physical attributes and prefer to increase the fertility of the topsoil in order to allow roots to extract enough nutrients from the small volume of soil that is accessible to them. Most approaches try to ameliorate both the physical and the chemical components of soil productivity. Decisions on which management practice to follow depend mostly on economic and social considerations, more specifically the availability of management inputs such as mulches, fertilizers, herbicides, or machinery.

Climate is the most important environmental factor to which the

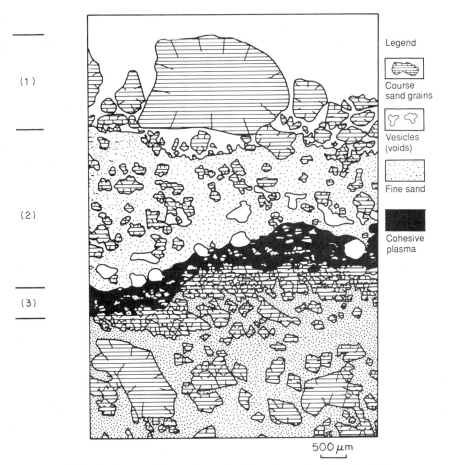

Figure 8.1 Microhorizons at the surface of subdesertic sand at Agadez, Niger. [*From Valentin (1985). C. Valentin, reprinted by permission.*]

technological "packages" have to adapt. If rainfall is high, then more materials are available for mulching and plants are stressed less by drought. The abundance of rain also decreases the risk of harmful competition for water between cover crops and the cash crop. If the climate is drier, then soil temperatures may become lethal to emerging seeds, especially in dry, sandy surface layers.

A few technological packages are described in the following sections. They vary because of differences in agroecological conditions. The methods adapted to the driest climates are described first. The more humid ecosystems follow. The practices range from systems used in Alfisols bordering deserts to the extremely leached Ultisols with perudic moisture regimes.

The management practices that farmers adopt also depend on social and economic conditions, such as the availability of land, labor, and capital. From their viewpoint the decisions on land improvement or maintenance techniques are based on cost-to-benefit considerations. New technologies to alleviate the constraints are only acceptable to the farmers if the costs and risks to introduce the technologies are less than the costs of opening new land that does not have the targeted limitations.

Deep plowing of sandy soils in Senegal.   Large areas in west Africa are covered with Alfisols that developed in sandy sediments or on sandstones. The climate is semiarid, characterized by a short rainy season with erratic rainfall. Timely land preparation is crucial to successful farming.

Nicou and Charreau (1980) recognized a textural threshold in the diagnosis of topsoils: Surface layers with less than 18 percent clay and that consist exclusively of kaolinitic minerals tend to be more affected by hardening than others.

Nicou (1974) found in trials in southern Senegal that the total porosity of Alfisols with less than 10 percent clay in the surface horizons is approximately 40 percent, which is a minimum value but does not fully account for the unsuitability for root growth (Fig. 8.2). Plowing requires a traction of 145 kg/dm^2 when the soil is dry, but only 30 to 36 kg/dm^2 when wet (Nicou, 1975). Bulk densities vary between 1.57 and 1.62 g/cm^3. Upon wetting there is a 5 percent increase in total porosity, which allows better root penetration and reasonable yields. Plowing to 20 cm depth when the moisture content is close to field capacity produces an additional 10 to 20 percent increase of up to 48 per-

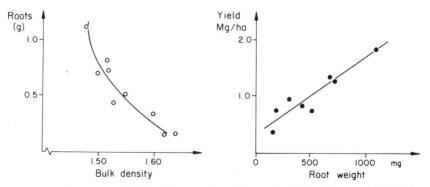

Figure 8.2   Influence of bulk density on root growth and its relation to groundnut yields. [*From Nicou (1974).*]

cent total porosity, resulting in higher production. This increase is explained by aggregate formation. However, the improved structure is very easily degraded and is lost by trampling, harrowing, and raindrop impact. Weeding by herbicides reduces the adverse effects of traffic in the fields.

Other benefits may be obtained by 20-cm-deep plowing: i.e., more roots may grow closer to water that is eventually stored in the subsoil. The disruption of crusts also increases the amounts of water that infiltrate. When the plowing takes place at the end of the previous rainy season, before the soil has dried completely, the hardening of the topsoil during the dry season is less intense, and the final seed bed preparation at the onset of the rains is easier and less time-consuming. Water acceptance from rain is also increased by the postharvest plowing. If weeds are eliminated and capillary voids broken, evapotranspiration from the subsoil is reduced and some carryover of stored water is possible. Therefore, postharvest plowing is recommended, provided the exposure of the surface does not increase the risks of wind or water erosion (Charreau, 1974).

Some soil properties influence the effects of deep plowing. The intensity of the hardening of the coarse-textured surface horizons increases with clay content. Slow desiccation produces stronger cohesion than abrupt drying; mulches may in fact aggravate the hardening process if soils are allowed to dry out slowly before tillage operations. Soil organic matter has favorable effects on soil consistency, and cultural practices that increase its content are recommended. The conservation of optimum physical conditions also depends on the kind of root systems of the crops that are grown in the rotation; cereals usually do better to maintain favorable surface soil conditions than groundnuts.

Alfisols in the semiarid tropics. The technological package proposed for medium-textured Alfisols was developed by ICRISAT at Hyderabad, India. The Indian peninsula is exposed to a harsh monsoon climate with very distinct and contrasting dry and rainy seasons. The mean annual rainfall is approximately 800 mm at Hyderabad. Many Alfisols are shallow or contain large quantities of gravel, thus reducing their water-holding capacity. The topsoils are often coarse-textured, and subsoils are heavier and less permeable than the sandy soils of Senegal discussed in the previous section. Clay minerals are mostly kaolinitic and retain few nutrients. Extreme rainfall densities often cause ponding and increased erosion hazards.

Land smoothing and the installation of field drains received high priority among the recommended management practices. They are needed because they reduce waterlogging of field depressions during the monsoon. Flat with graded or contour bunds as seed bed configuration is preferred. Raised beds to protect plants against flooding during heavy storms and to evacuate surplus water safely may have negative effects when furrows expose the less permeable argillic or kandic horizons and thus enhance runoff and erosion (El-Swaify, Singh, and Pathak, 1987). Based on experience in Africa, alternate graded, tied ridges, designed with the ties (the ties are small dams interrupting the furrows) lower than the ridges, seem the best solution for optimum water uptake in dry years and for minimizing the risk of gully erosion; ideally the raised-bed configuration should be combined with grassed drainage ways that evacuate surplus water during exceptional storms.

Tillage is considered necessary for creating favorable conditions for root proliferation and for increasing water acceptance. Shallow intercultivation and harrowing to break crusts and produce a dust mulch improve seed emergence and water conservation (El-Swaify, Singh, and Pathak, 1987).

Organic mulches reduce runoff and soil loss. They improve surface soil conditions in two ways (Perrier, 1987): First, they reduce raindrop impact. Second, they increase biological activity, particularly by termites that build channels and increase the number of continuous tubular pores in the topsoil. The formation of surface crusts in microdepressions or tied furrows is reduced by the mulch, and infiltration of water into the subsoil is increased, thus favoring seed emergence and the conservation of water.

Minimum tillage. Minimum tillage is the soil management package recommended on soils of west Africa by the International Institute of Tropical Agriculture (IITA) at Ibadan. At this station the topsoils are coarse-textured, with clay contents ranging between 10 and 25 percent. The profiles are classified as Alfisols. The parent rocks are mostly gneisses and quartzites, which belong to the precambrian Basement Complex. The natural vegetation is a moist semideciduous forest. The annual precipitation may average 1280 mm, with 3 to 4 months receiving less than 50 mm. The agroecological conditions are considerably more humid than in the regions discussed previously, and the water conservation aspects included in the technology are less important. Water shortages, however, may occur if crops are forced in the short second rainy season. Surface sealing and the rigidity of the topsoil are among the physical problems reported in the area (Moormann, Lal, and Juo, 1975).

The minimum tillage package includes the use of herbicides that kill the regrowth that invades the fields during the dry season. The killed weeds form a surface mulch that protects the soil against direct sunshine and raindrop impact. The crops are seeded at appropriate depth with minimum disturbance of soil and mulch, using special equipment. The mulch has a significant influence on soil temperatures. For example at 3:00 p.m. and 5 cm depth the temperatures are below 35°C in no-till mulched treatments, while plowed fields with their residues incorporated reach soil temperatures above 42°C. The combined effects of lower soil temperatures and reduced drought stress promote seed emergence (Lal, 1986) and also contribute to the conservation of soil organic matter.

No-tillage techniques without crop residues that are incorporated or used as surface mulch can be detrimental to soil structure, and water infiltration rates into the soil may actually decrease with time (Lal, 1978). After long periods of no-tillage, meliorative measures such as chiseling, controlling traffic, plowing at the end of the rains, cover crops, or fallows may be necessary (Lal, 1983).

The advantage resulting from no-tillage with residue mulch is that it reduces erosion, which is often one of the major constraints to sustained agriculture in Alfisols. Crop responses vary according to soil fertility. In fertile soils that have no nutrient deficiencies or that are adequately fertilized, roots can find sufficient nutrients in a small volume of soil to produce satisfactory yields. There is consequently less need to increase the number of macropores by plowing, and minimum tillage is usually appropriate. In poor soils that can only provide few nutrients per unit volume, the increased macroporosity resulting from plowing often produces higher yields than no-till practices. There are other aspects to consider when selecting methods of seed bed preparation. For example, tillage may stimulate the mineralization of organic matter and increase the supply of nitrates to the crops.

The initial physical conditions of the surface layers also determine the suitability of no-tillage. Not all undisturbed surface soil situations are ideal for root penetration; some soils have deteriorated under mismanagement. The most successful examples of minimum tillage are in areas that were recently cleared by conventional methods where bulk densities are low and decaying root systems provide continuous pores for easy root development.

Humid tropics Ultisols. The technological packages that were developed at Yurimaguas, Peru (Sanchez and Salinas, 1981) on acid Ultisols with light-textured topsoils face fertility constraints and aluminum toxicities rather than physical limitations. However, the maintenance of favorable physical conditions continues to be a prerequisite if fertilizers are to improve performance of field crops. Given the perudic moisture regime of the soils, mulches are more readily available than in the drier regions; since there is less competition for water, cover crops are more justifiable as soil management components. There is also a wider selection of crops. The year-long rains make trees the common pioneer plants in the regrowths that invade abandoned fields. In leached soils, trees are generally more efficient restorers of soil fertility than grasses.

At Yurimaguas, given the intense leaching of the soils, nutrient deficiencies appear a few years after removing the rain forest and cropping the land. Clearing methods are critical, and those that mechanically least disturb the surface layers are the most appropriate. In the traditional *slash and burn,* no-till planting is practiced.

To change from shifting cultivation that abandons fields as soon as nutrient deficiencies reduce yields to continuous cropping requires fertilization and liming. Tillage also is necessary. When heavy machinery is used, compaction is to be expected and the surface soil needs protection. Mulching, green manures, managed fallows, intercropping, and multiple cropping are among the practices proposed by Sanchez and Salinas (1981) to preserve adequate physical conditions in low-input technologies. These authors rightfully insisted that the soils should be kept covered as much as possible.

The purposes of mulching are to lower surface soil temperatures and increase soil moisture by limiting weed growth and by reducing crusting; both effects promote the proliferation of roots that are then able to explore a larger volume of soil. This strategy is particularly effective when there is either a nutrient deficiency or a shortage of water per unit volume of soil. Mulching therefore is most needed in dry years and in poor soils.

The incorporation of green manures in the soil has, besides the additions of nutrients contained in the plant residues, beneficial effects on water retention and the consistency of the topsoil. Fallow systems, either natural or managed, generally need less time in Ultisols to restore the fertility than in Oxisols. The guidelines for handling rotations that include tree fallows and determining their duration have been discussed in Chap. 3.

Some Ultisols under rain forests have been cleared by inappropriate methods. Surface soil conditions often deteriorate by improper use of heavy machinery. Even under optimum conditions, mechanized clearing without burning, such as in the experiment conducted at Yurimaguas, Peru, using a D-6C Caterpillar bulldozer results in higher bulk densities and lower water infiltration rates than in the traditional slash and burn method (Seubert, Sanchez, and Valverde, 1977). Topsoils with 10 percent clay may reach bulk densities of 1.67 g/cm^3. Some lands have to be abandoned after a few years of cropping (Alegre, Cassel, and Bandy, 1986) because of adverse surface soil conditions and the existence of a compacted zone between 15 and 45 cm depth. The subsoils can be reclaimed for continuous cropping by chiseling or subsoiling to 25 to 30 cm depth. Surface soil physical conditions (0 to 15 cm) can be improved by grass fallows (8 years in the Yurimaguas experiment) or tillage.

8.4.2 Chemical properties

Alfisols and Ultisols of tropical regions vary considerably in their chemical characteristics. Guidelines on the management of their fertility can only be given when restricting their range to the low-activity clay soils (CEC less than 24 cmol(+)/kg clay) and limiting the discussion to cations and phosphate. The principles that govern nitrogen supplies have been explained in Chap. 3.

Alfisols by definition have a higher base saturation than Ultisols, and for this reason, Alfisols are generally less deficient in Ca, Mg, and K than Ultisols. However, some caution should be exercised in making fertility interpretations. The high base saturation of Alfisols does not necessarily mean that they are "rich" in bases. Since many tropical Alfisols have a low cation exchange capacity and contain large amounts of gravel that act as a diluent, their volumetric nutrient content may be low.

Some examples may illustrate the order of magnitude of the volumetric nutrient content of several Alfisols at the International Institute of Tropical Agriculture at Ibadan, Nigeria (Moormann, Lal, and Juo, 1975) and of an Ultisol at Onne, Nigeria (Friesen, Juo, and Miller, 1982) that are given in Table 8.1. The Alfisols (at 0 to 100 cm) contain roughly 20 times more calcium than the Oxisol of Fig. 7.6 but still contain 10 times less calcium than the Mollisol shown in the same illustration. When tropical Alfisols are compared with volcanic ash soils in similar agroecological zones (Typic Haplustands or Eutric Hapludands), they have volumetric nutrient contents, except for potassium, that are roughly five-fold lower.

The differences in quality between Alfisols and Ultisols are not only a function of quantities of nutrients per unit volume. An equally important agronomic attribute that separates Alfisols and Ultisols is the severity of the aluminum toxicity that prevails in Ultisols.

Acidity and aluminum toxicities. Most Ultisols and Alfisols contain weatherable minerals. Under strong leaching conditions they release Al that produces acidity upon hydrolysis. At low pH the Al is present in the exchange complex and diffuses in the soil solution where it may cause toxicities.

The amounts of extractable Al are usually greater in Ultisols than in Oxisols. Abruña et al. (1975) found that the extractable Al ranged from 7 to 10 cmol(+)/kg in the Ultisols of Puerto Rico; in the Oxisols it only amounted to 1 cmol(+)/kg or less. On the other hand, the

TABLE 8.1   Volumetric Nutrient Content of Soils at Ibadan and Onne, Nigeria

Series	Depths, cm	Amounts*						
		C	N	Ca	Mg	K	Na	Al
Apomu								
	0–20	31	2324	1151	300	143	21	7
	0 100	44	3917	4400	564	332	64	19
Ekiti	0–20	28	2435	2061	390	250	15	5
	0–66	59	5093	5920	1034	675	56	13
Iwo	0–20	35	3566	2278	383	208	15	12
	0–100	86	8899	8317	1517	1326	70	78
Egbeda	0–20	30	1708	1975	352	179	15	3
	20–100	36	4363	5646	1045	390	132	38
Gambari	0–20	49	4653	3556	601	317	36	17
	20–100	21	2618	5394	1388	464	66	34
Iregun	0–20	15	1799	845	193	129	10	3
	0–100	61	7395	4836	1052	644	74	22
Matako	0–20	24	2987	336	247	88	22	7
	0–100	52	6467	1560	891	284	157	33
Onne	0–20	23	2541	—†	21	54	—	401
Ultisol	0–100	72	6890	—	50	148	—	2490

*Units of t/ha/m depth for C and of kg/ha/m depth for N, Ca, Mg, K, Na, and Al.
†Trace levels found.

amounts of extractable Al are greater in Ultisols than in Alfisols because the higher pH in Alfisols reduces the solubility of aluminum hydroxides (Fig 8.3).

An example of a relationship between pH and Al saturation in Puerto Rican soils, most of which are Ultisols, is shown in Fig. 8.3. (Abruña et al., 1975). *Aluminum saturation* is the aluminum extracted by 1 M KCl expressed as a percentage of the sum of bases exchanged with 1 M NH_4OAc at pH 7 plus the KCl-extractable Al. It is frequently used as a soil fertility diagnostic criterion.

The *critical levels* of exchangeable aluminum in the soil differ with the tolerance of the crops and with the *salt concentration* of the soil water. The latter often varies according to fertilization. For example, Friesen, Juo, and Miller (1982) found that the aluminum concentration in the soil solution of a fertilized Ultisol is six times that measured in the "same" but unfertilized soil (17.1 mg/L versus 2.7 mg/L). Reductions in maize yields were observed starting at soil solution concentrations of 4 mg/L, which corresponded to a 28 percent exchangeable aluminum saturation in the fertilized soil as compared to 40 percent for the unfertilized soil (Friesen, Miller, and Juo, 1980).

Liming. The first objective of liming Ultisols is to eliminate the aluminum toxicity that impairs root development and consequently reduces the nutrient and water uptake by the plants. This toxicity may

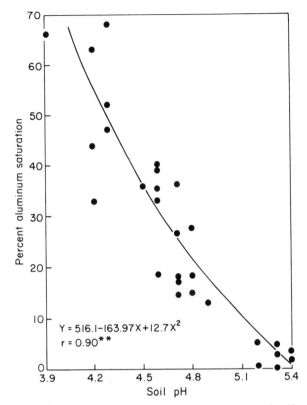

Figure 8.3 Changes in Al saturation percentages with pH.
[*From Abruña et al. (1974).*]

prevail in the virgin soil, or it may be produced or increased by the use of fertilizers. Secondary objectives may be to reach an optimum pH required by a particular crop, reduce phosphate fixation, or eliminate manganese toxicities. The target layer may be the topsoil with or without the subsoil. The time span may vary from a few years to decades, depending on the residual effects that are expected to benefit the crops included in the rotations or farming systems.

In most circumstances small farmers in tropical regions cannot afford large investments. Therefore the usual liming recommendations aim at applying just as much as is needed to *eliminate the aluminum toxicity* in the surface layers. The following formula (Kamprath, 1970; Sanchez and Salinas, 1981)[2] is often used:

---

[2] A $CaCO_3$ equivalent is the weight of $CaCO_3$ in a liming material, assuming that all the Ca and Mg are present as $CaCO_3$.

Tons $CaCO_3$ equivalents/ha $= 1.5 \times$ cmol($+$) Al/kg soil

Cochrane et al. (1980) proposed a formula that takes into account the tolerance of the crop to the aluminum toxicity as well as the initial percentage aluminum saturation in the soil. The formula is:

$$\text{Tons } CaCO_3 \text{ equivalents/ha} = \frac{1.8[Al - RAS(Al + Ca + Mg)]}{100}$$

where Al, Ca, and Mg are the cmol($+$)/kg of each exchangeable cation, and RAS is the desired percent aluminum saturation. They quoted from the literature values of tolerance of 70 percent for maize and 30 percent for soya beans, but higher percentages for upland rice, cassava, cowpeas, groundnuts, and grasses. Sorghum on the contrary is very sensitive to aluminum.

A parallel approach with the objective of eliminating Al toxicities would be to apply as much lime as is needed to raise the pH (H_2O) of the soil to 5.4. A laboratory pot experiment that keeps the soil in contact with variable amounts of lime during a 6-week period and monitors the changes in pH would be appropriate.

Most experiments on liming have been conducted on plots that receive adequate and complete fertilization. There is little evidence that liming alone can produce satisfactory crop responses in unfertilized nutrient-deficient Ultisols in the lowland tropics. However, in Ultisols that contain large amounts of organic matter, such as Haplohumults, 6 Mg/ha of lime alone doubled the control yields of 1000 kg/ha maize, while complete fertilizer treatments with lime reached yield levels between 3000 and 4000 kg/ha (Morrison et al., 1989).

In a 6-year liming experiment in Ultisols that received basal fertilizer applications. Friesen, Juo, and Miller (1982) found that relatively low rates of lime are adequate to sustain yields in a continuous maize and cowpea rotation system. Liming at a rate of 0.5 t/ha powdered $Ca(OH)_2$ maintained near maximum maize yields for 2 years after application. Sustained yields for 5 years or more are possible with a one-time application of 2 t/ha $Ca(OH)_2$. The critical level of exchangeable Al required for 90 percent maximum yield of maize is about 35 percent. For cowpeas the critical level of exchangeable Al ranged from 25 to 55 percent, depending upon rate of application of chemical fertilizer as well as the cowpea variety. The extractable Al saturation in all subsoil layers of all treatments 3 years after liming exceeded 40 percent and the soil pH (H_2O) was less than 4.3, indicating that lime is leached as neutral Ca salts and has little effect in ameliorating subsoil acidity.

Residual effects of liming were studied in Ultisols in northern Zambia (McKenzie et al., 1988). Crops still respond after 8.5 years. Lime applied at 8 t/ha reduces soluble Al at depths of 30 to 45 and 60 to 75 cm to ⅓ and ½, respectively, of that found in the control; smaller amounts of lime are less effective at these depths.

Leaching losses of Ca from the surface soil during the first 3 years of the Nigeria experiments reported above (Friesen, Juo, and Miller, 1982) are less than 0.5 t/ha of $Ca(OH)_2$ equivalents in the less than 2 t/ha lime treatments. An amount of 1.6 t/ha $Ca(OH)_2$ was leached from the 15-cm surface soil in the 5 t/ha treatment. The authors assumed that the mobility of Ca in this Ultisol is related to the presence of acidifying nitrogenous fertilizers. After 3 years nearly all the Ca applied is found within 90 cm depth. However, most of the Ca migrates with NO_3^- and Cl^- anions and does not influence pH or the aluminum saturation in the subsoil.

Side effects of liming include higher nitrification rates, and, consequently, stronger leaching of nitrates (Arora and Juo, 1982). Liming acid soils increases the availability of soil and fertilizer Mo and S; it does not directly improve soil P uptake. Liming may accelerate the decomposition of organic matter, which may release organic P, and liming increases the availability of *subsequently* added fertilizer P. Nodulation and *Rhizobium* growth are both pH-sensitive, and liming acid soils improves biological nitrogen fixation by legumes (Adams, 1978).

Overliming is a risk in all nutrient-deficient soils with little cation exchange capacity, particularly for acidophilic crops. For example, the tolerance to high soil acidity of two cassava cultivars was studied in a pot experiment with an Ultisol (pH 4.25) and eight levels of liming (0 to 6 t/ha) (Edwards and Kang, 1978). Fresh tuber yields are drastically decreased by additions above 1.6 t/ha for one variety and above 2.5 t/ha in another. The reduction in tuber yield at high lime levels is attributed to zinc deficiency. There are other examples of overliming of field crops (Spain, 1976) that illustrate the need for accurate lime requirement diagnosis.

Phosphate. The phosphate requirements of the Ultisols and Alfisols with *light-textured surface horizons* are not as demanding as those of Oxisols. The low clay and iron oxide contents that are due to eluviation do not generate strong sorption mechanisms for phosphates. For example, the phosphate sorption isotherm of the topsoil at Yurimaguas, Peru, which contains 10 percent clay, shows that 100 mg/kg sorbed P is sufficient to create a 0.2 mg/L P concentration in the soil solution (Sanchez and Salinas, 1981). In surface layers of Ultisols developed on Miocene to late Pleistocene terraces of the river Niger in Nigeria, the P-sorption capacities that correspond to an equilibrium concentration of 0.2 mg/L in the soil solution average 110 ± 25 mg/kg. The mean clay content of these topsoils is approximately 15 percent (Loganathan, Isirimah, and Nwachuku, 1987).

The possible mechanisms of P sorption in Ultisols do not differ from those that operate in other acid soils. Multiple regression analyses show that exchangeable (KCl extractable) Al (Al_{exch}), amorphous Al (Al_{am}), and amorphous Fe (Fe_{am}) control much of the variability in P sorption in the Ultisols of the Niger terraces. The regression equations that estimate P sorbed (expressed in mg/kg) at 0.2 mg/L in solution are (Loganathan, Isirimah, and Nwachuku, 1987):

$$P \text{ sorbed} = 0.224 \, Al_{exch} + 0.016 \, Al_{am} + 0.007 \, Fe_{am}$$

on the oldest terrace and

$$P \text{ sorbed} = 0.418 \, Al_{exch} + 0.049 \, Al_{am} + 0.004 \, Fe_{am}$$

on the late Pleistocene terrace.

In light-textured Ultisols developed from acid rocks such as granites, gneisses, or other rocks that are rich in quartz, the *total* P content may be the primary limiting factor. For example, in peninsular Malaysia the 0- to 15-cm layer of Ultisols have the lowest total P values (201 mg/kg) (Kalpage and Wong, 1977). In most cases small additions of fertilizer P may adequately correct the deficiency. For acid-tolerant crops, slowly soluble P sources may be used. For other uses P can be applied well before liming the soil, thus avoiding its precipitation as calcium phosphate and taking advantage of the soil acidity to increase the solubility of the P fertilizer.

Light-textured Alfisols share some of the properties of the Ultisols with regard to P. However, if the pH is higher than 5.2, Al plays a lesser role in P sorption than in Ultisols. Iron oxides and clays instead are the dominant sorbing phase (Juo and Fox, 1977).

The optimum P rate for dryland sorghum in an Alfisol in Nigeria (Ogunlela, 1988) is 11 kg/ha P. In a red sandy loam soil of Hyderabad an application of 30 kg/ha P is essential for sorghum intercropped with red gram (*Cajanus cajan*). Higher yields of sorghum are obtained with P at 50 kg/ha (Venkateswarlu, Das, and Rao, 1986). At Ibadan, Nigeria, cowpea seed yields are significantly increased by 13 kg/ha P. The critical Bray P-1 level (Bray, 1948) for growing cowpeas is 7 mg/L P. P significantly increases the N content and decreases the K, Zn, Ca, and Mn contents of cowpea leaves. It also raises the P content of the seeds (Kang and Nangju, 1983). Optimum yields of cowpeas are obtained with soil solution concentrations of P between 0.12 and 0.27 mg/L (Anyaduba and Adepetu, 1983). The P requirement of light-textured Alfisols for yield increases in maize varies between 26 and 52 kg/ha P (Kang and Osiname, 1979).

There may be residual effects of applied P. In experiments on

Egbeda soils in Nigeria one addition of 52 to 104 kg/ha P was sufficient for 2 to 3 maize crops (Kang and Osiname, 1979). The Alfisols and Ultisols that have no light-textured surface horizon and usually contain more than 40 percent clay react very much like Oxisols with respect to P sorption. The most typical soils are either derived from mafic rocks or limestones. In these soils iron oxides with a specific surface on the order of 300 m²/g provide the dominant area for P sorption. In west Africa the topsoils need between 190 and 430 mg/kg sorbed P and the subsoils need between 390 and 1160 mg/kg sorbed P to achieve a P concentration of 0.2 mg/L in the soil solution (Juo and Fox, 1977). Alfisols often contain as much calcium phosphates as aluminum phosphates, and in this respect they often differ from Ultisols. Liming Ultisols before the application of P fertilizers in order to neutralize the exchangeable Al may increase the availability of added P. However, such practices may have the opposite effect in Alfisols because of the formation of calcium phosphates.

References

Abruña, F., R. W. Pearson, and R. Perez-Escolar, "Lime response of corn and beans in typical Ultisols and Oxisols of Puerto Rico," in E. Bornemisza and A. Alvarado (eds.), *Soil Management in Tropical America, Proceedings of a Seminar,* North Carolina State University, Raleigh, 1975, pp. 261–299.

Adams, F., "Liming and fertilisation of Ultisols and Oxisols," in C. S. Andrew and E. J. Kamprath (eds.), *Mineral Nutrition of Legumes in Tropical and Subtropical Soils,* CSIRO, Australia, 1978, pp. 377–394.

Alegre, J. C., D. K. Cassel, and D. E. Bandy, "Reclamation of an Ultisol damaged by mechanical land clearing," *Soil Sci. Soc. Amer. J.* 50:1026–1031 (1986).

Anyaduba E. T. and J. A. Adepetu, "Predicting the phosphorus fertilization need of tropical soils: Significance of the relationship between critical soil solution P, requirement of cowpea, P sorption potential, and free iron content of the soil," *Beitrage zur Tropischen Landwirtschaft und Veterinarmedizin.* 21(1):21–30 (1983).

Arora, Y. and A. S. R. Juo, "Leaching of fertilizer ions in a kaolinitic Ultisol in the high rainfall tropics: Leaching of nitrate in field plots under cropping and bare fallow," *Soil Sci. Soc. Amer. J.* 46(6):1212–1218 (1982).

Bray, R. H. "Requirements for successful soil tests," *Soil Sci.* 66:83–89 (1948).

Brinkman, R, "Ferrolysis, a hydromorphic soil forming process," *Geoderma* 3:(3):199–206 (1970).

Charreau, C., *Soils of Tropical Dry and Dry-Wet Climatic Areas of West-Africa and Their Use and Management,* Agronomy Mimeo 74-26, Department of Agronomy, Cornell University, Ithaca, NY, 1974.

Cochrane, T. T., J. G. Salinas, and P. A. Sanchez, "An equation for liming acid mineral soils to compensate crop aluminum tolerance," *Trop. Agric.* 57:133–140 (1980).

Edwards, D. G. and B. T. Kang, "Tolerance of Cassava (*Manihot Esculento* Crantz) to High Soil Acidity," *Field Crops Res.* 1(4):337–346 (1978).

El-Swaify, S. A. and P. Pathak, "Physical and conservation constraints and management components of SAT Alfisols," in *Alfisols in the Semi-Arid Tropics.* ICRISAT Center, Patancheru, India, 1987, pp. 33–48.

FAO, *FAO-Unesco Soil Map of the World, Vol 1, Legend.* Unesco-Paris, 1964.

FAO, *Soil Survey in Irrigation Investigations,* FAO Soils Bulletin #42, FAO, Rome, 1979.

FAO, *Guidelines: Land Evaluation for Rainfed Agriculture*, FAO Soils Bulletin #52, FAO, Rome, 1983.

FAO, *FAO-Unesco Soil Map of the World, Revised Legend*, World Soil Resources Report 60, FAO, Rome, 1988.

Friesen, D. K., M. H. Miller, and A. S. R. Juo, Liming and lime-phosphorus-zinc interactions in two nigerian Ultisols: II. Effects on maize root and shoot growth," *Soil Sci. Soc. Amer. J.* 44(6):1227–1232 (1980).

Friesen, D. K., A. S. R. Juo, and M. H. Miller, "Residual effects of lime and leaching of calcium in a kaolinitic Ultisol in the high rainfall tropics," *Soil Sci. Soc. Amer. J* 46(6):1184–1189 (1982).

Hoogmoed, W. B., "Crusting and sealing problems on west African soils," in F. Callebaut, D. Gabriels, and M. De Boodt (eds.), *Assessment of Soil Surface Sealing and Crusting*, State University of Ghent, Ghent, Belgium, 1985, pp. 48–55.

Juo, A. S. R. and R. L. Fox, "Phosphate sorption characteristics of some benchmark soils of West Africa," *Soil Sci.* 124(6):370–376 (1977).

Kalpage, F. S. C. and H. H. Wong, "Forms of native phosphorus and their relationships to soil parameters in acid tropical soils," in K. T. Joseph (ed.), *Clamatrops Conference on Classification and Management of Tropical Soils*, Malaysian Society of Soil Science, Kuala Lumpur, Malaysia, 1977, pp. 122–138.

Kamprath, E. J., "Exchangeable aluminum as a criterion for liming leached mineral soils," *Soil Sci. Soc. Amer. J.* 34:252–254 (1970).

Kang, B. T. and O. A. Osiname, "Phosphorus response of maize grown on Alfisols of southern Nigeria," *Agronomy J.* 71(5):873–877 (1979).

Kang B. T. and D. Nangju, "Phosphorus response of cowpea (*Vigna unguiculata* L. Walp)," *Trop. Grain Legume Bulletin* 27:11–16 (1983).

Lal, R., "Influence of tillage methods and residue mulches on soil structure and infiltration rate," in W. W. Emerson, R. B. Bond, and A. R. Dexter (eds.), *Modification of Soil Structure*, John Wiley & Sons, 1978, London; New York, pp. 393–402.

Lal, R., "Soil conditions and tillage methods in the tropics," in I. O. Akobundu and A. E. Deutsch (eds.), *No-Tillage Crop Production in the Tropics*, Oregon State University. Corvallis, OR, 1983, pp. 217–235.

Lal, R., "Effects of eight tillage treatments on a tropical Alfisol: Maize growth and yield," *J. Sci. Food Agric.* 37(11):1073–1082 (1986).

Loganathan, P., N. O. Isirimah, and D. A. Nwachuku, "Phosphorus sorption by Ultisols and Inceptisols of the Niger delta in southern Nigeria," *Soil Sci.* 144(5):330–338.

Magai, R. N., "Wetland soils of Zambia," in *Wetland Soils: Characterization, Classification, and Utilization*, International Rice Research Institute, Los Baños, Philippines, 1985, pp. 489–498.

McKenzie, R. C., D. C. Penney, L. W. Hodgins, B. S. Aulakh, and H. Ukrainetz, "The effects of liming on an Ultisol in northern Zambia," *Comm. Soil Sci. Plant Anal.* 19: 1355–1369 (1988).

Moormann, F. R., R. Lal, and A. S. R. Juo, *The Soils of IITA*. IITA Technical Bulletin No. 3, Communications and Information Service, IITA, Ibadan, Nigeria, 1975.

Morrison, R. J., R. Naidu, P. Gangaiya, and Y. W. Sing, "Amelioration of soil acidity and the impact on the productivity of some highly weathered Fijian soils," in *Soil Management and Smallholder Development in the Pacific Islands*, IBSRAM Proceedings #8, Bangkok, Thailand, 1989.

Nicou, R., "Contribution à l'étude et à l'amélioration de la porosité des sols sableux et sablo-argileux de la zone tropicale sèche. Conséquences agronomiques," *Agronomie Tropicale* 29(11):1100–1127 (1974).

Nicou, R., "Le problème de la prise en masse à la dessication des sols sableux et sablo-argileux de la zone tropicale sèche," *Agronomie Tropicale* 30(4):325–343 (1975).

Nicou, R. and C. Charreau, "Mechanical impedance related to land preparation as a constraint to food production in the tropics (with special reference to fine sandy soils in west Africa)," in *Priorities for Alleviating Soil-Related Constraints to Food Production in the Tropics*, International Rice Research Institute, Los Baños, Philippines, 1980, pp. 371–388.

Ogunlela, V. B., "Growth and yield responses of dryland grain sorghum to nitrogen and phosphorus fertilization in a ferruginous tropical soil (Haplustalf)," *Fertilizer Res.* 17(2):125–135 (1988).

Perrier, E. R., "An evaluation of soil-water management on an Alfisol in the semi-arid tropics of Burkina Faso," in *Alfisols in the Semi-Arid Tropics,* ICRISAT Center, Patancheru, India, 1987, pp. 59–65.

Sanchez, P. A. and J. G. Salinas, "Low-input technology for managing Oxisols and Ultisols in tropical America," *Adv. Agronomy* 34:279–406 (1981).

Seubert, C. E., P. A. Sanchez, and C. Valverde, "Effects of land clearing methods on soil properties of an Ultisol and crop performance in the Amazon jungle of Peru," *Trop. Agric. (Trinidad)* 54(4):307–321 (1977).

Soil Survey Staff, *Soil Taxonomy, A Basic System of Soil Classification for Making and Interpreting Soil Surveys,* Agriculture Handbook No. 436, Soil Conservation Service, U.S.D.A., Washington, D.C., 1975.

Soil Survey Staff, *Keys to Soil Taxonomy,* SMSS Technical Monograph #6, International Soils, Agronomy Department, Cornell University, Ithaca, NY, 1987.

Somasiri, S., "Wet Alfisols with special reference to Sri Lanka," in S. Banta and C. V. Mendoza (eds.), *Wetland Soils: Characterization, Classification, and Utilization.* International Rice Research Institute, Los Baños, Philippines, 1985, pp. 421–438.

Spain, J. M., "Field studies on tolerance of plant species and cultivars to acid soil conditions in Colombia," in M. G. Wright (ed.), *Plant Adaptation to Mineral Stress in Problem Soils,* Cornell University, Ithaca, NY, 1976, pp. 213–222.

Valentin, C., "Surface crusting of arid and sandy soils," in F. Callebaut, D. Gabriels, and M. De Boodt (eds.), *Assessment of Soil Surface Sealing and Crusting,* State University of Ghent, Ghent, Belgium, 1985, pp. 40–47.

Venkateswarlu, J., S. K. Das, and U. M. B. Rao, "Phosphorus management for castor-sorghum + red gram intercropping rotation in semi-arid Alfisols," *Indian Soc. Soil Sci.* 34(4):799–802 (1986).

Wilding, P. and J. A. Rehage, "Pedogenesis of soils with aquic moisture regimes," in *Wetland Soils: Characterization, Classification, and Utilization,* International Rice Research Institute, Los Baños, Philippines, 1985, pp. 139–157.

9

Vertisols

9.1 Definition

The most striking characteristics of Vertisols are their high contents in smectites that shrink and swell following pronounced seasonal changes in moisture. These processes create typical soil structure and crack patterns. When Vertisols are wet, the soils are sticky, cohesive, plastic, and practically impermeable. When they are dry, they may become extremely hard and massive or they may break into blocks or prisms that are separated by wide cracks occupying large parts of their volume.

The criteria used in the detailed definition of Vertisols include the following three properties (Soil Survey Staff, 1987):

1. The soils are at least 50 cm deep and contain to that depth more than 30 percent clay, *and*
2. At certain periods in most years, the soils have open cracks that are at least 1 cm wide at 50 cm depth, *and*
3. The soils have one or more of the following characteristics:
   a. Gilgai, *or*
   b. At some depth between 25 and 100 cm, slickensides close enough to intersect *or* tilted (10 to 60 degrees), wedge-shaped aggregates.

The tropical Vertisols are subdivided at the suborder level into *Uderts, Usterts,* and *Torrerts,* depending on the length of time the cracks remain open during the year. These subdivisions parallel the definitions of suborders in other taxa on the basis of udic, ustic, or aridic moisture regimes.

The soils that are grouped in the Vertisols order have always caught the attention of farmers and have received many local and regional names. In the tropics the following names are the most common (Dudal and Eswaran, 1988); *Badobes* (Sudan); Black Clays, Black

Cracking Clays, Black Cotton Soils, Dary Clay Soils, and *Dian Pere* (Francophone, west Africa); *Firki* (Nigeria); *Grumusols* and *Karail* (India); *Makande* (Malawi); *Mbuga* (Tanzania); *Mourcis* (Mali); *Regurs* (India); *Sonsocuite* (Nicaragua); *Teen Suda* (Sudan); *Tirs* (North Africa); and Tropical Black Earth.

The term "Vertisols" is also used in other classification systems such as FAO-Unesco (FAO, 1988) in which it covers a concept that is very similar to the one used in *Soil Taxonomy* (Soil Survey Staff, 1987). An earlier name was *Grumusols* (Oakes and Thorp, 1951).

Although the concept of Vertisols is narrowly defined in terms of their physical soil properties, their chemical characteristics vary widely. Some contain calcic or gypsic horizons or both; others are free of them, although most would be fully saturated with bases. There are, however, some rare examples of acid soils that meet the requirements of Vertisols.

9.2   Profile Descriptions and Analyses

The descriptions and analyses of two Vertisols from Africa are included in App. B. They have ustic moisture and isohyperthermic temperature regimes and contain more than 50 percent clay. The subsoils have calcium carbonate nodules and the base saturation is 100 percent. The highest pH values measured in water reach 8.5.

9.3   Occurrence

The largest areas of Vertisols are located in Australia (70 million hectares), India (79 million hectares), and Sudan (50 million hectares). Smaller areas that are larger than 10 million hectares occur in Chad and Ethiopia. Vertisols occupy approximately 4 percent of the land area in the tropics (Dudal and Eswaran, 1988).

The occurrence of Vertisols depends on specific soil-forming processes. The first process to consider is the accumulation of 2:1 clays, generally montmorillonitic, that are either the product of authigenic clay formation or the result of selective sedimentation from slowly flowing water or lakes.

Smectites synthesize in environments in which soil solutions contain adequate amounts of silica and bases, particularly magnesium. Favorable conditions, for example, exist close to weathering basalts or dissolving limestones. Other frequent sites in dry climates are the lower portions of slopes that receive alkaline seepage waters from uplands.

The second process necessary for the formation of Vertisols is the periodic shrinking and swelling that produces slickensides or gilgai microrelief (Fig. 9.1). The semiarid tropics and subtropics with

Figure 9.1   Cracking of heavy clay soil in Sudan.

marked dry and rainy seasons are therefore the typical climatic zones where most Vertisols occur. Closer to the deserts, the soil materials are seldom sufficiently moistened to swell significantly during the rainy season, and in the humid tropics the dry season is too short to form the cracks.

According to Jewitt et al. (1979) and Blokhuis (1982), the morphology of Vertisols changes according to the length of the rainy season. They observed that the longer the rainy season: (1) the deeper and wider the cracks, (2) the more pronounced the gilgai microrelief, (3) the more distinct the wedge-shaped soil structure, and (4) the shallower the surface mulch of fine aggregates.

9.4   Genesis

The formation of slickensides and gilgai microrelief (Fig. 9.2) has been explained by several models. All of them include shrink-swell phenomena (Wilding and Tessier, 1988).

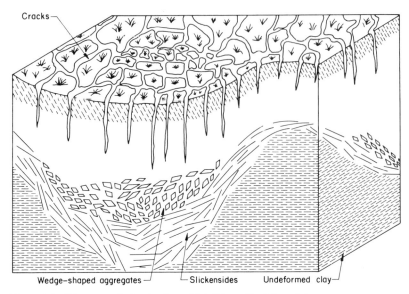

Cracks

Wedge–shaped aggregates ─╯   ╰─Slickensides   Undeformed clay─╯

Figure 9.2 Diagram of gilgai microrelief. [*From Dudal and Eswaran* (1988). *Soil Management Support Services, reprinted by permission.*]

The volume changes due to moisture depend on the composition of the soil materials. The components that reduce the extensibility are, for example, organic matter, carbonates, sesquioxides, and silica, all of which bind or cement particles. Low-activity clays also limit swelling and shrinking phenomena. Interlocking of skeleton grains tends to make materials more rigid. The potential for swelling on the other hand increases with the total and fine clay content.

Most shrink-swell phenomena in Vertisols are controlled by the losses or gains in *interparticle* water of the materials (normal shrinkage), rather than by volume changes of the clay crystals due to displacement of lattice water (structural shrinkage). In the *normal* process the volume changes of the material are almost equal to the gains or losses in water. For example, the dimensions of shrinking narrowly correlate with the amount of interparticle water lost between 33- and 2000-kPa tensions (McGarry and Malafant, 1987).

The shrink-swell potential between 1- and 100-kPa water tensions decreases in the following order: sodium smectites > calcium and magnesium smectites > calcium, magnesium, and potassium illites > kaolinites (Wilding and Tessier, 1988). According to Ahmad (1983), a large number of fine cracks is characteristic of carbonate-free, sodium-dominated clays, while calcium clays are associated with fewer but wider cracks.

Figure 9.3   Aspect of a slickenside in a Vertisol.

9.4.1 Slickensides

Slickensides are polished and grooved surfaces that form when one soil mass slides past another (Fig. 9.3). They can be viewed as the product of failures of soil materials along shear planes. The failures occur whenever the swelling pressures exceed the shear strengths (Wilding, 1985). The amount of swelling of a material is a function of its extensibility per unit moisture and the quantity of added water. The *pressures* in a soil layer depend on the specific swelling, the amount of materials that filled the cracks during dry periods,[1] and the forces of confinement.

Swelling pressures have two components: vertical and horizontal. In the topsoil the vertical pressures can be relieved by upward movement of the surface. Similarly, horizontal swelling in the surface layers does not produce much pressure because little in-filling of cracks occurs at shallow depth. Therefore, no large pressures are generated and only a few slickensides are formed in the surface horizons.

In the subsoil the weight of the overlying soil partly confines the vertical movements, and pressures produce shear planes at angles that depend on the ratio between the vertical and the horizontal com-

---

[1]For "normal" shrink and swell, the volume increases during wetting equal the volume losses during drying, and no pressure develops.

ponents of the swelling forces (Wilding and Tessier, 1988). For these reasons most subsoil slickensides are tilted; exceptions to this may be due to changes in soil composition such as the presence of carbonates that reduce the extensibility of the materials.

9.4.2 Gilgai microrelief

Gilgai microrelief consists of either a succession of enclosed microbasins and microknolls in nearly level areas or of microvalleys and microridges that run up and down the slope (Soil Survey Staff, 1975). The vertical interval between the depressions and the mounds or shelves varies generally from 10 to 50 cm; the microdepressions may be separated by distances that on average range from 3 to 20 m (Probert et al., 1987).

The same shrink-swell processes that are responsible for the formation of slickensides are active in the formation of gilgai. According to Wilding and Tessier (1988), gilgai formation is a consequence of the microvariability in soil moisture that locally produces differences in swelling pressures. Soil materials are pushed upwards along oblique shear planes from the points that receive the most water to the drier surrounding areas, thus creating small differences in elevation. This mechanism is subsequently enhanced by the inflow of water into the microlows that causes stronger shrink-swell phenomena as well as more intensive leaching of carbonates. Continuing wetting and drying cycles over long periods produce stronger differences in elevation between the microknolls and the microbasins, and result in the maximum expression of a microrelief that is in equilibrium with erosion.

9.5 Management Properties

Vertisols differ markedly in their suitability for land use. Some of them cannot be used for agriculture, while others are exceptionally suited for crop production. The actual use of Vertisols also depends on the technology that is available to farmers. For example, in Africa most Vertisols are under some sort of livestock farming, grazing, or related animal production system (Kanwar and Virmani, 1987). On the Indian peninsula they are intensively used and are considered one of the best agricultural soils of that subcontinent.

It is generally accepted that physical properties are more critical in controlling the qualities of Vertisols than chemical properties. This is because the most serious problem is the preparation of a seed bed that provides adequate water and air to plants at the beginning and during the growing season.

9.5.1   Physical properties

Surface structure. The structure of the topsoil is one of the most critical attributes in the management of Vertisols. When the soils are dry their structure may almost totally consist of very coarse prisms that upon wetting *may or may not* disintegrate into a loose mulch of fine or medium, angular or subangular blocks that create excellent conditions for seed germination. The soils that have this property are said to be self-mulching or self-loosening (Willcocks, 1987). They are the better soils among the Vertisols. Those that lack this quality, on the contrary, are extremely difficult to cultivate and are often left idle or are only used for grazing. One of the usual soil management approaches is to select Vertisols that have self-mulching properties for agriculture and keep this condition as permanent as possible. The ideal approach is to avoid the dispersion and detachment of clay particles from the aggregates during tillage operations or heavy rains and maintain the stability of the structure in the surface mulch.

According to Probert et al. (1987) three factors have an effect on the type of surface structure:

1. *Weather.* Coughlan (1984) observed that the slower the wetting and the higher the temperature of drying, the larger the number of small aggregates. Ahmad (1984) found that slow and frequent wetting of air-dry soil produces the best surface structures.

2. *Climate.* Rainfall correlates with surface conditions in that the wettest areas generally have the poorest structures; where rains reach 760 mm/year, hard, massive surface conditions prevail (Kanyanda, 1985).

3. *Exchangeable cations.* The exchangeable cations, particularly calcium and sodium, influence the stability of the clay. There is no general agreement, however, on critical levels for the *exchangeable sodium percentage* (ESP). The African Vertisols have good surface structures at ESP values between 8 and 16 that are sometimes associated with high magnesium percentages. These cationic proportions are considered too high by Australian soil scientists.

Soil characteristics other than those previously mentioned influence soil structures. Kanyanda (1985) reported that free iron oxides contribute to the quality of surface structures in Vertisols of Zimbabwe. This may explain why basaltic Vertisols generally are less affected by adverse physical topsoil conditions than alluvial soils derived from more acidic rocks. Critical levels of organic matter contents in surface layers are given: 3 percent organic matter (Ahmad, 1984) and 1.2 percent organic carbon (Kanyanda, 1985). The effects are enhanced by calcium in the exchange complex. High sodium and silt contents on the contrary tend to favor the formation of surface crusts. The ability of Vertisols to maintain a suitable surface structure varies with the stability of the aggregates that form the mulch. However, there is no agreement about how much stability is achieved. Coughlan, McGarry, and Smith (1986) reported that slaking (breakdown of small aggregates) only occurs when dry soil is wetted rapidly. It is a function of the rate at which water is applied and taken up by the aggregates. Dispersion is enhanced by extremely low initial moisture conditions of the aggregates (Lal and Greenland, 1978); it most commonly takes place when mechanical forces are applied on moist soil, such as raindrop impact or tillage.

There are no satisfactory laboratory tests to predict the stability of the aggregates in the surface mulch. Field observations such as the presence of crusts, sand separations, and clay "curls" in depressions may be diagnostic of poor physical conditions. Soils showing ponding and turbid water after light rains are most likely to have physical problems (Coughlan, McGarry, and Smith, 1986).

Water infiltration.   The qualities of Vertisols depend on their ability to conduct water. They have very high initial infiltration rates when dry (10 cm/h), but have extremely low saturated hydraulic conductivities when wet (as low as 0.2 mm/h). The large variations are due to the presence or absence of desiccation cracks and the clayey textures of the soil materials. Controlling cracks is critical in the use of Vertisols, and this is why their management has often been called the management of cracks.

The rain density at the beginning of the rainy season, e.g., a monsoon, defines whether the water penetrates into the subsoil or not. Initial heavy rains on dry, cracked soil let the water flow into deep horizons that adsorb the moisture; light rains in the beginning of the growing season, on the contrary, tend to wet the surface soil that swells and closes the top of the cracks, thus preventing water from reaching the subsoil.

Water retention.   The water-holding capacities of Vertisols are usually high. According to observations by Russell (1980) a 2-m-deep profile can hold more than 250 mm plant-available water. Ahmad (1988) mentions wilting point percentages between 25 and 30 (g/g) and field capacities between 60 and 80 percent by weight. The actual availability of that water to crops essentially depends on the depth of cracking, which creates planes of weakness and allows plant roots to penetrate into the subsoil.

The large water-holding capacities and the high water-acceptance rates of dry Vertisols extend the length of growing seasons considerably beyond what is normally expected in semiarid environments. Krantz et al. (1978) for example indicated that in the same area Vertisols can support crops for 26 weeks while Alfisols are only able to supply water for 17 weeks.

Variable volume soils like Vertisols often keep water at the same tension and availability to plants over a wider range of moisture contents than soils with a rigid framework. Pores in fact may become wider when the soils lose water and thus compensate for the increased retention caused by finer capillaries.

Chemical properties. The cation exchange capacity of Vertisols varies between 40 and 70 cmol(+)/kg of soil. Calcium and magnesium usually dominate the exchange complex. Potassium contents are generally adequate for crop production. Many Vertisols contain large amounts of free calcium carbonates or gypsum. A few are acidic and contain considerable amounts of exchangeable aluminum. Zinc is often unavailable to plants in calcium-saturated Vertisols.

According to Le Mare (1987), Vertisols that are continuously cultivated contain little organic carbon; percentages between 0.3 and 0.9 percent are reported for India. Nitrogen fertilization is normally necessary for high yields. Ammonia volatilization is the main cause of nitrogen losses. In waterlogged soils, e.g., rice paddies, denitrification is rapid, particularly under intermittent flooding (Craswell and Vlek, 1982).

Phosphates in Vertisols are usually present as calcium phosphates. At the pH of most Vertisols they control the P concentration in the soil solution (Le Mare, 1987). Soils derived from basalts are generally less deficient in P than others.

Erodibility. Vertisols are highly erodible because they have very slow infiltration rates and tend to form surface crusts that favor runoff. Most of them occur on long slopes that enhance the risks of surface water concentrations and may cause rill erosion and cut gullies into farmland. To set an order of magnitude, some data reported by El-Swaify et al. (1985) are given in Table 9.1.

Most soil losses occur during major storms (rainfall > 90 mm) that may reach intensities between 10 and 35 mm/h and produce more than 75 percent of the seasonal runoff and soil loss (Pathak et al.,

TABLE 9.1   Annual Runoff and Soil Losses from Vertisols under Traditional Monsoon Fallow

Year	Rainfall, mm	Runoff, mm	Soil loss Mg/ha
1976	710	238	9.20
1977	586	53	1.68
1978	1117	410	9.69
1979	682	202	9.47
1980	688	166	4.58

SOURCE: El-Swaify et al. (1985).

1985). The recurrence interval of such storms is usually less than 5 years, and seed bed preparation techniques alone are inadequate to cope with them.

9.6   Management Practices

Management strategies depend on the climatic conditions, the kinds of crops to be grown, the amount of water stored in the subsoil at the onset of the growing season, and the possibility of using irrigation water. The availability of mechanized power also increases the number of options that are open to farmers for the preparation of seed beds.

9.6.1   Rainfed post-rainy season cropping

This is the most common agricultural use of Vertisols by small farmers who do not have access to irrigation water or the resources to build drainage systems to evacuate surplus water from the land during the rainy season. An example is India where traditionally Vertisols are not cropped during the monsoon because at that time trafficability on the heavy clays is too poor. Instead, land is left in fallow to store water for the crops to be grown during the post-rainy season. During the wet period fallows are harrowed to control weeds and reduce transpiration losses. The water-holding capacity and the depth of the soils are the critical attributes in this type of cropping system. The risks of erosion during the fallow are one of the disadvantages of this type of land use (see Table 9.1).

9.6.2   Rainy season cropping

In this approach the land preparation techniques aim at one or more of the following objectives:

1. Capture as much precipitation as possible and keep it available for plant growth

2. Avoid surface ponding, runoff, and erosion

3. Protect the soil against direct raindrop impact

4. Maintain a favorably structured surface mulch

Most systems (Fig 9.4) use beds, ridges, or furrows. The beds or ridges place plant roots above the level of occasional flooding, and the furrows either serve as drainage ways or, when tied, increase the infiltration of water into the soil.

One of the systems is the *broad bed and furrow* (BBF) system, which has been developed in Hyderabad, India, by ICRISAT and is recommended for areas receiving 750 to 1250 mm/year rain. This amount of precipitation is characteristic of the more humid parts of the semiarid tropics. The BBF land preparation is described in the following sections together with some variants that are imposed by climatic conditions different from those that prevail in the area of Hyderabad.

Surface water management. The BBF system (Fig. 9.5) includes wide beds separated by furrows that drain into grassed waterways. The beds are 1 m wide. The furrows extend over 50 cm and are 15 cm deep and their recommended length at Hyderabad is 100 m (El-Swaify et al., 1985). Optimum performance is obtained on slopes of 0.4 to 0.8 percent (Kampen, 1982). The land preparation design can accommodate different row-spacing requirements. The furrows are used for traffic.

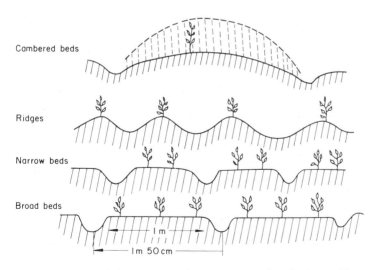

Figure 9.4   Microrelief patterns used to improve surface drainage. [*From Ahn (1988). IBSRAM, reprinted by permission.*]

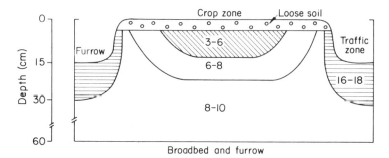

Figure 9.5   Penetration resistance zones (kg/cm) under broad bed and furrow system. [*From ICRISAT (1982)*.]

Land smoothing and installation of drains to avoid short-term waterlogging is necessary. Drainage ways are such that during extraordinarily heavy rains the excessive amounts of surface water do not overflow the land and do not cause rill erosion. Eventually the channels are to be protected by grass covers.

Supplemental irrigation can be obtained from small tanks or reservoirs that harvest water during heavy rains. They reduce the risks of crop failure in areas where rainfall is erratic or scarce. At certain locations only the topsoil needs some additional water to start or maintain a crop since the water content at lower depths is often adequate. Small tanks can achieve this purpose and avoid crop failures at the beginning of the rainy season.

Land preparation.   Under the climatic conditions of Hyderabad the best time for conducting primary tillage is at the end of the monsoon, immediately following the harvest of the previous crop. The moisture content in the surface layers at this time is optimum for plowing, and weeds are prevented from setting seeds. The plowed soil covers the desiccation cracks and reduces evaporation losses of residual moisture from the subsoil. The final land preparation, bed reforming, and cultivation is carried out at the onset of the rains, after the dry season. It is important to note that even after the long dry season the deep Vertisols in this area conserve available water below 30 cm in the subsoil.

In areas where the subsoil has to be recharged, the seed bed preparation should not begin before the subsoil moisture has been replenished through the cracks and, if possible, not before weeds have germinated. Tillage before the rains risks creating a suspended moistened seed bed above a dry soil in which crops may be exposed to lethal drought during dry spells. In irrigation agriculture, subsoil moistening can only be achieved when the soil is dry (Willcocks, 1987).

In most instances deep plowing is not recommended. Vertisols do not need major tillage operations, except for weed control. On the other hand, moisture conditions are optimum for plowing only for short periods each year. Therefore, in mechanized agriculture, high-speed shallow cultivation is most appropriate. In traditional farming, the availability of labor at the right time is critical. Deep plowing with heavy machinery includes the risk of bringing to the surface large hard clods that resist the formation of a suitable seed bed. Furthermore, the energy requirements for deep tillage are either unavailable or too costly (Willcocks, 1987).

The maintenance of a surface mulch of small, stable aggregates is not only important for root penetration but also for better water acceptance from rain by the soil and for the avoidance of ponding and surface runoff. The surface mulch contributes to water conservation and lengthens the growing season. A thick surface plow layer allows dry sowing techniques. For example, for sorghum, maize, or pigeonpea, seeds can be placed at 7 to 10 cm depth. Good soil contact and the gradual moistening of 10 cm of topsoil prevents premature germination.

Mitchell (1987) reported that in Malawi farmers sow cotton in small hollows (0.3 m long, 0.25 m wide, and 0.1 m deep), placing the seed in them uncovered. Local runoff during heavy rains covers the seeds, which then germinate after only 15 mm of rain instead of 50 mm for planting on the flat.

The critical question in maintaining an open surface structure is whether the structure will resist raindrop impact and persist until a plant canopy will protect the soil. The topsoil should also remain sufficiently permeable when the subsoil needs to be refilled with water.

According to Lal and Greenland (1978), sealing and dispersion cause the most damage when rains hit very dry topsoils. The risks of sealing and crusting may be diminished by maintaining some moisture in the surface layer at the end of the dry season before the rainy season starts, i.e., by sprinkler irrigation.

9.6.3 Irrigation agriculture

The aridity of many climates where Vertisols occur has called for the application of irrigation on land that is flat and occupies a lowland position reachable by water supplies. The costs of lifting water should not be prohibitive.

However, Vertisols do pose problems for irrigation. The cracks may

cause tremendous losses of water from the irrigation canals that have
to be lined, kept permanently wet, or protected against the swelling
forces of the surrounding clay sediments. In the case of salinity, deep
Vertisols are difficult to drain when excess salts have to be removed
by leaching. In large-scale projects where tillage operations with
heavy machinery are necessary, soils may suffer serious damage when
not plowed at the right moisture content.

The structure and the saturated hydraulic conductivity of saline-
alkaline Vertisols in Barbados are markedly improved by applications
of 10 Mg/ha gypsum, resulting in a six-fold increase of the original
0.010 cm/h hydraulic conductivity and better penetration into the soil
by the roots of sugarcane. It was also reported that the desalinization
occurred by upward movement of the salts by capillary forces, their
accumulation on the surface, and their removal by the flushing action
of subsequent rain and surface runoff (Ahmad and Webster, 1988).

In spite of many constraints to irrigation agriculture, there are several successful irrigation schemes for Vertisols in the tropics. A good
example is the Gezira in Sudan (Jewitt et al., 1979) on Usterts that
conserved excellent surface structures even after 60 years of one annual crop per year.

Puddling of soaked surface horizons in Vertisols does not present
major problems, and many Usterts and Uderts are used in rice-based
cropping systems (De Datta, Raymundo, and Greenland, 1988). The
regeneration of soil structure for upland crops that are included in a
paddy rice rotation and the lack of adequate drainage to evacuate
toxic substances that are produced by the anaerobic soil conditions are
among the major soil management issues.

9.6.4 Crop adaptation and special land treatments

Many Vertisols have a reputation of being excellent for cotton and are
still called *Black Cotton Soils* in some countries. On the other hand,
tree crops are seldom recommended because tree roots find it difficult
to establish themselves in the subsoil without being damaged by the
shrinking and swelling phenomena.

The two examples given above illustrate the importance of crop adaptation. Other approaches, however, prefer to use some special land
preparation techniques to alleviate the constraints imposed by the
soil.

Some crops tolerate flooding better than others. Taro (*Colocasia
esculenta* var. *esculenta*), sweet potatoes (*Ipomea batatas*), and yam
(*Dioscorea* spp.) grow well on beds built in poorly drained soils that

remain wet most of the time (Ahmad, 1988). *Gramineae* differ in their resistance to flooding. Among them sugarcane is the best suited because it can be planted in wet soil and because it ratoons and does not need seed bed preparation every growing season. In Uderts, however, ripening of the crop may be difficult, and drainage by pumping may be required to bring the crop to maturity.

Trees can only be planted in Uderts and Usterts on high-cambered beds with deep drains that evacuate water rapidly. Deep, large planting holes filled with lighter textured soil or mixed with compost are necessary to give young plants a good start. Coconut (*Cocos nucifera*) and oil palm (*Elaeis guineensis*) are among the trees that suffer least from temporary standing water on Vertisols (Ahmad, 1988). In some shallow Vertisols that rest on sands or other permeable sediments that contain water, narrow bore holes filled with sand may be sufficient to let tree roots reach the subsoil and grow satisfactorily.

Unimproved pastures are the most widespread use of Vertisols where equipment for seed bed preparation is unavailable or where the microvariability in crop response due to the pronounced gilgai relief cannot be overcome.

There are some special treatments mentioned in the literature to restore the soil structure for upland crops after many years of cultivation, for example, in sugar cane plantations. *Flood fallowing* (Ahmad, 1988) is one of them. Under this system the land is flooded for 6 to 9 months. Gases produced by fermentation and the redistribution of iron oxides improve the rooting conditions in the heavy clay topsoils. For subsoil drainage of Uderts or Vertisols with an aquic moisture regime, mole drainage is usually recommended (Ahmad, 1988).

9.6.5 Soil fertility management

Nitrogen and phosphorus are among the most important nutrients that are considered in the fertilization of Vertisols. The variability in fertilizer requirements, however, is high and probably reflects differences in cropping intensities and parent materials. The zinc deficiencies are related to the pH of the soils. The few acid Vertisols usually contain high amounts of exchangeable aluminum and need to be limed.

The response to nitrogen fertilization depends on the soil moisture conditions during the growing season. Extremely dry conditions that reduce crop growth lower the efficiency of nitrogen and induce losses by volatilization. In high-rainfall areas losses by denitrification are probable. The methods of application and the type of nitrogen sources also determine the effects of nitrogen fertilizers. Deep placement is best in dry years; in wet years split banding is preferred. Split applications have the advantage that farmers at midseason can adjust the amounts of fertilizers to the levels of precipitation. The time of fertilizer applications should allow full utilization of the nitrates released by soil organic matter decomposition at the beginning of the rainy season; dry sowing will decrease nitrogen losses in seasons with heavy, early monsoon rains. The optimum time seems to be just before the period showing a 70 percent chance of rain (El-Swaify et al., 1985).

Phosphorus is the second nutrient in Vertisols that may limit crop production. Vertisols derived from basalts seldom respond to phosphate fertilizers. In other cases, when higher yields are obtained by nitrogen fertilizers or by irrigation, the levels of phosphate applications have to be adapted. Placement of phosphate in or near the plant rows increases its efficiency (Katyal, Hong, and Vlek, 1987).

In India, Vertisols are considered deficient at less than 5 ppm Olsen-extractable P; a positive response to P fertilization may be expected when the water-soluble P content is lower than 0.5 ppm. No consistent results have been obtained in experiments testing the efficiency of superphosphates as compared to nitrophosphates (Katyal, Hong, and Vlek, 1987).

The results of chemical analyses performed on crushed samples of Vertisols may give misleading estimates of nutrient availabilities in Vertisols. The interior of many aggregates is often inaccessible to roots. Therefore the interpretations of soil tests need calibrations that are specific to Vertisols and are often only applicable to a narrow range of conditions. Potassium fertilization is usually not required. Minor element deficiencies most frequently include zinc and sulfur deficiencies.

References

Ahmad, N., "Vertisols," in L. P. Wilding, N. E. Smeck, and G. F. Hall (eds), *Pedogenesis and Soil Taxonomy. II. The Soil Orders,* Elsevier, Amsterdam, 1983, pp. 91–123.

Ahmad, N., "Tropical clay soils, their use and management," *Outlook Agric.* 13:87–95 (1984).

Ahmad, N., "Management of Vertisols in the humid tropics," in L. P. Wilding and R. Puentes (eds.), *Vertisols: Their Distribution, Properties, Classification and Management,* SMSS Technical Monograph #18, Texas A&M University, College Station, TX, 1988, pp. 97–116.

Ahmad, N. and J. L. Webster, "Improvement and management of salt affected Vertisols (Scotland clays) using irrigation and organic and inorganic soil amendments in Barbados, West Indies," in *Classification, Management and Use Potential of Swell-Shrink Soils,* Oxford & IBH Publishing Co., New Delhi, India, 1988, pp. 232–233.

Ahn, P. M., "Some observations on Vertisol properties relevant to management," in *Site Selection and Characterization.* Ibsram Technical Notes No. 1, Bangkok, Thailand, pp. 11–20.

Blokhuis, W. A., "Morphology and genesis of Vertisols," in *Vertisols and Rice Soils of the Tropics*, Symposia Papers II, Trans. 12th Int. Congr. Soil Sci., New Delhi, India, 1982, pp. 23–45.

Coughlan, K. J., "The structure of Vertisols," in J. W. McGarity, E. H. Hoult, and H. B. So (eds.), *The Properties and Utilization of Cracking Clay Soils. Reviews in Rural Science 5*, University of New England, Armidale, Australia, 1984, pp. 87–96.

Coughlan, K. J., D. McGarry, and G. D. Smith, "The physical and mechanical characterization of Vertisols," in *IBSRAM (ed.), Management of Vertisols under Semiarid Conditions*, IBSRAM Proceedings No. 6, Bangkok, Thailand, 1986, pp. 89–105.

Craswell, E. T. and P. L. G. Vlek, "Nitrogen management for submerged rice soils," in *Vertisols and Rice Soils in the Tropics*, Symposia Papers II, Trans. 12th Int. Congr. Soil Sci., New Delhi, India, 1982, pp. 193–212.

De Datta, S. K., M. E. Raymundo, and D. J. Greenland, "Managing wet Vertisols in rice-based cropping systems," in L. P. Wilding and R. Puentes (eds.), *Vertisols: Their Distribution, Properties, Classification and Management*, SMSS Technical Monograph #18, Texas A&M University, College Station, TX, 1988, pp. 117–127.

Dudal, R. and H. Eswaran, "Distribution, properties and classification of Vertisols," in L. P. Wilding and R. Puentes (eds.), *Vertisols: Their Distribution, Properties, Classification and Management*, SMSS Technical Monograph #18, Texas A&M University, College Station, TX, 1988, pp. 1–22.

El-Swaify, S. A. and E. W. Dangler, *Erodibilities of Selected Tropical Soils in Relation to Structural and Hydrologic Parameters*, SCSA Special Publ. No. 21, SCSA, Ankeny, IA, 1977, pp. 105–114.

El-Swaify, S. A., P. Pathak, T. J. Rego, and S. Singh, "Soil management for optimized productivity under rainfed conditions in the semi-arid tropics," *Adv. Soil Sci.* 1:1–64 (1985).

FAO, *FAO-Unesco Soil Map of the World. Revised legend*, World Soil Resources Report 60, FAO, Rome, 1988.

ICRISAT, *Icrisat Annual Report 1981*, Patancheru, India, 1982.

Jewitt, T. N., R. D. Law, and K. H. Virgo, "Vertisol soils of the tropics and subtropics: Their management and use," *Outlook Agric.* 10(1):33–40 (1979).

Kampen, J., *An Approach to Improved Productivity on Deep Vertisols*, ICRISAT Info. Bull. II, Patancheru, India, 1982.

Kanyanda, C. W., "Properties, management and classification of Vertisols in Zimbabwe," in *Fifth Meeting of the Eastern African Sub-Committee for Soil Correlation and Land Evaluation, Wad Medani, Sudan, 1983*, World Soil Resources Report No. 56, FAO, Rome,1985, pp. 94–109.

Kanwar, J. S. and S. M. Virmani, "Management of Vertisols for improved crop production in the semi-arid tropics," in IBSRAM *Management of Vertisols under Semiarid Conditions*, Proceedings No. 6, Bangkok, Thailand, 1987, pp. 157–172.

Katyal, J. C., C. W. Hong, and P. L. G. Vlek, "Fertilizer management in Vertisols," in IBSRAM *Management of Vertisols under Semiarid Conditions*, Proceedings No. 6, Bangkok, Thailand, 1987, pp. 247–266.

Krantz, B. A., J. Kampen, and M. B. Russell, "Soil management differences of Alfisols and Vertisols in the semiarid tropics," in M. Stelly (ed.), *Diversity of Soils in the Tropics*, ASA Special Publication #34, Soil Science Society of America, Madison, WI, 1978, pp. 77–96.

Lal, R. and D. J. Greenland, "Effect of soil conditioners and initial water potential of a Vertisol on infiltration and heat of wetting," in W. W. Emerson, R. D. Bond, and A. R. Dexter, (eds.), *Modification of Soil Structure*, John Wiley & Sons, Chichester, 1978, pp. 191–198.

Le Mare, P. H., "Chemical fertility characteristics of Vertisols," in IBSRAM, (ed.), *Management of Vertisols under Semiarid Conditions*, IBSRAM Proceedings No. 6, Bangkok, Thailand, 1987, pp. 125–137.

McGarry, D. and K. W. F. Malafant, "The analysis of volume change in unconfined units of soils," *Soil Sci. Soc. Amer. J.* 51(2):290–297 (1987).

Mitchell, A. J. B., "Management problems of cotton on Vertisols in the lower Shire valley of Malawi," in IBSRAM. *Management of Vertisols under Semiarid Conditions,* Proceedings No. 6, Bangkok, Thailand, 1987, pp. 221–229.

Northcote, K. H., *A Factual Key for the Recognition of Australian Soils,* Rellim Techn. Publica., Glenside, S. Africa, 1971.

Oakes, H. and J. Thorp, "Dark-clay soils of warm regions variously called Rendzina, Black Cotton Soils, Regur, and Tirs," *Soil Sci. Soc. Amer. Proc.* 15:347–354 (1951).

Pathak, P., S. M. Miranda, and S. A. El-Swaify, "Improved rainfed farming for semi-arid tropics implications for soil and water conservation," in S. A. El-Swaify, W. C. Moldehauer, and A. Lo (eds.), *Soil Erosion and Conservation,* Soil Conservation Society of America, Ankeny, IA, 1985, pp. 338–354.

Probert, M. E., I. F. Fergus, B. J. Bridge, D. McGarry, C. H. Thompson, and J. S. Russell, *The Properties and Management of Vertisols,* C.A.B. International, Wallingford, UK, and IBSRAM, Bangkok, Thailand, 1987.

Russell, M. B., "Profile moisture dynamics of soil in Vertisols and Alfisols," *Proc. Intl. Work. Agrocl.* ICRISAT, Hyderabad, India, 1980, pp. 75–78.

Soil Survey Staff, *Soil Taxonomy. A Basic System of Soil Classification for Making and Interpreting Soil Surveys,* Agriculture Handbook No. 436, Soil Conservation Service. U.S.D.A., Washington, D.C., 1975.

Soil Survey Staff, *Keys to Soil Taxonomy,* SMSS Technical Monograph #6, International Soils, Agronomy Department, Cornell University, Ithaca, NY, 1987.

Virmani, S. M., M. V. K. Sivakumar, and S. J. Reddy, "Climatological features of the SAT in relation to the farming systems research program," *Proc. Intl. Work. Agrocl.* ICRISAT, Hyderabad, India, 1980, pp. 106–135.

Yule, D. F. and J. T. Ritchie, "Soil shrinkage relationships of Texas Vertisols. I. Small cores," *Soil Sci. Soc. Amer. J.* 44:1285–1290 (1980).

Wilding, L. P., "Genesis of Vertisols," in *Proceedings of the 5th International Soil Classification Workshop,* Soil Survey Administration, Khartoum, Sudan, 1985, pp. 47–62.

Wilding, L. P. and D. Tessier, "Genesis of Vertisols. Shrink-swell phenomena," in L. P. Wilding and R. Puentes (eds.), *Vertisols: Their Distribution, Properties, Classification and Management,* SMSS Technical Monograph #18, Texas A&M University, College Station, TX, 1988, pp. 55–81.

Willcocks,T. J., "Avenues for the improvement of cultural practices on Vertisols," in IBSRAM (ed.), *Management of Vertisols under Semiarid Conditions,* IBSRAM Proceedings No. 6, Bangkok, Thailand, 1987, pp. 207–220.

10

Andisols

The Andisol order is a new soil order that was added in 1990 to the 10 orders of *Soil Taxonomy* (Soil Survey Staff, 1987, 1990). Most soils that are now called Andisols previously belonged to the suborder of *Andepts* in the order Inceptisols. Their names contained the radical "andepts," i.e., Eutrandepts, Dystrandepts, Hydrandepts, and Vitrandepts. The radicals "and," "ando," or "andi" come from the Japanese word *ando* meaning *dark* soil, the color being the typifying characteristic of volcanic ash soils.

There are at present probably only a few soil resource inventories that use the new terminology and for that reason this chapter adds as often as possible references to the names that were used in the 1975 version of *Soil Taxonomy*. Most of the differentiating criteria that were introduced in 1990 originated from research in temperate regions. The criteria may not have been tested as thoroughly in the tropics as they have been at high latitudes, and some of the new criteria may not be entirely satisfactory for the diagnosis of tropical volcanic ash soils.

10.1 Definitions

10.1.1 Profile descriptions and analyses

Three Andisols are included in App. B. They contrast with all other soils by their large organic matter contents; thick, dark topsoils; and low bulk densities. The particle size analyses have been carried out with acid dispersing agents.

10.1.2 Andic soil properties

The key attributes of Andisols are the *andic soil properties*. They either reflect the presence of volcanic ejecta such as ash, pumice, cinders, or lava in the soil, or they denote the properties of amorphous clays that characterize volcanic ashes that weather rapidly in humid or perhumid climates. These clay minerals have been described in Chap. 2 as the products of the *amorphous clay formation environment.* According to Wada (1985a), Andisols in temperate regions can reach maturity in 2000 or 3000 years. The identification of andic soil properties is based on several chemical dissolution techniques.

Acid oxalate extraction.   The acid oxalate extraction procedure is performed by shaking in the dark 1 g of soil in 100 ml of 0.2 M ammonium oxalate solution (pH 3) for 4 h. The extraction is assumed to dissolve selectively "active" aluminum and iron components that are present in amorphous materials as well as associated or independent amorphous silica. The method extracts allophane, imogolite, aluminum plus iron humus complexes, and amorphous or poorly crystallized oxides or hydroxides. The extraction is also assumed not to dissolve gibbsite, goethite, hematite, or phyllosilicates. The intent is clearly to measure the quantities of amorphous materials in the soil; the acid oxalate extraction is at present considered the most precise chemical method for achieving this. However, in principle, it cannot be expected that chemical methods are able to perfectly distinguish degrees of crystallinity, and some caution is to be exercised in the interpretation of the analytical data.

Phosphate retention.   *Phosphate retention* is defined as the percent phosphate retained by 1 g of soil after equilibration with a KH_2PO_4 solution. Phosphate retention is determined after shaking overnight 5 g of soil in 25 ml of a 1% KH_2PO_4 solution (pH 4.6). The criterion is diagnostic for the dominance of active aluminum in amorphous clay minerals that synthesize in rapidly weathering volcanic ashes. The method has been selected for taxonomic identification purposes and the critical levels do not necessarily coincide with soil fertility criteria; the concentration of 1 percent KH_2PO_4 is higher than current P contents in soil solutions, and in some cases the method probably overestimates the positive charges on the soil colloids under field conditions.

Diagnostic criteria.   The identification procedures for Andisols use chemical norms, one of which corresponds to the acid oxalate extractable aluminum plus one-half of the acid oxalate extractable iron. It is called *oxalfe* in this text. The other criterion is the phosphate retention and is given the acronym *pret*. Andic soil properties are only recognized in materials that have less than 25 percent organic carbon.

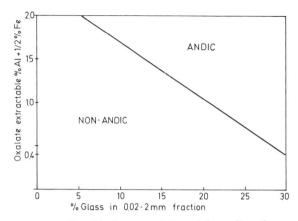

Figure 10.1 Graphical representation of the andic soil property for criterion 2c. [*From Soil Survey Staff (1990)*.]

The requirements for andic soil properties include one or both of the following (Soil Survey Staff, 1990):

1. The fine earth fraction[1] has all of the following properties:
   a. Oxalfe is at least 2.0 percent, *and*
   b. The bulk density measured at 33 kPa water retention is 0.90 g/cm^3 or less, *and*
   c. Pret is greater than 85 percent.
2. Thirty percent of the 0.02- to 2.0-mm fraction in the fine earth has a pret that is greater than 25 percent and meets *one* of the following requirements:
   a. The fine earth has oxalfe of at least 0.40 percent, and there is at least 30 percent volcanic glass in the 0.02- to 2.0-mm fraction, *or*
   b. The fine earth has oxalfe of at least 2.0 percent, and there is at least 5 percent volcanic glass in the 0.02- to 2.0-mm fraction, *or*
   c. The fine earth has oxalfe between 0.40 and 2.0 percent, and there is at least a reciprocal amount of volcanic glass, ranging from between 30 to 5 percent, in the 0.02- to 2.0-mm fraction. Figure 10.1 shows the domain of the andic soil properties in this particular instance.

10.1.3   Andisols

Andisols are mineral soils that have andic soil properties in at least 35 cm of the top 60 cm of the mineral soil or in 60 percent of the total soil

---

[1]This fraction consists of the less than 2-mm particle size fraction.

thickness if there is a lithic or paralithic contact at a depth of 60 cm or less (Soil Survey Staff, 1990). Most of these soils were previously called Andepts.

Andisols include a wide range of soils that not only differ by the properties of the subsoils that occur under the andic layer but also by large variations in soil moisture regimes, weathering stages, and leaching conditions. The emphasis in this chapter is placed on profiles that have andic properties throughout.

It should be noted that not all soils that formed from volcanic materials are classified as Andisols. Some of them have developed into more "mature" soils such as Mollisols, Alfisols, Ultisols, or Oxisols. Andisols in fact are transitional soils between fresh volcanic ejecta (Entisols) and one of the previously mentioned soil orders.

Subdivisions

Suborders. Andic soil properties actually group two kinds of soils that have developed in volcanic ejecta. The most recent ones, the *Vitrands,* still contain appreciable amounts of unweathered volcanic materials and not enough time has elapsed to produce appreciable amounts of amorphous clay minerals. The second kind consists of soils that are dominated by allophanes, imogolite, and other *x*-ray amorphous materials that are the typical weathering products of volcanic glass. The representative tropical soils of the second group are subdivided at the suborder level into *Ustands* and *Udands* on the basis of their soil moisture regime.

A simplified key for identifying the well-drained tropical Andisols into the suborders that are recognized in *Soil Taxonomy* (Soil Survey Staff, 1990) is as follows:

1. Andisols that have 1500 kPa water retention of less than 15 percent on air dry samples and less than 30 percent on undried samples throughout a thickness of 35 cm or more within 60 cm of the mineral soils surface   ...VITRANDS

2. Other Andisols that have an ustic moisture regime   ...USTANDS

3. Other Andisols   ...UDANDS

Great groups. Tropical Udands are subdivided into several great groups on the basis of specific differentiating characteristics. For example, the placic horizon is a criterion to distinguish a *Placudand* great group. They are typical for high-elevation isomesic perhumid regions underlain by rhyolitic ashes.

In the absence of placic horizons the cool perudic profiles may have such large accumulations of organic matter that the water retention at 1500 kPa of undried samples exceeds 100 percent by weight, a property that is diagnostic for *Hydrudands*. Some horizons in Hydrudands may have bulk densities below 0.5 g/cm³. Their topsoils often present hydrophobic properties or dry into a soil material made of irreversibly hardened aggregates. The Udands of the tropics that do not have these special horizons commonly belong to *Hapludands*, a term that translates into "simple" Udands.

Ustands seldom show strong accumulations of organic matter and never contain placic horizons. The most frequent "special" horizon development in tropical Ustands is the result of incomplete leaching of silica that accumulates in a duripan at some depth in the profile. The duripan is defined in Sec. 4.2.4. It is often impervious to water and cannot be penetrated by roots. It may drastically modify the management characteristics of Andisols, particularly when it occurs at shallow depth. The Ustands with a duripan are called *Durustands*; the others are *Haplustands*.

The description of Andisols in this chapter and the discussion of their management properties concentrate almost exclusively on Hapludands and Haplustands. The other great groups would require a level of detail that is not compatible with the objectives of this book.

Equivalents. The equivalents of Andisols are *Andosols* in the FAO-Unesco classification; in previous versions of *Soil Taxonomy* (Soil Survey Staff, 1987) most of them were Andepts, a suborder of Inceptisols. The French C.P.C.S. (1967) classification groups them in a *Classe des Andosols*.

Several local names were given to Andisols. They are *Trumaos* (Chile); Black Dust Soils or High Mountain Soils (Indonesia); Soapy Hills (West Indies); Humic Allophane Soils and *Kuroboku* Soils (Japan); and Yellow Brown Loams (New Zealand).

10.2 Occurrences

The geographical distribution of Andisols is linked to recent eruptions of volcanoes. Volcanic eruptions most frequently occur in active tectonic zones, the most important of which is the circum-Pacific ring that borders the western coasts of the Americas and includes the Philippines and Indonesia and other islands in the Pacific Ocean. The West Indies are a part of that ring. The rift valleys in Africa also contain regions that were recently affected by volcanic eruptions.

An inventory of the area covered by Andisols in the developing world (Dudal, Higgins,and Pecrot, 1983) expressed in million hectares gives the following figures: 5.4 for Africa, 7.3 for southeast Asia, 30.4 for South America, and 13.5 for Central America. The total represents only approximately 1 percent of the land area, or 56.8 million hectares. This is a rather low proportion of the total land surface. However, the Andisols are among the most densely populated areas in the tropics, and they are intensively used. Densities of 391 inhabitants, mostly farmers, per square kilometer are known, for example, in Rwanda, Africa, on soils with isothermic temperature regimes (Mizota and Chapelle, 1988). Some Andisols have an extremely high agricultural potential. However, others often support very fragile ecosystems that require constant attention and protection. Andisols have been reported with any kind of moisture regime, although they most frequently occur in humid and perhumid climates.

10.3   Properties of Andisols

10.3.1   Udands and Ustands

Udands and Ustands by definition only differ by their moisture regimes. There are some properties that are associated with these climatic characteristics.

Morphology. The most striking characteristics of Hapludands and Haplustands are attributable to the dominance of amorphous materials in their clay fraction. These minerals typically form by weathering of porous volcanic ash sediments under humid climatic conditions. Aluminum, iron, and silica are released by the rapid decomposition of primary minerals, and aluminum, in combination with silica, and iron precipitate as amorphous gels rather than slowly crystallize into lattice clays. In most circumstances, strong water percolation intensifies leaching that continuously removes the excess of silica. In addition, high rainfall rates favor the production of large amounts of organic matter, the mineralization of which is retarded by the amorphous clays. The clays that synthesize are essentially flocculated allophanes, precipitates of iron oxyhydroxides, and eventually opaline silica. In surface layers aluminum humus complexes may prevail. The high organic matter accumulations are probably due to aluminum that is present in toxic amounts; Fox (1980) mentioned annual decomposition rates as low as 0.5 percent in Andisols of Colombia, South America.

The morphological characteristics of Hapludands are the dark colors of the topsoils and their porous crumb structure, friable nonsticky consistence, and gradual or diffuse transitions to subsoil horizons, unless they are influenced by lithological discontinuities related to different depositional ash layers. The A horizons of Andisols are usually thicker and darker than the A horizons of soils in the same region that are not developed from volcanic ash.

The subsoils are generally brown and are thixotropic.[2] The latter attribute is apparent by a genuine type of consistence that has been qualified as having a soapy, smeary, slippery, greasy, or unctuous feeling (Mizota and van Reeuwijk, 1989). In the case of Andisols the crushing destroys voids filled with free water, and the soil releases water in quantities that may reach the liquid limit.

The structure is crumb and very porous. There are no evidences of clay illuviation. The low bulk density of the soil fragments is also a useful criterion to recognize Andisols in the field. It is often said that the morphological properties of Andisols are the result of *transformation* processes rather than the consequences of *translocations* of soil constituents to form illuvial horizons.

Volcanic ash soils and other soils respond similarly to moisture and temperature regimes; good examples of the influences of climate on soil morphology are the rainfall and temperature gradients that are related to altitude. The coolest locations in climatic toposequences are frequently typified by soils with the largest organic matter contents, darkest A horizons, and lowest bulk densities. The extremes that have perudic[3] moisture regimes and approach isomesic soil temperature regimes may be characterized by soils with placic or spodic horizons that often place them taxonomically in the Spodosol order of *Soil Taxonomy* (Soil Survey Staff, 1990).

Mineralogy. The colloidal fractions of Andisols are mixtures that contain predominantly noncrystalline minerals with small amounts of layer lattice clays. They also include aluminum and iron humus complexes. However, there are no clear-cut analytical methods to measure precisely their relative amounts. Several names have been used to qualitatively indicate their presence: *x*-ray amorphous oxides and oxyhydroxides, allophane, poorly ordered inorganic materials, and inorganic gel materials (Smith and Mitchell, 1987).

Amorphous clays. The most commonly identifiable constituents of amorphous clays in volcanic ash soils are:

1. *Allophanes.* These are hydrous aluminosilicates with short-range order and a predominance of Si-O-Al bonds (Van Olphen, 1971). Their molecular structure consists of incomplete 1:1 phyllosilicate layers that contain aluminum both in octahedral and tetrahedral

---

[2]Thixotropic means that a solid may undergo reversible sol-gel transformations, for example, liquefy by crushing a fragment between thumb and fingers or by other mechanical disturbances.

[3]A perudic moisture regime is a regime in which the monthly precipitations always exceed the monthly potential evapotranspirations.

positions. The defects produce hollow spherules of 3.5 to 5 nm in diameter. Openings in the sheets allow small molecules to enter the allophane structures. Allophane 1 has six Al-OH or $AlOH_2^+$ against one Si-OH reactive group; in Allophane 2, the ratio is six to two. There are several kinds that differ by their SiO_2 to Al_2O_3 molar ratios, which range from 0.8 to 2.0, or by their Al to Si atomic ratios, which range from 1.0 to 2.5. Allophanes with SiO_2 to Al_2O_3 molar ratios of between 1.7 and 2.0 disperse in alkaline solutions (pH 10); those with lower ratios deflocculate in acid media (pH 4). The measured specific surface of allophanes ranges between 700 and 1100 m^2/g. Allophanes are dissolved by hot 0.5 M NaOH or 0.15 to 0.2 M oxalate–oxalic acid (pH 3.0 to 3.5) (Wada, 1985a).

2. *Allophanelike constituents.* These are similar to allophane but are difficult to identify or have more defective structures.

3. *Imogolite.* These are threads of 10 to 30 nm in diameter, consisting of fine tubes that are 2.0 nm thick. The walls are approximately 0.7 nm thick. The external surfaces of the tubes contain aluminum hydroxide groups similar to those of gibbsite. The formula of imogolite is $1.1SiO_2 \cdot Al_2O_3 \cdot 2.3–2.8H_2O(+)$; the measured specific surface area varies between 900 and 1000 m^2/g (Wada, 1985a).

4. *Aluminum and iron humus complexes.* These complexes are extractable by sodium pyrophosphates ($Na_4P_2O_7$); this extraction is also used to isolate translocated aluminum and iron humus complexes from spodic horizons.

5. *Opaline silica.* This occurs as amorphous fine spheres.

6. *Amorphous iron oxides (ferrihydrite) and aluminum hydroxide gels.* In some Andisols that are derived from basaltic ashes, the ferrihydrite content may reach 35 percent, and it is assumed that these large amounts are responsible for the irreversible drying of the soil into hard aggregates (Parfitt, Childs, and Eden, 1988).

The formation of aluminum humus complexes is said to compete with the synthesis of allophane; allophane is often absent in surface horizons that are rich in organic matter in which aluminum humus complexes dominate. This trend is most commonly observed in soils of temperate regions (Wada, 1985a; Mizota and van Reeuwijk, 1989) but is also reported in Andisols of the tropics (Mizota and Chapelle, 1988).

Layer lattice clays. Gibbsite is a frequent constituent of Andisols in humid and perhumid climates. Halloysite is usually the first phyllosilicate to appear in profiles that are transforming into more mature soils. In the tropics halloysite forms preferentially in lower rainfall regions or climates that include a dry season (Colmet-Daage,

1969). Halloysite tubes have an Si-O-Si outer surface that is less reactive to phosphates than the Al-OH-Al outer sheet exposed by
allophane. Curly smectite has been mentioned in volcanic ash soils in
parts of Kenya and Tanzania that are characterized by a marked dry
season (Mizota and van Reeuwijk, 1989).

Physicochemical properties: Surface charges. The absence of well-
defined crystallized minerals in the clay fractions or their very low
proportions in the mineral assemblage typifies Andisols as genuine
"variable charge" soil. There is practically no permanent charge in
modal Andisols, although no soil is absolutely pure with respect to the
nature of the surface charges.

   In variable charge systems the charges of the soil colloids are
largely dependent on the pH *and* the electrolyte concentration of the
soil solution. The charges are generated by various soil constituents
that act rather independently of each other. They either produce positive or negative charges, depending on the soil solution pH and the
isoelectric point of the soil constituents. The reactions that generate
charges by proton exchange at both sides of the isoelectric point are
illustrated in Fig. 10.2; pH_0 is the pH at which the solid neither generates a positive nor a negative surface charge by protonation or
deprotonation processes.

   The pH_0 of gibbsite is 5.0 to 5.2; for amorphous aluminum hydroxides it is 7.0 to 9.0; and for iron oxyhydroxides it is approximately 8.5.
Phyllosilicates have low pH_0, generally at pH values well below the
current soil pH; the same is true for amorphous silica (pH_0 = 2) and
organic matter. The charge densities that develop on soil surfaces
when the soil pH moves away from the pH_0 value differ with the soil
constituent as well as with the concentration of electrolytes in the soil
solution.

   The electrolyte concentration is important for two reasons: for interpreting soil fertility and for understanding the evolution of exchange
properties during soil development. The influence of electrolyte concentration on the charge characteristics of Andisols is illustrated in
Fig. 10.3. The concentrations of the NH_4Cl solutions in which the ion
exchange capacities were measured included 0.1 *M* (labeled 1 in the
figure), 0.02 *M* (labeled 2) and 0.005 *M* (labeled 3). The first example
(905) is a sample that has a low Si to Al ratio (0.5) and contains es-

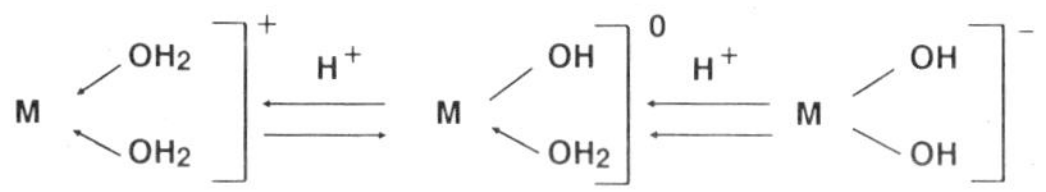

$$M\begin{array}{c}{\diagup OH_2}\\{\diagdown OH_2}\end{array}\Bigg]^+ \quad \xrightleftharpoons{H^+} \quad M\begin{array}{c}{\diagup OH}\\{\diagdown OH_2}\end{array}\Bigg]^0 \quad \xrightleftharpoons{H^+} \quad M\begin{array}{c}{\diagup OH}\\{\diagdown OH}\end{array}\Bigg]^-$$

Figure 10.2   Generation of charges by proton exchange.

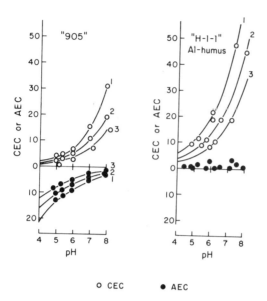

○ CEC   ● AEC

Figure 10.3 Examples of electric charge characteristics of Andisols, where CEC = cation exchange capacity and AEC = anion exchange capacity. [*From Wada* (1980). *New Zealand Society of Soil Science, reprinted by permission.*]

sentially allophanes and imogolite; the second sample (H-1-1) is an aluminum humus complex. Only sample 905 develops considerable positive charges at low pH.

The variations in exchange properties that are due to analytical procedures have to be taken into account for soil fertility interpretations, i.e., base saturation and cation exchange capacity. The routine method that uses 1 M NH$_4$OAc at pH 7 considerably overestimates the field CEC (*cation exchange capacity*) in Andisols and consequently underestimates the base saturation calculated on that basis. Table 10.1 gives an example for a Typic Hapludand in Indonesia.

Table 10.1 also shows that an increase from 0.002 to 1 M in the electrolyte concentration of the analytical solution results in a five-fold increase in the estimation of the cation exchange capacity of the soil. The base saturation in the first horizon decreases from 91.5 percent to 25.3, 21.3, and 19.5, respectively, when moving from methods 1 to 4. The analytical results are clearly laboratory artifacts. The recommendation would be to use dilute solutions that approximate the field electrolyte concentrations of the soil water for soil fertility purposes. Gillman and Fox (1980) proposed a method using unbuffered 0.002 M BaCl$_2$ that has an ionic strength close to that of the soil (0.006 M).

The cation exchange capacity of Andisols of humid environments decreases markedly if the samples are oven-dried before the analytical

TABLE 10.1   Cation Exchange Capacities (CEC) Measured at Different
Concentrations and pH Values on Samples of a Typic Hapludand

Depth, cm	Organic C, %	Clay, %	pH (H2O)	Sum of cations, cmol/kg	(1)*	(2)†	(3)‡	(4)§
40	4.13	20.9	5.9	7.0	7.65	27.7	32.8	35.9
105	4.13	19.2	6.1	6.1	5.40	22.7	37.7	41.4
207	2.89	39.2	6.0	5.3	5.90	29.1	39.5	44.3

*Ca^{2+} adsorbed from 0.002 M $CaCl_2$ solution; units of cmol(+)/kg.
†CEC by unbuffered 1 M NH_4Cl; units of cmol(+)/kg.
‡CEC by 1 M NH_4OAc at pH 7; units of cmol(+)/kg.
§CEC by 1 M NH_4OAc at pH 8.2; units of cmol(+)/kg.
SOURCE: Sutanto et al. (1988).

determination. The cause of this reduction is assumed to be shrinking
of the specific surface of the colloidal material during drying; the
changes in CEC tend to be irreversible in materials with low silica to
sesquioxide ratios, but some exchange capacity can be recovered upon
rewetting younger, less desilicated soils (Uehara and Gillman, 1981).

Chemical properties.   The variability in the chemical composition of
Andisols is such that no useful general statements can be made about
their use without distinguishing more detailed subdivisions of the order. These have been established at great group and subgroup level in
Soil Taxonomy (Soil Survey Staff, 1990) and are generally based on
nutrient contents, aluminum contents, base saturation, and organic
carbon. It is beyond the scope of this chapter to enter into a discussion
of all subgroups of Andisols, and a compromise has been made by limiting the description of the chemical composition of Andisols to a few
of their subgroups.
    Several soil-forming factors act on the chemical properties of
Andisols during their development:

1. The original composition of the volcanic ashes

2. The soil climatic regimes that control the weathering intensities on
   the one end and the rates of leaching and desilication on the other
   end

3. The accumulation of organic matter

The development sequence is to be considered in a geological time
scale.

Original composition of volcanic ash.   Table 10.2, extracted from Heiken
and Wohletz (1985), exemplifies orders of magnitude of chemical compositions of volcanic ashes. According to these data basaltic

TABLE 10.2   Chemical Composition of Common Igneous Rocks

Name	SiO$_2$	Al$_2$O$_3$	Fe$_2$O$_3$	FeO	CaO	K$_2$O
			Weight percentage			
Rhyolitic	73.7	13.4	1.25	0.75	1.13	5.35
Dacitic	63.6	16.7	2.24	3.00	5.53	1.40
Andesitic	57.7	17.0	2.4	3.6	7.5	1.6
Basaltic	50.8	14.1	2.88	9.05	10.4	0.82

SOURCE: *Heiken and Wohletz (1985)*.

mineralogies are more apt to create calcium and magnesium rich soil solutions than felsic types. They also produce environments that contain higher amounts of iron oxyhydroxides and oxides. Rhyolitic ashes on the other hand develop into calcium and iron poor soils from which cationic nutrients such as potassium leach out very rapidly. They also favor organic matter translocations into the subsoil that often lead to the formation of spodic or placic horizons.

In ashes with high molar ratios of silicon to iron and aluminum, and at the same time large iron to aluminum ratios, weathering produces more silicon-iron oxides than aluminum-iron oxides (Wielemaker and Wakatsuki, 1984) and the amorphous clay fractions may not show the same reactivity with phosphates that characterizes soils derived from less silicic volcanic ashes (Wielemaker, 1984).

Desilication and leaching.   During the initial stages of soil development, volcanic ash soils are characterized by cation-rich environments and by amorphous materials with high silica to sesquioxide ratios. These young stages, even in humid environments, for a short time supply sufficient nutrients for sustained crop production. In drier climates and in profiles with ustic moisture regimes, these incipient stages last longer, and this is why the high nutrient availability is considered typical for Ustands. A good example of a Haplustand is given in Table 10.3, where analytical results show striking differences in composition between the Haplustand and the Typic Hydrudand. In Table 10.6 the Typic Haplustand accounts for a volumetric nutrient content of approximately 35, 12, and 2 Mg/ha/1 m depth for calcium, magnesium, and potassium, respectively.

In earlier versions of *Soil Taxonomy* the base-saturated volcanic ash soils are classified as *Eutrandepts* and are recognized as very productive soils. In the Udand suborder they are now identified as *eutric* subgroups on the basis of the presence of a profile section thicker than 15 cm that contains more than 25.0 cmol(+) exchangeable cations per kilogram fine earth. The Eutric Fulvudand of Table 10.6 is an example. It contains nutrient amounts that are comparable to the Typic Haplustand mentioned in the previous paragraph; having no serious

TABLE 10.3   Chemical Properties of Two Andisols of Hawaii

Soil	Depth, cm	Organic C, %	Total analysis, %		Exchangeable bases, cmol/kg			pH (H2O)	pH (KCl)
			SiO$_2$	Al$_2$O$_3$	Ca	Mg	K		
Typic Haplustand	50	3.29	51.1	13.9	17.2	9.1	0.7	5.4	4.5
	78	1.97	44.1	15.9	22.8	11.7	0.7	5.8	4.8
	90	0.90	37.9	20.2	33.7	19.2	0.5	6.6	5.6
	133	0.60	35.8	22.2	35.8	23.8	0.3	6.7	5.6
Typic Hydrudand	40	5.30	12.9	24.3	2.0	1.8	0.1	5.8	5.6
	65	3.23	9.84	34.2	2.4	0.3	Tr.	6.3	6.4
	83	2.28	9.42	35.4	1.4	0.3	Tr.	6.4	6.4
	128	3.12	11.9	32.6	1.7	1.0	Tr.	6.4	6.4

SOURCE: *Uehara and Gillman (1981)*.

water limitations, the Eutric Fulvudand is a soil capable of sustaining intensive agricultural production. Close equivalents of these well-saturated Andisols in the FAO-Unesco classification would be called *Mollic Andosols.*

As soil development continues and when a large proportion of the weatherable minerals is consumed, silica and nutrients become deficient because the leaching losses are not compensated by supplies produced by weathering of fresh volcanic ash. The aluminum to silicon ratio of the amorphous materials increases and active aluminum starts to control several important agronomic properties—i.e., the phosphate retention and the organic matter accumulation. With higher rainfall, soil fertility conditions deteriorate faster and the soil solutions become more depleted of plant nutrients. The representative profiles at such stages are known as *Dystrandepts* in *Keys to Soil Taxonomy* (Soil Survey Staff, 1987). They are now called *acrudoxic, alic,* or *typic* subgroups of various great groups of Udands (Soil Survey Staff, 1990).

The end terms in this evolutionary sequence are the *Typic* and *Acrudoxic Hydrudands* in which the organic matter accumulation reaches a maximum. Many Hydrudands correspond to the *Hydrandepts* of earlier classifications. Their diagnostic criteria are given in Sec. 10.1.3.

The chemical properties of the two Andisols that are given in Table 10.3 are representative of the extremes in the sequence. They illustrate the large variability that exists in the soil order. This diversity justifies the use of chemical methods in diagnostic procedures, both for soil genesis and soil fertility purposes. Some of the chemical methods are briefly discussed in the following sections.

1. *Delta-pH.* This procedures is used to estimate the proportion of positive and negative charges on variable charge colloids. Uehara

and Gillman (1981) consider that delta-pH, which equals pH (KCl) minus pH (H2O), values above -0.5 are indicative of soils that are dominated by variable charge minerals. Positive values imply that the soils have a net positive charge (Mekaru and Uehara, 1972). The subsurface horizons of the Typic Hydrudand in Table 10.3 have delta-pH values of 0 or $+0.1$ that are characteristic for cation-depleted soils with high phosphate-retention capacity.

2. *Isoelectric point.* The isoelectric point (pH_0) of amorphous clays varies with their silica to aluminum ratio. The isoelectric point value shifts to higher pH values as the proportion of aluminum increases during desilication. The sign of the charges that develop on the soil colloids depends on the position of the soil pH with respect to the isoelectric point, i.e., it is the algebraic sum of the soil pH minus the isoelectric point ($pH_{soil} - pH_0$). The isoelectric point of the Haplustand in Table 10.3 is presumably low (approximately 3.5) but is high for the Hydrudand (approximately 6.4). The two subsoils have the same soil pH (around 6.5) but develop charges of opposite signs: On the one hand, the Haplustand is negatively charged and has a high CEC that is able to retain large amounts of cations, while the Hydrudand, on the other hand, in spite of a larger organic matter content, fails to do so. The use of pH_0 in the interpretation given above assumes that only one kind of soil colloid is present.

3. *Point of zero net charge (PZNC).* The concept of the PZNC was explained previously in the sections on ion exchange reactions of oxic horizons. PZNC provides a better explanation of the physico-chemical behavior of soils that contain mixtures of different colloids than the use of a single pH_0 value. It would, for example, be wrong to assume that a soil with a high PZNC and a soil pH close to it has no charges at all. A zero net charge just means that the algebraic sum of all the charges equals zero. They may or may not completely neutralize each other. The potential for developing positive charges that could be responsible for phosphate retention may be present. For example, in variable charge systems such as Andisol an increase in the electrolyte concentration of the soil solution (i.e., by fertilization) may result in the development of effective cation and anion exchange capacities.

4. *Phosphate retention.* The phosphate retention has been mentioned on several occasions in the previous sections. It varies with the SiO_2 to Al_2O_3 ratio of the soil colloids. In materials that have high SiO_2 proportions, the density of surface hydroxyls linked to aluminum is too low to generate strong phosphate retention. The younger members of the Andisol development sequence therefore

do not produce the severe phosphate deficiency symptoms in crops that the older members are able to generate.

5. *Isoelectric weathering:* The end members of the Andisol evolutionary sequence are excellent examples of isoelectric weathering, a term proposed by Mattson (1932) and used by Uehara and Gillman (1981) for transformations that take place during soil formation, particularly by weathering and leaching processes. As Andisols reach the ultimate stages in their development, the secondary amorphous aluminosilicates that they contain tend to have lower silica to sesquioxide ratios with progressively higher isoelectric points. It is also observed that as the soil solution becomes more dilute by *leaching,* the pH of the soil migrates to that isoelectric point. The evolutionary sequence thus ends with a soil that has practically no charges, very little adsorbed cations, and a relatively neutral pH. They are generally recognized as acrustoxic subgroups of Andisols.

Cation selectivity. Allophanic soils exhibit low potassium selectivity, and potassium deficiencies are usually the first ones to appear among the cations. The low potassium affinity of Dystrandepts (Udands) is related to the selective adsorption of calcium ions by organic matter and allophane (Delvaux et al., 1989).

Phosphates. The active aluminum in allophane, imogolite, and the aluminum humus complexes is highly reactive with anions such as phosphates, sulfates, and silicates. The affinity of the amorphous minerals for phosphates is very strong because of the *density* of active aluminum on the colloidal fractions. The capacity of the soil material to retain large amounts of phosphate is also the result of the very high *specific surface* of the amorphous minerals. Uehara and Gillman (1981) further mentioned the ability of fully hydrated gels to deform, partially liquefy, and trap or encapsulate phosphates in voids that are not connected with the soil solution (occlusion of phosphates).

Physical properties

Bulk densities. The low bulk densities of Udands and Ustands are one of their most essential characteristics. They are important in the interpretation of pore space; availability of water, air, and nutrients; workability; and water intake capacity of the soils. It is obvious that the depth of the soil with respect to barriers that cannot be penetrated by roots or that are impervious to water also plays a role in the agronomic properties that have just been mentioned.

The low bulk densities are partly due to the high organic matter content and to the small particle density (2.55 g/cm^3) of volcanic glass. Alvarado and Bornemisza (1985) reported particle densities that vary between 2.1 and 2.4 g/cm^3 in Andisols of Costa Rica. Table 10.4 includes typical data for several perudic Andepts in Papua, New Guinea (Allbrook and Radcliffe, 1987). Usually less than 30 percent of the volume in Udands is occupied by solids; this may cause problems in the retention of plant nutrients and is often one of the reasons why they are rapidly lost by leaching. The open structure is also at the origin of the large pore space and the low bulk densities.

The high total porosity is not necessarily an indication of good aeration. The subsoils of the profiles in Tables 10.4 and 10.5 have less than 10 percent pores that are larger than 30 μm. Under some circumstances the lack of transmission pores creates anaerobic conditions that have to be corrected by deep plowing, particularly for crops that are sensitive to waterlogging. It is not known whether the shortage of O$_2$ could cause denitrification of nitrates in the subsoils.

Structure. The microstructure of Udands is described as very porous, ranging from very spongy to fine crumbly (Stoops, 1983). Several types of aggregates are distinguished, the smallest ones being micropeds about 50 μm in diameter that may coalesce into larger structural units. Loosely packed aggregates are considered the most typical microstructures observed in Udands. The absence of evidences of clay movements is considered a sign that the aggregates are stable.

The drying of Hydrudands may result in the formation of hard aggregates that do not easily rewet. Some Andisols lose cohesion upon drying and transform into dust that is difficult to protect against wind erosion. The name *Trumao* given to Andisols in Chile actually means dust.

Water retention. The water contents retained at various tensions measured on the same undried samples as those reported in Table 10.4 are given in Table 10.5. In all horizons 1 cm of soil on an average retains

TABLE 10.4   Physical Characteristics of Some Andisols in Papua, New Guinea

Soil	Horizon	Depth, cm	Organic C, %	Clay,%	Bulk density, g/cm^3	Total porosity
Hapludand	A	0–15	9.3	40	0.38	82
	B	19–75	3.9	77	0.70	73
Hydrudand	A	0–21	12.2	42	0.54	78
	2B2	62–147	1.7	50	0.75	74
Hydrudand	A	0–19	10.5	20	0.49	76
	2Bw	19–47	7.2	66	0.45	83

SOURCE: *Allbrook and Radcliffe (1987).*

TABLE 10.5   Water Contents Measured on Undisturbed Field Moist Samples by a
Pressure Plate Apparatus

Soil	Horizon	Depth, cm	\multicolumn{5}{c}{Moisture retained (g/100 g soil) at tensions of}
			10 kPa	33 kPa	100 kPa	1500 kPa	1500 kPa†
Hapuland	A	0–15	187	165	154	101	33
	B	19–75	103	97	92	59	24
Hydrudand	A	0–21	130	122	116	82	31
	2B2	62–147	93	88	84	62	25
Hydrudand	A	0–19	110	98	84	58	29
	2B$_w$	19–47	155	147	140	125	25

†Water content measured on oven-dry sample.
SOURCE: Allbrook and Radcliffe (1987).

approximately 6 to 7 mm of water at 33 kPa or, in other terms, at field
capacity. In the A horizons the available water held between 33 and
1500 kPa varies between 2.1 and 2.6 mm/cm of soil. In the B horizons
the variation is wider and ranges between 1 and 2.5 mm/cm of soil.
Compared to other soils in the tropics such as Oxisols, these figures
are exceptionally high.

Warkentin and Maeda (1980) found that water in Andisols is not
held on clay surfaces but rather in small voids, the volume of which is
a direct function of the allophane content. The organic matter increases the available water-holding capacity. Table 10.5 shows that
oven drying of allophanic soils drastically changes their water retention at 1500 kPa. These changes may be irreversible and are attributed by some to the shrinking of the specific surface. Warkentin and
Maeda (1980) stated that upon drying the field capacity decreases
more than the wilting percentage.

Water acceptance. Andisols are known for their ability to adsorb water readily during heavy rainstorms. Briones (1982) reported that applications as large as 38 mm/h of water do not cause ponding in a series of experiments on Hydrudands (Hydric Dystrandepts). Although
this amount is higher than the infiltration rates observed in many
other soils, it is stated that, in spite of the excellent physical conditions, runoff and subsequent erosion cannot be avoided because the intensity of rain in the tropics frequently exceeds the infiltration capacity of the soil.

10.3.2 Vitrands

The major characteristic of Vitrands is their sandy or gravelly particle
size class and the dominant presence of unweathered volcaniclastic
materials. The texture is such that they cannot retain more than 15

percent water at 1500 kPa, measured on air-dried samples, or 30 percent measured on undried samples.

Their morphology, mineralogy, and physical properties depend on the kinds of eruptions, distance from the craters, and mode of transportation and sedimentation. Many vitrands are stratified and contain layers that are products of different eruptions. The heterogeneity of the parent materials is often a useful field diagnostic feature.

10.4   Management of Andisols

10.4.1   Udands and Ustands

Interpretation of analytical data. Many land use and soil management decisions are made on the basis of the information provided by routine analytical procedures. In almost all cases the chemical and physical properties of the soils are reported on a weight percentage basis calculated from analyses that are performed on air-dried samples.

The rather unique properties of Udands and Ustands make special interpretive procedures compulsory. The most important in this respect is the computation of volume percentage, using bulk densities to recalculate the data. It is also recommended to carry out the analyses on field moist samples, especially in the case of soils that have a udic or perudic moisture regime. The root systems of the crops react to amounts of nutrients per unit volume rather than per unit mass. This is particularly relevant when bulk densities are extraordinarily low as in many Udands and Ustands. The variable charge characteristics of Andisols also call for special attention when base saturation data have to be interpreted.

Examples of the significance of such transformations have been given by Kimble and Nettleton (1985): 2% oxalic acid extractable Al per 100 g soil that has a bulk density of 0.9 g/cm^3 corresponds to 630 kg Al/ha/1 cm depth or 126 t/ha in the top 20 cm. If the soil had a bulk density of 0.5 g/cm^3, this amount would be reduced to 70 t/ha. The computation could also be used for P requirements, and in the example given above, the P requirement would be reduced to 55 percent of the amount calculated on a weight percentage basis.

Table 10.6 gives data on nutrient quantities on a volume basis in a number of Andisols. It is obvious that there are large soil fertility differences between the subgroups. For example, the calcium content in the 20-cm top layer varies from 9 t/ha in the Ustands and the eutric subgroups of Udands (Eutrandepts) to less than 200 kg/ha in acrudoxic subgroups of Hydrudands (Hydrandepts) that have suffered the strongest leaching losses.

TABLE 10.6   Volumetric Nutrient Contents in Some Andisols of the Tropics

Depths, cm	C	N	Ca	Mg	K	Na	Al
				Amounts[a]			
Typic Haplustand (Typic Eutrandept) in Hawaii[b]							
0–20	57	4350	5986	1924	476	134	0
20–100	144	11,605	29,514	9811	1605	757	0
Eutric Fulvudand (Udic Eutrandept) in Mutura, Rwanda[c]							
0–20	107	11,110	9322	890	116	13	0
20–100	305	32,163	27,997	4676	222	116	0
Pachic Fulvudand (Typic Dystrandept) in Nariño, Colombia[d]							
0–20	153	n.d.[e]	115	58	38	0	276
20–100	456	n.d.	996	333	112	92	311
Hydric Thaptic Fulvudand (Hydric Dystrandept) in Naga City, Luzon, Philippines[d]							
0–20	134	n.d.	293	44	95	27	33
20–100	161	n.d.	2083	504	324	132	12
Typic Dystrandept in Costa Rica[f]							
0–20	208	n.d.	1142	620	399	n.d.	n.d.
20–100	342	n.d.	5429	1036	363	n.d.	n.d.
Acrudoxic Hydrudand (Typic Hydrandept), isomesic, in Costa Rica[g]							
0–20	137	n.d.	182	222	119	n.d.	68
20–100	421	n.d.	737	855	390	n.d.	137
Thaptic Hydrudand (Typic Hydrandept), Hilo Series, in Hawaii[d]							
0–20	96	n.d.	1214	126	58	0	0
20–100	93	n.d.	612	19	9	0	3

[a]The concentration of C is in units of t/ha, while the concentration for the other elements is in kg/ha/m depth
[b]Soil Survey Investigation Report No. 29, 1976.
[c]Based on data from IVth International Soil Classification Workshop, Part 2: Field Trip and Background Soil Data. Abos-Agcd, Brussels, 1983.
[d]Mizota and van Reeuwijk (1989).
[e]n.d., not determined.
[f]Alvarado and Buol (1975).
[g]Alvarado and Bornemisza (1985).

Nutrient retention. Allophane, imogolite, and aluminum humus complexes are weak retainers of NH_4^+ and potassium. Leaching problems are the most severe in desilicated Andisols in which the weatherable mineral reserves have been depleted.

Calcium, magnesium, and potassium theoretically can be protected against excessive losses by developing negative surface charges and thus creating additional cation exchange capacity. This can be achieved either by raising the pH, increasing the electrolyte concentration in the soil solution, or lowering the isoelectric points or the PZNC. Liming to raise the pH of strongly buffered Andisols to optimum values is seldom economically feasible. Added fertilizers, the second option, may deprotonate soil colloids and acidify the soil solution. The third alternative, to lower the PZNC, has received the most attention from soil scientists. Silicates and phosphates block positive charges on the clay and thus increase the net negative charge. Organic matter may also mask the effects of positive adsorption sites. Phosphates and silicates are expensive or not available, and organic matter buildup is difficult in warm climates. Since phosphates are needed at any rate in "mature" Andisols, phosphate applications usually offer the best solution for improving the retention of cations. Figure 10.4 illustrates the effects of phosphate applications on the retention of cations in an Hydrudand of Hawaii. In order to avoid excessive losses of NH_4^+ in paddy rice cultivation, the addition of $(NH_4)_2HPO_4$ as a nitrogen fertilizer has been recommended (Wada, 1985*b*).

Liming.   Ustands and eutric subgroups of Udands (Eutrandepts) seldom need additions of calcium to satisfy crop requirements. This is not always the case for Udands. However, most soils of this suborder are strongly buffered, and large amounts of lime are needed to raise the pH to values higher than 6 (Uehara and Gillman, 1981). Fox (1980) reported the pH values at several depths in an Acrudoxic Hydrudand in Hawaii that receives 4500 mm/year of rain. The measurements were made 5 years after the application of 21.3 and 48.0 t/ha lime. In the 0- to 15-cm layer the pH increased from 4.8 to 6.1 and 6.8; between

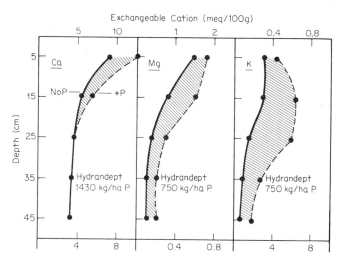

Figure 10.4   Increases in exchangeable Ca, Mg, and K retained by a Hydrudand fertilized using concentrated superphosphate for 8 years. [*From Gillman and Fox (1980). Soil Science Society of America, reprinted by permission.*]

30 and 45 cm the pH increased from 5.4 to 5.8 and 6.3; at 1 m the effect was still noticeable with a pH change from 5.8 to 6.0 and 6.3. The last pH values relate to the application of 48.0 t/ha lime.

Given the high quantities of lime needed to raise the pH in soils that contain appreciable amounts of variable charge clays, it is usually sufficient to select a pH that keeps aluminum insoluble (5.5), counteracts the effects of acidifying fertilizers, and, if possible, generates some cation exchange capacity (Uehara and Gillman, 1981).

In certain cases the liming of Andisols that contain large quantities of aluminum humus complexes may enhance the decomposition of organic matter, releasing aluminum and thus decreasing the availability of phosphates in these soils (Le Mare and Leon, 1989).

Phosphate. The most severe nutrient deficiency in Udands and Ustands, except for some *eutric* subgroups of Udands and some Ustands, is P. It is difficult to devise a satisfactory test to determine the P requirements of Andisols. On the one hand, crops differ in their needs for P. For example cassava, yams, and sweet potatoes are plant species that have low external P requirements.[4] On the other hand, soils vary in their capacity to maintain a given concentration of P in the soil solution; Fox and Kamprath (1970) derived this capacity from an interpretation of P sorption curves measured in the laboratory. However, as stated by Fox (1980), these diagnostic techniques do not take into account the amounts of P that can be supplied by the decomposition of soil organic matter.

The Andisols differ in their P requirements. For example, to reach an external P requirement of 0.2 mg/L, a Typic Haplustand (Eutrandept) needs 200 mg sorbed P/kg of soil; a Hydric Hapludand (Hydric Dystrandept) needs 1200 mg/kg; and a Typic Hydrudand (Typic Hydrandept) needs 2900 mg/kg. The latter amount corresponds to approximately 4 t/ha of P (Fox and Searle, 1978).

Several techniques have been tried to correct the P deficiencies in Andisols with large P sorption capacities: broadcast fertilizer applications to quench the P needs of the soil, application in bands to reduce the large initial investment costs, and use of silicates to diminish the number of P sorption sites on the colloids. Most of these management practices are capital-intensive and are seldom available to small farmers. They have to rely on crop adaptation to cope with severe P shortages or on manuring to build up P reserves in the soil organic matter. In Sec. 1.1.2, the effects of *guie* (soil burning) practiced by Ethiopian farmers to increase the availability of P in volcanic ash soils are described.

---

[4]The external P requirement is the concentration of P in the soil solution required by a crop for optimum yields.

According to Silva (1984), the modified Truog method of Ayres and Hagihara (1952) gives the most satisfactory test of P deficiencies in Andisols. The extraction uses a 0.02 M H_2SO_4 plus 0.03% $(NH_4)_2SO_4$ solution applied on undried samples. No crop response is to be expected at or above the critical value of 25 ppm modified Truog extractable P. Silva (1984) also devised a method to calculate the amounts of P to be added to the soil to reach the 25-ppm critical level. For the Hydric Dystrandept used in the experiment in the Philippines this amount was 189 μg P/g of soil. In the case of a 20-cm-deep topsoil with 0.7 g/cm^3 bulk density this amount corresponds to 265 kg P/ha, or 1214 kg P_2O_5/ha.

Similar orders of magnitude on P fertilizer requirements are reported in the literature. For example, for maximum potato yields in Costa Rica Alvarado and Bornemisza (1985) mentioned 480 kg P_2O_5/ha in a Typic Dystrandept and 860 kg/ha in an Hydric Dystrandept.

The effectiveness of the P fertilization depends on the method of application. Briones (1982) reported that at low doses (20 to 40 kg P/ha), placement may produce substantial yield increases that approach those of broadcast applications that use four times as much P. At high doses, broadcast application is superior to the fertilizer placement technique.

Workability. The workability of Udands and Ustands is considered easier than in other soils not derived from volcanic ash, particularly for plowing and subsoiling. The mechanical power needed is estimated at 20 to 40 horsepower (Alvarado and Bornemisza, 1985). The Hydrudands may have low trafficability, and soil may stick to the plowshare during tillage operations.

Moorman and Van Breemen (1978) reported that the stability of the open structure in Udands very often retards the formation of a plow pan that is needed to reduce water losses in paddy fields. Water percolations that exceed 50 to 60 mm/day have been mentioned whereas 10 mm/day is considered adequate. Japanese farmers prevent excessive permeabilities by adding up to 10 t/ha of bentonite to their paddy fields (Wada, 1985*b*).

Soil conservation, landslides, and mud flows. Andisols, because of their open structure and large rainwater acceptance, have a reputation for being more resistant to water erosion than other soils. However, demographic pressure and steep slopes very often lead to severe losses of soil. On the other hand, the thixotropic properties of some Andisols make them very susceptible to landslides that can have catastrophic consequences both for the land and the people who live on or close to them. Excessive confidence in the slope stability of Andisols is probably due to the fact that Andisol regions are often characterized by free-standing road cuts 30 m high and 70° slopes.

Massive soil movements usually occur after prolonged heavy rains. Soils that remain wet the whole year are the most subject to land slumps or landslides. Deforestation or other disturbances often bring about a decrease in the stability of the slopes that may suddenly collapse, soil masses sliding over the lithological discontinuities that caused the accumulation of excessive amounts of water above the contact. Durudands often lose the complete soil layer above the duripan, exposing after the slump the duripan directly at the surface.

10.4.2 Vitrands

The management properties of Vitrands vary markedly with climate. Their low water-holding capacities make actual rainfall the most critical growth factor for crop production. Deep-rooting perennials suffer less from short dry spells in the growing season than annual crops. The next land quality to consider for Vitrands that are located at high elevations is usually temperature that may be too low for many crops.

Nutrient contents of Vitrands are given in Table 10.7. It has been compiled from analyses of soils in the high Andes of Ecuador (Del Posso and Luzuriaga, 1984). When compared to the 250, 120, and 100 kg/ha/1-m depth for Ca, Mg, and K, respectively, found in the Oxisol of Zambia (see Fig. 7.6), the Vitrands of Table 10.7 contain sufficient amounts of nutrients.

TABLE 10.7   Volumetric Nutrient Content of Some Vitrands

	Amounts*					
Depths, cm	C	N	Ca	Mg	K	Na	Al

Typic Udivitrand (Vitrandept), Ecuador, 3400 m, isomesic, ash, paramo vegetation, and profile ECU-03

Depths, cm	C	N	Ca	Mg	K	Na	Al
0–20	110	6902	1233	441	137	58	90
20–100	208	n.d.†	3455	1869	344	256	205

Typic Udivitrand (Vitrandept), Ecuador, 3200 m, isomesic, ash and unweathered pumice, wheat and barley, and profile ECU-01

Depths, cm	C	N	Ca	Mg	K	Na	Al
0–20	65	7867	2892	268	383	27	0
20–100	160	n.d.	10,820	1346	964	168	0

Typic Udivitrand (Vitrandept), Ecuador, 3420 m, isomesic, ash and unweathered pumice, cropland, and profile ECU-09

Depths, cm	C	N	Ca	Mg	K	Na	Al
0–20	70	7029	3347	468	524	18	2
20–100	304	19,023	8637	1293	1016	159	99

Typic Udivitrand (Vitrandept), Ecuador, 1500 m, isothermic, ash and unweathered pumice, pasture, and profile ECU-11

Depths, cm	C	N	Ca	Mg	K	Na	Al
0–20	109	8791	3967	347	353	50	14
20–100	228	n.d.	4547	281	443	209	143

Typic Ustivitrand (Vitrandept), Ecuador, 3200 m, isomesic, ash over tuff, grassland, severe wind erosion, and profile ECU-02

Depths, cm	C	N	Ca	Mg	K	Na	Al
0–20	17	1400	2576	646	328	62	0
20–100	51	n.d.	11,245	3266	2777	341	0

*C is expressed in t/ha/m depth, while the other elements are expressed in kg/ha/m depth.
†n.d., not determined.
SOURCE: *Del Posso and Luzuriaga (1984)*.

References

Allbrook, R. F., and D. J. Radcliffe, "Some physical properties of Andepts from the southern highlands, Papua, New Guinea," *Geoderma* 41:107–121 (1987).

Alvarado, A. and S. Buol, "Toposequence relationships of Dystrandepts in Costa Rica," *Soil Sci. Soc. Amer. Proc.* 39(5):932–937 (1975).

Alvarado, A. and E. Bornemisze, "Management and classification of Andisols of Costa Rica," in F. H. Beinroth et al. (eds.), *Proceedings 6th International Classification Workshop, Chile and Ecuador. Part 1: Papers,* Sociedad Chilena Ciencia del Suelo, Santiago, Chile, 1985, pp. 69–96.

Ayres, A. A. and H. H. Hagihara, "Available phosphorus in Hawaiian soil profiles," *Hawaiian Planters' Record* 44:81–99 (1952).

Briones, A. A., "Characteristics and fertilization of Andepts in the Philippines," *Tropical Agriculture Research Series,* Tropical Agriculture Research Center, Tsukuba, Japan, 1982, 15:251–264.

Colmet-Daage, F., "Nature of the clay of some volcanic ash soils of the Antilles, Ecuador and Nicaragua," in *Panel on Volcanic Ash Soils in Latin America,* CATIE (Centro de Agricultura Tropical para la Investigación y Enseñanza), Turrialba, Costa Rica, 1969, B.2.1.–B.2.11.

C.P.C.S. (Commission de pédologie et de cartographie des sols), *Classification des Sols, Edition 1967,* Laboratoire de Géologie-Pédologie de l'E.N.S.A., Grignon, France, 1967.

Del Posso, G. and C. Luzuriaga, *Guide Book for Field Trip. Sixth Int. Soil Classification*

Workshop. Tour Guide. Part 2: Ecuador, SMSS, U.S.D.A., Soil Conservation Service, Washington, D.C., 1984.

Delvaux, B., J. E. Dufey, L. Vielvoye, and A. J. Herbillon, "Potassium exchange behavior in a weathering sequence of volcanic ash soils," *Soil Sci. Soc. Amer. J.* 53:1679–1684 (1989).

Dudal, R., G. M. Higgins, and A. Pecrot, "Utilization of soil resource inventories in agricultural development," *Proceedings Fourth International Soil Classification Workshop, Rwanda,* ABOS, Brussels, Belgium, 1983, pp. 6–17.

Fox, R. L., "Soils with variable charge: agronomic and fertility aspects," in B. K. G. Theng (ed.), *Soils with Variable Charge,* New Zealand Society of Soil Science, Soil Bureau, Lower Hutt, New Zealand, 1980, pp. 195–224.

Fox, R. L. and E. J. Kamprath, "Phosphorus sorption isotherms for evaluating the phosphate requirements of soils," *Soil Sci. Soc. Amer. Proc.* 34:902–907 (1970).

Fox, R. L. and P. G. E. Searle, "Phosphate adsorption by soils of the tropics," in M. Stelley (ed.), *Diversity of Soils in the Tropics,* Amer. Soc. of Agronomy, Special Publication No. 34, 1978.

Gillman, G. P. and R. L. Fox, "Increases in the cation exchange capacity of variable charge soils following superphosphate applications," *Soil Sci. Soc. Amer. J.* 44:934–938 (1980).

Heiken, G. and K. Wohletz, *Volcanic Ash,* University of California Press, 1985.

Kimble, J. M. and W. D. Nettleton, "Analytical characterization of Andepts and Andisols," in F. H. Beinroth et al. (eds.), *Proceedings 6th International Classification Workshop, Chile and Ecuador: Part 1: Papers,* Sociedad Chilena Ciencia del Suelo, Santiago, Chile, 1985, pp. 211–233.

Knox, E. G. and F. Maldonado, "Suelos de cenizas volcánicas, excursión al volcán Irazú," in *Panel Sobre Suelos Derivados de Cenizas Volcánicas de América Latina,* Turrialba, Costa Rica, 1969, pp. 1–12.

Leamy, M. L., *Icomand Circular Letter No. 10,* New Zealand Soil Bureau, Lower Hutt, New Zealand, February, 1988.

Le Mare, P. H. and A. Leon, "The effects of lime on adsorption and desorption of phosphates in five Colombian soils," *J. Soil Sci.* 40:59–69 (1989).

Mattson, S., "The laws of soil colloidal behavior: amphoteric reactions and isoelectric weathering," *Soil Sci.* 34:209–240 (1932).

Mekaru, T. and G. Uehara, "Anion adsorption in ferruginous tropical soils," *Soil Sci. Soc. Amer. Proc.* 36:296–300 (1972).

Mizota, C. and J. Chapelle, "Characterization of some Andepts and andic soils in Rwanda, Central Africa," *Geoderma,* 41:193–209 (1988).

Mizota, C., I. Kawasaki, and T. Wakatsuki, "Clay mineralogy and chemistry of seven pedons formed in volcanic ash, Tanzania," *Geoderma* 43:131–141 (1988).

Mizota, C. and L. P. van Reeuwijk, *Clay Mineralogy and Chemistry of Soils Formed in Volcanic Material in Diverse Climatic Regions,* Soil Monograph 2, ISRIC, Wageningen, The Netherlands, 1989.

Moorman, F. R. and N. Van Breemen, *Rice: Soil, Water, Land,* International Rice Research Institute, Los Baños, Philippines, 1978.

Parfitt, R. L., C. W. Childs, and D. N. Eden, "Ferrihydrite and allophane in four Andepts from Hawaii and implications for their classification," *Geoderma* 41:223–241 (1988).

Silva, J. A., (ed.), *Soil-Based Agrotechnology Transfer,* Benchmark Soils Project, College of Tropical Agriculture and Human Resources, University of Hawaii, 1984.

Soil Survey Staff, *Keys to Soil Taxonomy,* SMSS Monograph #6, International Soils, Agronomy Department, Cornell University, Ithaca, NY, 1987.

Soil Survey Staff, *Keys to Soil Taxonomy,* 4th ed., SMSS Monograph #19, Blacksburg, VA, 1990.

Smith, B. F. L., and B. D. Mitchell, "Characterization of poorly ordered minerals by selective chemical methods," in M. J. Wilson (ed.), *A Handbook of Determination Methods in Clay Mineralogy,* Blackie & Son Ltd., Glasgow, 1987, pp. 275–294.

Sutanto, R., F. De Coninck, and M. Doube, *Mineralogy, Charge Properties and Classification of Soils on Volcanic Materials and Limestone in Central Java (Indonesia),* Geological Institute, State University of Ghent, Belgium, 1988.

Uehara, G. and G. Gillman, *The Mineralogy, Chemistry and Physics of Tropical Soils with Variable Charge Clays,* Westview Tropical Agriculture Series, No. 4, Westview Press, Boulder, CO, 1981.

Van Olphen, H., "Amorphous clay materials," *Science,* 171:90–91 (1971).

Wada, K., "Mineralogical characteristics of Andisols," in B. K. G. Theng (ed.), *Soils with Variable Charge,* New Zealand Society of Soil Science, Lower Hutt, New Zealand, 1980, pp. 87–107.

Wada, K., "The distinctive properties of Andosols," *Adv. Soil Sci.* 2:173–228 (1985a).

Wada, K., "Properties of Andisols important to paddy rice," in F. H. Beinroth et al. (eds.), *Proceedings 6th International Classification Workshop, Chile and Ecuador. Part 1: Papers,* Sociedad Chilena Ciencia del Suelo, Santiago, Chile, 1985b, pp. 97–107.

Warkentin, B. P. and T. Maeda, "Physical and mechanical characteristics of Andisols," in B. K. G. Theng (ed.), *Soils with Variable Charge,* New Zealand Society of Soil Science, Soil Bureau, Lower Hutt, New Zealand, 1980, pp. 195–224.

Wielemaker, W. G., "The importance of variable charge due to amorphous siliceous clay constituents in some soils from Kenya," *Geoderma* 32:9–20 (1984).

Wielemaker, W. G. and T. Wakatsuki, "Properties, weathering and classification of some soils formed from peralkaline ash in Kenya," *Geoderma* 32:21–44 (1984).

11

Inceptisols

Inceptisols are soils that are in the incipient stages of formation toward *mature* soil orders but have not yet fully developed their diagnostic properties. Conceptually Inceptisols fill an area between soils
with minimum or not appreciable development (Entisols) and soils of
various orders that have been accepted by pedologists as carrying the
marks of well-defined soil-forming processes.

There is a wide variety of Inceptisols because there are many direc-
tions to soil development. However, only a few of them are important
in the tropics. The selection made in this text includes the great soil
groups *Aquepts* and *Tropepts*.

The Inceptisols have none of the sets of properties that identify
Andisols, Vertisols, Spodosols, Oxisols, Ultisols, or Alfisols. There are
nevertheless a large number of horizons that are used in the diagnosis
of Inceptisols. They are included in the key to the soil orders of Chap.
6 and most of them are explained in the glossary. In the humid tropics
the histic, umbric, sulfuric, and cambic horizons are important; in
semiarid regions the calcic, petrocalcic, and gypsic are the most common.

Soils that are equivalent to Inceptisols in the FAO-Unesco classification (FAO, 1988) are spread over several classes. The poorly drained
soils would mostly belong to the *Gleysols* except for the profiles with
plinthite that are grouped with the *Plinthosols*. Many well-drained
Inceptisols of dry areas correspond to *Calcisols* or *Gypsisols*. The remaining Inceptisols meet the requirements of *Cambisols*.

11.1 Profile Description and Analysis

Only one pedon of the Inceptisols has been included in App. B. The
example is a soil formed in alluvium on a river terrace in Thailand. It
was assumed in the field that the profile has a textural B horizon, but

the clay increase between A and B was not sufficient to identify an argillic horizon. The micromorphological examination furthermore did not detect any evidence of clay translocations. The maximum development of a blocky structure justifies the recognition of a cambic horizon (Soil Survey Division, 1983).

The high silt content sets the soil apart from more strongly weathered materials. X-ray analysis of the clays reveals the presence of micas and montmorillonite. The very fine sands contain between 10 and 20 percent potash feldspars. The 100 percent base saturation reflects the immaturity of the soil or the nonaggressiveness of the climate; the water-holding capacity of more than 10 percent by volume adds to the quality of the soil.

11.2 Diagnostic Horizons

11.2.1 The cambic horizon

Definition. Cambic horizons are noncemented subsurface horizons in which the marks of original rock structures or thin bedding have been obliterated in at least one-half of their volume. They have textures of very fine sand or finer. They also have one or more horizons that have been altered by one or more of the following three processes:

1. *Removal of carbonates.* This results in the formation of a horizon that contains less carbonates *than an underlying* k horizon. This kind of cambic horizon is usually brown and overlies a whitish material that is coated with lime. It occurs mostly in the semiarid tropics that border deserts.

2. *Weathering.* Weathering of primary minerals releases free iron oxides and/or produces clays that give subsurface horizons stronger chromas or redder hues *than the underlying* horizons. In carbonate-free parent materials the formation of soil structure alone is considered an alteration that is sufficiently strong to accept the presence of a cambic horizon. However, the development may not be such that the horizon would meet the requirements of an argillic or a kandic horizon. It may have neither the mineralogical nor the chemical properties of oxic horizons.

3. *Reduction and removal of iron oxides.* This results in the formation of dominant gray colors in soils with an aquic moisture regime. This horizon is sometimes covered by a histic or umbric epipedon in which organic matter accumulates because of seasonal anaerobic conditions.

The cambic horizons should be thick enough so that their lower boundaries are deeper than 25 cm.

The three sets of processes are the basis for detailed definitions of cambic horizons (Soil Survey Staff, 1990). Except for the third kind of horizon, they always correspond to horizons that either present a maximum in the development of colors, grades of blocky structures, or clay contents. For these reasons, in the past they have been called *color (B)'s* and *structural B's*.

Genesis. The development of a soil structure in a carbonate-free material is recognized by *Soil Taxonomy* (Soil Survey Staff, 1990) to be sufficient to consider a horizon to be a cambic horizon, *provided* it does not at the same time meet the requirements of an argillic, kandic, or oxic horizon. The nonilluvial "structural" B horizon frequently displays the morphological features of an argillic horizon and has often been erroneously identified with it. There is no sharp boundary between their morphology, and much confusion has clouded the precision of soil classification systems that use them as diagnostic horizons. Certain schools are inclined to consider all subsurface horizons that have maximal structural development as argillic horizons; others, on the contrary, reject clay illuviation horizons in strongly weathered materials and prefer to classify them as cambic horizons.

Moniz and Buol (1982) hypothesized that in strongly weathered materials blocky structures form by compression induced by alternating saturation with water and desiccation. The water supply on slopes would come from subsurface lateral water flows. Subsequent desiccation would initiate the formation of cracks that shape blocky soil aggregates that, at the next wetting cycle, are marked by pressure faces. Fluctuations of the groundwater in lowlands are assumed to produce the same effects.

Cambic horizons may contain clay skins on ped surfaces that are similar to the illuviation cutans observed in argillic horizons. However, it is difficult to determine where the translocated clay particles come from. In one wetting and drying cycle in the laboratory, Chauvel and Pedro (1978) produced oriented clay films in a strongly weathered material derived from a red *Sol ferrallitique*. The identification of the argillic horizon therefore requires a clay increase between horizons as an additional differentiating characteristic.

Many toposequences in the tropics that have Oxisols on the highest surface include soils with *structural B horizons* or cambic horizons on the slopes that lead to the rivers dissecting the landscape. The sediments exposed by the incision of the valleys are usually less weathered than the Oxisols that characterize the plateaus. This is often reflected in the mineralogy of the sand, silt, or clay fractions, and may be the result of admixtures of fresh materials exposed by the downcutting rivers. Lateral groundwater seepage or subsurface water flows that carry silicon or other elements may also be at the origin of distinct clay formation environments that favor the synthesis of 2:1 clays that are prone to shrinking and swelling, easily dispersible, and tend to contribute effectively to the formation of cambic and argillic horizons.

11.2.2 The sulfuric horizon

Soil Taxonomy (Soil Survey Staff, 1990) defined the sulfuric horizon as follows:

> The sulfuric (L. *sulfur*) is composed either of mineral or organic soil material that has both a pH < 3.5 (1:1 in water) or jarosite mottles (the color of fresh straw that has a hue of 2.5Y or yellower and chroma of 6 or more).
>
> A sulfuric horizon forms as a result of artificial drainage and oxidation of sulfide-rich mineral or organic materials. Such a horizon is highly toxic to plants and virtually free of living roots.

Not all horizons that contain jarosite have pHs below 3.5. In the central plain of Thailand the pH of freeze-dried samples reconstituted in 1:1 soil to water mixtures is 3.64 ± 0.2; the 1 M KCl extractable aluminum content is 9.7 ± 3.8 cmol(+)/kg (Osborne, 1984).

The acidity in sulfuric horizons is produced by the oxidation of pyrite (FeS_2) according to the following chemical reactions:

$$FeS_2 + \tfrac{7}{2}O_2 + H_2O \rightarrow Fe^{2+} + 2SO_4^{2-} + 2H^+$$

$$6Fe^{2+} + 15H_2O + 2O_2 \rightarrow 6Fe(OH)_3 + 12H^+$$

If the acidity is not neutralized by carbonates, jarosite is formed:

$$3Fe(OH)_3 + K^+ + 2SO_4^{2-} + 3H^+ \rightarrow KFe_3(SO_4)_2(OH)_6 + 3H_2O$$

When no more SO_4^{2-} is present, the jarosite may hydrolyze and produce iron hydroxides such as goethite.

11.3 Subdivisions

A simplified key to suborders, extracted from *Soil Taxonomy* (Soil Survey Staff, 1990), that accommodates the most common Inceptisols of the tropics is as follows:

1. Soils that have an aquic moisture regime and have one or more of the following:
   a. A histic epipedon, *or*
   b. A sulfuric horizon with its upper boundary within 50 cm of the mineral soil surface, *or*
   c. Ped surfaces or a matrix that have dominant chromas of 2 or less if there are mottles, or 1 or less if there are no mottles. ...AQUEPTS

Aquepts are subdivided at the great group level into:

1. Aquepts that have a sulfuric horizon within 50 cm depth
...SULFAQUEPTS
2. Other Aquepts that have a horizon that is dominantly plinthite
...PLINTHAQUEPTS
3. Other Aquepts   ...TROPAQUEPTS
4. Other Inceptisols that have an isomesic or warmer isotemperature regime   ...TROPEPTS

Tropepts are further subdivided at the great group level into the following taxa:

1. Tropepts that have a sombric horizon   ...SOMBRITROPEPTS
2. Other Tropepts that have at least 12 kg organic carbon per square meter and per meter depth and have a base saturation of less than 50 percent (by NH_4OH at pH 7) in some horizon between 25 and 100 cm depth   ...HUMITROPEPTS
3. Other Tropepts that have an ustic moisture regime and have at least 50 base saturation (by NH_4OH at pH 7) in all horizons between 25 and 100 cm depth   ...USTROPEPTS
4. Other Tropepts that have at least 50 percent base saturation (by NH_4OH at pH 7) in all horizons between 25 and 100 cm depth   ...EUTROPEPTS
5. Other Tropepts   ...DYSTROPEPTS

11.4 Occurrence

The Inceptisols of tropical regions are soils of recent geomorphic surfaces that may either be erosional or aggradational. Given the high intensity of weathering and biological activity in the humid tropics, many young alluvial deposits rapidly develop into Inceptisols. Fresh rocks that are exposed at shallow depth and recent colluvia derived from them produce cambic horizons in a relatively short time. Landscape facets that in temperate environments would be characterized by Entisols, or nonsoils, are generally characterized by Inceptisols in the humid tropics.

The mountain ranges and the subsiding zones of the orogenic belts are the structural regions of the world that contain the largest areas of Inceptisols. The Andes Cordillera, Central America, the Caribbean, southeast Asia, parts of the Orinoco, and the Ganges valleys belong to these regions. According to Moormann and Van Breemen (1978), Inceptisols are the most important single soil order in rice-growing areas.

11.5 Management Properties and Appraisal

It is difficult to give general recommendations on the use of Inceptisols. There are several reasons for this: One of them is the diversity in their diagnostic properties, and another is their *incipient* development so that the influences of the parent materials are not yet masked. More than in other soil orders, the rock typos, weathering stages of sediments, textures, and drainage classes are predominant criteria to assess the management qualities of Inceptisols.

11.5.1 Aquepts

Sulfaquepts. The Sulfaquepts generally occur in the deltas of the major tropical rivers. Examples are the Ganges, Mekong, and Senegal. The 1983 estimate of the area under rice mangrove–swamp cultivation in west Africa is 214,000 ha (WARDA, 1983).

Sulfaquepts are soils with an aquic moisture regime that have a sulfuric horizon starting at less than 50 cm depth in the profile. The major restrictions for agriculture in these soils are the strong acidity and the aluminum toxicities that result from the oxidation of pyrite after sulfidic materials have been exposed to air. When used for paddy rice, Fe^{2+} toxicities may also be damaging, particularly at the beginning of the flooding (Prade et al., 1988).

The adverse soil fertility conditions in Sulfaquepts can be corrected in several ways. However, none of them is easy to implement. The pH can be raised either by flooding or liming or the sulfates can be leached out of the rooting zone of the crops to be grown. In many cases additional difficulties arise because of soil salinity in areas that are influenced by the tides; the land has to be protected against them.

The most effective way to manage acid sulfate soils is to *prevent* the massive oxidation of pyrite. They should only be drained very progressively, thus reducing the rate of the formation of sulfuric materials and the production of sulfates. The control of the groundwater level is the key to maintaining reasonable growing conditions for paddy rice. In areas that are influenced by sea water, this type of land management requires fresh water and protection against the tides.

In land that has already been damaged by drainage, *flooding* may restore extremely low pH values to acceptable ones. Bloomfield and Coulter (1973) reported several experiments where pH values close to neutrality were reached after 1 or 2 months saturation with water. The reclamation procedures may use sea water to reduce the acidity, but fresh water should be available to remove the salts before crops can be grown.

The roots of oil palms do not grow in artificially drained soil layers that have been acidified by the oxidation of pyrite to pH values between 2.5 and 3.0. When the materials are resaturated with water and subsequently kept under flooded conditions, the root development into the affected soil section is comparable to that observed in freely drained nonacid soils (Poon and Bloomfield, 1977).

Leaching of the soluble salts that are formed during oxidation offers another reclamation alternative; however, it raises technical problems as well as environmental impact issues because the acidity may spread over large neighboring areas. It is estimated that a minimum of 10 years is needed to arrive at acceptable growing conditions if only rainwater (1200 mm/year) is used; laboratory experiments indicate that the quantities of water necessary to reclaim the soil may amount to from 5 to 15 times the weight of the soil (Bloomfield and Coulter, 1973). The costs of the investments may be reduced by limiting the volume of soil to be treated to the root zone of the crops or by building raised beds and using furrows for drainage. *Liming* may also contribute to the reduction of acidity, but the requirements are usually not economically feasible.

Brackish water fish ponds that are established in acid sulfate soils can be improved by removing the acidity produced by the oxidation of pyrites in the top 15 cm of the pond bottoms. The procedure includes 4 to 5 times repeated sequences of drying, tilling, flooding, and draining of the pond bottom as well as leaching of the dikes (Brinkman and Singh, 1982).

Plinthaquepts.   The Plinthaquepts are poorly drained soils that have plinthite forming a continuous phase or constituting more than one-half of the matrix in the upper 125 cm of the soil. They occur in flat areas that have fluctuating water tables within that depth. They are often associated with Plinthaquults, a great group of Ultisols (see Sec. 8.3.2).

Most of the Plinthaquepts are acidic and nutrient deficient and often contain large amounts of exchangeable aluminum. For these reasons they are seldom worth draining and are rarely used for agricultural purposes.

Tropaquepts.   The Tropaquepts are poorly or very poorly drained soils in which the groundwater periodically stands close to the surface. They have predominantly gray colors in the top horizons and are mottled in deeper layers. Their parent materials are mostly alluvial deposits. Tropaquepts may contain plinthite but by definition never as a continuous phase or in quantities that would fill more than one-half of their soil volume.

Tropaquepts is not synonymous with "tropical wetland soils"; they only form a part of them. Organic soils, Plinthaquepts, Sulfaquepts, and the aquic subgroups of Tropepts are often associated with them in the wetlands. Some poorly drained Ultisols may also characterize parts of them. The very poorly drained Tropaquepts or those that are characterized by an isothermic or cooler soil temperature regime usually have a histic epipedon. Most poorly drained or moderately well drained isohyperthermic soils have an ochric epipedon. The water table fluctuations in Tropaquepts may be very pronounced and may reach several meters. Some profiles may be entirely at the wilting point for several months in the dry season but flooded, either by rain or river overflow, during the monsoon season.

Many Tropaquepts are actually in the active weathering phase of soil formation and contain large amounts of exchangeable aluminum if they are acid. For example, an Aeric Tropaquept formed in colluvium from shales in southern Nigeria has the chemical properties listed in Table 11.1 (Okusami et al., 1985).

Most Tropaquepts are acid. For this reason they seldom have the qualities that make land attractive for development of farming systems based on upland crops because drainage and soil fertility constraints are difficult to overcome. Paddy rice offers the best land use alternative since the flooding temporarily alleviates the soil acidity problem and reduces the aluminum toxicity. Phosphate fertilization is often necessary to obtain optimum yields.

The Tropaquepts in areas that have a long dry season are often used for extensive grazing during the time that adjacent uplands cannot be used because of excessive drought. This is mostly the case in lowland areas that have very contrasting wet and dry seasons such as in monsoon climates.

11.5.2 Tropepts

It is useful to recognize in this suborder two great soil groups that characterize the *cool* tropics by their high organic matter contents: the Humitropepts and the Sombritropepts. They almost always have

TABLE 11.1 Chemical Properties of an Aeric Tropaquept of Southern Nigeria

| Horizon | Depth, cm | Organic C, % | Exchangeable cations (cmol(+)/kg soil) | | | | | pH (H$_2$O) | pH (KCl) |
			Ca	Mg	Na	K	Al		
A11	12	1.52	0.80	2.19	0.10	0.10	0.5	4.6	3.8
A12g	33	0.32	0.50	0.68	0.07	0.02	1.4	4.7	3.5
2B1g	45	0.30	0.21	0.32	0.07	0.02	2.2	4.7	3.5
2B21g	63	0.57	0.32	0.90	0.21	0.05	6.8	4.7	3.3
2B22g	120	0.34	0.67	1.73	0.32	0.07	8.0	4.9	3.3
2B23g	160	0.18	0.57	11.2	0.83	0.13	8.1	4.8	3.1
2Cg	180	0.13	7.25	29.2	1.66	0.37	13.8	4.7	2.8

SOURCE: *Oskusami et al. (1985).*

isothermic soil temperature regimes. The other Tropepts are typical for young erosion surfaces in the lowland tropics and are subdivided into Ustropepts, Eutropepts, and Dystropepts.

Humitropepts. The Humitropepts are by definition acid and usually have a udic moisture regime. They have many properties in common with the Dystropepts except for their higher organic matter contents. This is not necessarily an indication of better nitrogen availability. The colder temperatures lower the mineralization rates of organic matter, and their frequent acid reaction and phosphate deficiencies reduce biological "nitrogen fixation."

The isothermic temperature regime of most Humitropepts makes them suitable for a group of crops that are different from those adapted to the warmer isohyperthermic soils of the lowland tropics. Tea and *Arabica* coffee are typical tree crops on Humitropepts. Potatoes (*Solanum tuberosum*) are also grown in them.

Sombritropepts. There are only a few detailed descriptions of Sombritropepts, and their land qualities have seldom been thoroughly reviewed. Their agronomic properties probably approach those of the Humitropepts, except that the Sombritropepts usually have a cooler, isomesic temperature regime.

Frankart (1989) considered that the high exchangeable aluminum contents, the low available phosphate levels, and the deficiencies in cations are the major limiting factors. The Humic Sombriudox that are associated with them are reported to produce approximately 270 kg/ha beans or 300 kg/ha sorghum. These low yields are probably also characteristic for the Sombritropepts and Humitropepts. Yields are increased by the application of manure (35 Mg/ha), and the recommended optimum liming schedule is 4 Mg/ha, which would have a positive effect on yields for 4 years. Chemical fertilizers without liming have not resulted in significant yield increases.

Dystropepts. The Dystropepts are characteristic soils of recent erosion surfaces in the warm humid tropics. By definition they have at least one subhorizon between 25 and 100 cm that has a base saturation of less than 50 percent (by NH_4OH at pH 7). Their epipedons are generally ochric, and their cambic horizons are commonly formed by the development of stronger, blocky structures or more pronounced colors than the underlying materials. They still contain weatherable minerals that in humid environments release nutrients *and* exchangeable aluminum. The latter may negatively influence the root development of sensitive crops, and this is particularly the case in soils with a perudic moisture regime. Their silt (0 to 20 μm) to clay ratios are generally higher than 0.10, which gives Dystropepts somewhat higher water-retention capacities than the Oxisols they are often associated with.

The major limitation that restricts the use of Dystropepts is topography. Many areas with steep slopes are unsuitable for any kind of use that exposes the soil surface to direct raindrop impact and runoff. The limitation imposed by slope may even be more stringent when mechanized agriculture is envisaged. The suitability of Dystropepts for growing crops is therefore often determined by the gradients and shapes of the slopes on which they occur. Concave sites are always preferred to convex slopes, which are best kept under the protection of trees. If the Dystropepts occur on alluvial flats, there are no topographical limitations, but the depth of the groundwater may interfere with root development.

The other land qualities of the Dystropepts that are important for land use are strongly parent material dependent. Most Dystropepts originated from felsic rocks or sedimentary rocks that are rich in quartz or have been strongly leached. As a result, in the long run aluminum toxicities become the major limitation for crops that have their optimum pH close to neutral. On the other hand, acidophilic perennials such as coffee or tea usually do well, provided the organic matter in the topsoils has not been removed by erosion.

Dystropepts are appraised by farmers by comparing them with soils that occur in association with them. They are usually preferred to Oxisols derived from the same parent materials because of better nutrient availability; if lime is available, its application yields better responses. Experience has shown that if Dystropepts are not exposed to erosion, they are in a more favorable position than Oxisols to support sustainable farming systems that have sufficient resources to use fertilizers and lime to reduce the nutrient deficiencies and eliminate the aluminum and manganese toxicities.

Eutropepts. By definition, Eutropepts have a udic moisture regime and a base saturation of at least 50 percent (by NH_4OAc at pH 7). They contain weatherable minerals or silicate clays that are not entirely kaolinitic. Eutropepts seldom present the disadvantages of the heavy clay textures of Vertisols. Therefore, Eutropepts are endowed by a set of favorable attributes that are seldom found in soils of the tropics: satisfactory water supply and adequate nutrient content. This is probably the reason why only a few of them have been reviewed in the literature and examined by soil management specialists who generally have to concentrate their research efforts on *problem* soils.

The most common parent materials of Eutropepts are derived from mafic rocks or sedimentary rocks that are rich in calcium-bearing minerals. Limestones are also frequently at the origin of the Eutropepts. Many well-drained or moderately well-drained soils in young alluvial valleys belong to the Eutropepts, and in this case the flat topography adds to the quality of the land.

Ustropepts. Ustropepts are Tropepts that have an ustic moisture regime and have a base saturation (by NH_4OAc) of 50 percent or more in all subhorizons between depths of 25 and 100 cm. Given the wide range of water availability in ustic pedons and the presence of dry seasons that may vary from 3 to 9 months, drought stress may be one of the most common limitations that affects the use of Ustropepts. The profiles that are seasonally influenced by groundwater may temporarily escape this restriction and are usually very fertile soils; examples are the aquic and the fluventic Ustropepts.

A description and an analysis of a Fluventic Ustropept is given in App. B. The physical and the chemical soil characteristics do not show any attributes that would seriously limit plant growth. In a context of an evaluation of soils of the tropics, the cation exchange capacity of the clay, the base saturation, and the pH point at favorable soil fertility conditions. The physical quality indicators are the fine silt content and the adequate water-holding capacity.

References

Bloomfield, C. and J. K. Coulter, "Genesis and management of acid sulfate soils," *Avd. Agronomy* 25:265–326 (1973).

Brinkman, R. and V. P. Singh, "Rapid reclamation of fish ponds on acid sulfate soils," in H. Dost and N. Van Breemen (eds.), *Proceedings of the Bangkok Symposium on Acid Sulphate Soils,* ILRI Publication, Wageningen, The Netherlands, 1982, 31:318–330.

Chauvel, A. and G. Pedro, "Génèse de sols beiges (ferrugineux tropicaux lessivés) par transformation des sols rouges (ferrallitiques) de Casamance (Sénégal). Modalités de leur propagation," Cah. ORSTOM. Ser. *Pédologie* 16:231–249 (1978).

FAO (Food and Agriculture Organization of the United Nations), *FAO-Unesco Soil Map of the World. Revised Legend,* World Soil Resources Report 60, FAO, Rome, 1988.

Frankart, R., Les sols acides des écosystèmes humifères d'altitude du Rwanda et du Burundi. Aspects pédo-agronomiques, *Bull. Séanc. Acad. r. Sci. Outre-Mer* 35(4):495–531 (1989).

Moniz, A. C. and S. W. Buol, "Formation of an Oxisol-Ultisol transition in São Paulo, Brazil: I. Double-water flow model of soil development," *Soil Sci. Soc. Amer. J.* 46: 1228–1233 (1982).

Moormann, F. R. and N. Van Breemen, *Rice: Soil, Water, Land.* International Rice Research Institute, Los Baños, Philippines, 1978.

Okusami T. A., R. H. Rust, and A. S. R. Juo, "Characteristics and classification of some soils formed on post-cretaceous sediments in Southern Nigeria," *Soil Sci.* 140(2):110–119 (1985).

Osborne, J. F., "Soil analysis as a guide to acidity classes in mature acid sulphate soils of the southern central plain of Thailand," *Proceedings of the 5th ASEAN Soil Conference,* Department of Land Development, Bangkok, Thailand, 1984, C3.1–C3.15.

Poon, Y. C. and C. Bloomfield, "The amelioration of acid sulphate soil with respect to oil palm," *Trop. Agric.* 54(4):289–302 (1977).

Prade, K., J. C. G. Ottow, and V. Jacq, "Excessive iron uptake (iron toxicity) by wetland rice (*Oryza Sativa L.*) on an acid sulphate soil in Casamance/Senegal," in J. Dost (ed.), *Selected Papers of the Dakar Symposium on Acid Sulphate Soils,* ILRI Publication, Wageningen, The Netherlands, 1988, 44:150–162.

Soil Survey Division, "Characteristics of some soils in the central plain, northeastern and southeastern regions of Thailand," *Tour Guide of Fourth International Forum on Soil Taxonomy,* Land Development Department, Bangkok, Thailand, 1983.

Soil Survey Staff, *Keys to Soil Taxonomy,* SMSS Technical Monograph #19, Blacksburg, VA, 1990.

WARDA, *Quinquennial Review Report,* Regional Mangrove Swamp Rice Research Institute, Rokupr, Sierra Leone, 1983.

Spodosols

12.1 Definition

Spodosols are mineral soils that have a spodic horizon whose upper boundary is within 2 m of the surface. They also include soils that have a placic horizon that meets all the requirements of a spodic horizon except for thickness and the index of accumulation, defined later in this chapter. The placic horizon is only diagnostic of Spodosols if it rests on a spodic horizon.

Tropical Spodosols occur in two distinct ecological situations. The first one coincides with deep deposits of quartz sands in the lowland tropics; the second one is characterized by a cool isomesic temperature regime that prevails at high elevations in areas that have often been influenced by volcanic ash. The cool tropical Spodosols are typical for mountain slopes that are hit by clouds producing orogenic rains.

Spodic horizons only form in soils with udic or perudic moisture regimes. They may, however, be present in soils of the tropics with a pronounced dry season, for example, under savannah woodland (Brammer, 1973). In many cases spodic horizons in dry climates are considered to be fossil (Schwartz, 1988).

The equivalent of Spodosols in other classification systems is *Podzols*. This correlation should not be extended to soils that are called *Podzolic* soils, which are soils characterized by a horizon of illuvial clay accumulation and are therefore more comparable to Alfisols and Ultisols.

12.1.1 Profile description and analysis

One pedon is included in App. B. It represents the Spodosols that form in sands of humid tropical lowlands. The small amounts of nutrients and limited water-holding capacity characterize the soil. The spodic horizon is essentially made of an accumulation of organic substances.

12.1.2 Diagnostic horizons

The spodic horizon. The spodic horizon is a subsurface horizon that should meet one or more of the following requirements below a depth of 12.5 cm or below any A_p that is present (Soil Survey Staff, 1987):

1. Has a continuously cemented subhorizon that is more than 2.5 cm thick. The cementation is due to some combination of organic matter with iron or aluminum or both, *or*
2. has a sandy or coarse-loamy particle-size class and the sand grains are covered with cracked coatings or there are distinct dark pellets of coarse silt size or larger or both, *or*
3. has one or more subhorizons in which:
   a. There is either one:
      If there is 0.1% or more extractable Fe, then (Fe + Al)/clay% is at least 0.2,
      If there is less than 0.1 extractable Fe, then (Al + C)/% clay is at least 0.2.[1], and
   b. The ratio between the percentages of pyrophosphate-extractable and dithionite-citrate extractable FE + AL is at least 0.5, *and*
   c. The combined index of accumulation[2] of amorphous material must be 65 or more.

The definition intends to group subsurface horizons that are the result of the accumulation of translocated organic matter. This mobile organic matter is able to carry with it small but variable amounts of Fe and Al. The Fe and Al that is extracted by pyrophosphate at pH 10 during the diagnostic procedures is assumed to represent the transported Fe and Al complexed by the organic matter and accumulated in the spodic horizon. The amounts of Fe and Al that move with the organic C depend on the original Fe and Al content that were present in the parent materials; spodic horizons that developed in pure quartz sands, for example, practically do not contain Fe or Al.

The development of spodic horizons is conditioned by the production of large amounts of organic substances. Schwartz et al. (1986) estimated that the *ortstein* formation on the Bateke Plateau in Africa de-

---

[1]All extractions use pyrophosphate at pH 10.

[2]The index is obtained by subtracting half of the percent clay from the CEC at pH 8.2 and multiplying the remainder by the thickness of the subhorizon in cm.

veloped by annual organic matter incremental additions of 230 kg/ha for 10,000 years. These organic substances should be mobile, a condition that is dependent on their hydrophilic character as colloids and on the absence of elements in the soil solution that produce their flocculation. Large amounts of polyvalent cations such as Ca, Al, and Fe would reduce their stability as dispersed colloids. Given these prerequisites, the most favorable conditions for the production of mobile organic matter exist in humid climates with soils that are often saturated with water in which a lack of oxygen retards the decomposition of organic matter and that form in sediments poor in Ca, Fe, and Al. For all these reasons the Spodosols most frequently occur in poor quartz sands that are periodically saturated with water, under forest vegetations that produce much organic matter, and in permeable sediments in which water can freely move. Other environments that allow spodic horizons to develop exist in weathering products of rhyolitic and dacitic volcanic ashes in perhumid cool climatic conditions.

However, the mobility of organic matter is not sufficient for the formation of spodic horizons. The mobile complexes need to be stopped and precipitated at a certain level in the profile. Andriesse (1969) argued that the accumulation occurs in a zone of abrupt moisture changes that are caused either by distinct particle size differences in the sand fractions or by the presence of a water table. The immobilization may also be the result of changes in physicochemical conditions (De Coninck, 1980). If the suspended and solubilized organic matter is not retained in the soil by flocculation, it percolates to the groundwater and joins the rivers that are then pigmented and often called *black waters*.

There are actually several kinds of spodic horizons. Some of them are cemented and correspond to *ortstein* (German) or *alios* (French). Others that are not cemented are distinguished on the basis of the iron to carbon ratios in the horizon. If the free iron (iron extracted by dithionite-citrate) to carbon ratio (both elemental) is six or more, the spodic horizon is diagnostic for the suborder *Ferrods*. If the ratio is less than 0.2, the spodic horizon characterizes the *Humods*. In all other cases, except for the poorly drained profiles, the Spodosols belong to the *Orthods*.

The composition of the organic matter that cements spodic horizons formed in sandy materials in the state of Pahang in peninsular Malaysia is given in Table 12.1. The amounts of organic carbon that can be extracted by pyrophosphate and the percentages of fulvic acids are smaller than in spodic horizons of temperate regions (De Coninck, 1980).

TABLE 12.1   Chemical Composition of Spodic Horizons in Sandy Soils of Malaysia

Series	Horizon	Total organic C, %	Extractable organic C, %*	Fulvic acid†	Humic acid†	Humin†	Fe_2O_3‡
Jambu	B21h	0.45	65	25	40	35	2.2
	B22ir	1.02	35	25	10	65	8.4
Rudua	B21h	1.59	63	40	23	37	3.7
	B22ir	0.25	60	46	14	39	11.5
Rudua	B21h	0.77	88	61	27	12	2.1
	B23ir	0.30	58	53	5	42	16.5

*Organic carbon extracted with pyrophosphate.
†Expressed as percent of total organic carbon.
‡Fe_2O_3 extracted with dithionite-citrate, 0/00.
SOURCE: De Coninck (1980).

The placic horizon. The placic horizon is a thin, black to dark reddish pan cemented by iron, iron and manganese, or iron organic matter complexes. Its thickness on average varies between 2 and 10 mm. The placic horizon is a barrier to water and roots. The organic carbon content of placic horizons distinguishes it from ironstone sheets that form at lithological discontinuities within the zone of fluctuations of water table levels (Soil Survey Staff, 1987). The placic horizon in tropical regions most frequently occurs in soils with an isomesic, udic, or perudic climatic regime that contains allophane.

12.2   Occurrences

Tropical Spodosols have been mentioned as early as in the 1920s in the soil science literature describing several locations on almost every continent. Klinge (1966, 1968) estimated that there are approximately 77,000 km^2 lowland Spodosols in the tropics or 2 percent of its surface. According to Driessen and Dudal (1989), the tropical Spodosols do not total more than 10 million hectares.

12.2.1   Spodosols of the lowland tropics

The Spodosols of the lowland tropics form in sandy regoliths. They usually occur in the lowest parts of the landscapes, in valley bottoms or in depressions. Sandy deposits with spodic horizons also occur along coastlines bordering the sea.

Spodosols may be saturated with water during the rainy season but may be completely dry during the dry season; in some Spodosols the groundwater cannot be reached at any time. According to Schwartz (1988), the annual rainfall in regions where Spodosols occur varies between 700 and 7200 mm and the length of the dry season may reach 7 months. Some spodic horizons are fossil. For example, in the People's

Republic of Congo on the Bateke plateau they were formed more than 30,000 years ago in soil climatic conditions that were different from the present ones (Schwartz, 1988).

Table 12.2 gives analytical data on a Spodosol on the Bateke plateau in the People's Republic of Congo. The B2m horizons are cemented into *ortstein*. They have very high carbon to nitrogen ratios in the spodic horizons but extremely low iron content. The organic carbon in the spodic horizon is assumed to have accumulated under the influence of a fluctuating water table in sands covered by a forest vegetation. *Ortstein* is considered the roof that covers the groundwater in the sands, but the bottom of a hanging water table is the E horizon that rests on it. More than 90 percent of the carbon in the *ortstein* is extractable by pyrophosphate. The cementation is due to aluminofulvic complexes (Schwartz et al., 1986).

The Spodosols in the lowland rain forests are usually located on the lower footslopes of recent valleys where colluvial sands have accumulated as sediments that may be several meters thick. The quartz grains are assumed to have lost part of their iron oxide coatings during transportation. Klinge (1965) described several soil catenas in the Amazon basin where Spodosols occupy the lowest portions of the valley slopes bordering the alluvial flats of the rivers. Some spodic horizons are complete in that they include a top subhorizon (B_h) with organic matter accumulation that rests on a B_s subhorizon of organic matter–sesquioxide complexes. It is not always clear what part a temporary water table plays in the accumulation of the organic matter in these spodic horizons. River waters that come out of heath forests on Spodosols in the Amazon are always black waters (Klinge, 1965).

Not all white sands are the result of colluviation in depressions or valleys. Chelation of iron by organic compounds in soil water that percolates through very permeable sediments removes the iron coatings from the quartz grains and creates the white albic horizon that may be very thick. Bleackley and Khan (1963) saw evidence of this process in

TABLE 12.2   Chemical Composition of a Spodosol on the Bateke Plateau

Horizon	Depth, cm	Organic C, %	C/N	Dithionite extractable Fe, 0/00	Acid oxalate extractable Fe, 0/00
A11	0–15	2.3	31.3	0.2	0.1
E	55–88	0.2	3.8	0.2	0.2
B21h	110	0.9	18.9	0.5	0.2
B22mh	120	17.8	65.3	0.2	0.2
B23mh	200	9.4	52.2	0.2	0.0
B23mh	260	1.0	32.7	0.2	0.1

SOURCE: *Schwartz (1988)*.

the brown colors and the humic substances in solution that precipitate with hydrochloric acid at pH 2 in the groundwaters that drain from such areas. Most bleached E horizons in tropical Spodosols, or the white quartz sands in which spodic horizons develop, are extremely deep. Thicknesses of 3 to 5 m have been reported (Thompson and Hubble, 1977).

12.2.2 High-altitude occurrences

One of the earliest descriptions of tropical highland Podzols was written by Jenny (1948) and dealt with a mountain profile in Colombia that was called a "giant podzol." The qualifier "giant" referred to the unusually thick eluviated E horizon above the placic and spodic horizons. In subsequent publications on tropical Spodosols the term "giant" continues to be used with reference to the albic, bleached horizon above the spodic horizon.

According to Barshad and Rojas-Cruz (1950), the profile formed in two parent materials: an upper semiconsolidated tuff of dacitic mineralogy that contained fresh volcanic glass of relatively recent origin and a lower stratum composed of weathering products of kaolinitic shaly rock.

The horizon formation only took place in the top layer and includes A1 horizons of organic matter accumulation above a bleached albic horizon (E) that rests on the illuvial Bm horizon. The latter formed at the contact with the underlying stratum (2Cl). The Bm horizon is cemented into what the authors call *ortstein*; in terms used by *Soil Taxonomy* (Soil Survey Staff, 1987) it corresponds to a placic horizon. The chemical composition of the major soil horizons are given in Table 12.3. The free Fe_2O_3 and Al_2O_3 contents were determined by acid oxalate extraction. It is assumed that the top layer contains large amounts of amorphous silica and amorphous iron oxides.

TABLE 12.3   Chemical Composition of a Giant Podzol in Colombia

Horizon	Depth, cm	Organic C, %	Free Fe_2O_3, %	Free Al_2O_3, %	Gibbsite, %
A11	18	11.45	1.7	11.3	0.3
A12	32	6.09	1.8	19.2	0.7
E1	45	4.33	0.7	18.2	1.0
E2	56	2.84	0.7	15.4	2.0
Bm	60	5.02	45.7	6.3	7.0
2Cl	80	1.64	8.4	6.3	5.5

SOURCE: *Barshad and Rojas-Cruz (1950).*

12.3   Management Properties

Spodosols lack most of the qualities needed to support profitable agriculture that can compete with other land types. The nutrient content of sandy Spodosols is extremely poor, and the deeper the albic horizons, the more severe is the shortage of nutrients. P, Ca, N, and K are deficient, and many crops that grow on deep sandy Spodosols lack Cu, Zn, B, and Mo.

The Spodosols with an ustic moisture regime suffer from periodic drought and are difficult to reclaim even for reforestation. The most prominent nutrient deficiency for subtropical conifers is P (Thompson and Hubble, 1977).

On the east coast of peninsular Malaysia Spodosols receive more than 2000 mm annual rainfall, and coconuts, mangos, and cashews are grown successfully around the houses. For larger plantations, yields are very low. The best results with cashews (*Anacardium occidentale* L.) are obtained where a cemented *ortstein* horizon occurs between 50 and 100 cm depth; it retards the percolation of water and creates better conditions for organic matter accumulation. When the spodic horizon is not cemented or deeper, the yields decrease. Tobacco is grown in soils where the spodic horizon occurs within 50 cm of the surface. As for all other annual crops, split applications of fertilizers are necessary (Kamal et al., 1984).

The Spodosols of the cool tropics that develop in volcanic ejecta may have adequate amounts of organic matter but are usually extremely desaturated and difficult to supply with adequate amounts of lime and fertilizers. The cold and humid climate conditions make them inhospitable. When grasslands have replaced the mountain forests, some sheep grazing is practiced. Among the annual crops, heavily fertilized potatoes can be grown. Given the steep relief of the mountains that are covered by these soil types, the land is best kept protected by forest vegetation.

References

Andriesse, J. P., "The development of the podzol morphology in the tropical lowlands of Sarawak (Malaysia)," *Geoderma* 3:261–279 (1969).

Barshad, I. and L. A. Rojas-Cruz, "A pedologic study of a podzol soil profile from the equatorial region of Colombia, South America," *Soil Sci.* 70:221–236 (1950).

Bleackley, D. and E. J. A. Kahn, "White-sand areas of the Berbice formation," *J. Soil Sci.* 14–51 (1963).

Brammer, H., "Podzols in Zambia," *Geoderma* 10:249–250 (1973).

De Coninck, F., "Major mechanisms in formation of spodic horizons," *Geoderma* 24:101–128 (1980).

Driessen, P. M. and R. Dudal, *Lecture Notes on the Geography, Formation, Properties and Use of the Major Soils of the World*, Agricultural University Wageningen,

Katholieke Universiteit, Leuven, Belgium, 1989.

Jenny, H., "Great soil groups in the equatorial regions of Colombia, South America," *Soil Sci.* 66:5–28 (1948).

Kamal, J. M., O. Yaacob, A. Husin, and S. Paramananthan, "Fertility status of sandy beach ridges in peninsular Malaysia," *Proceedings 5th ASEAN Soil Conference*, Department of Land Development, Bangkok, Thailand, 1984, H8.1–H8.10.

Klinge, H. C., "Podzol soils in the Amazon basin," *J. Soil Sci.* 16:95–103 (1965).

Klinge, H. C., "Verbreitung tropischer Tieflandpodsolo," *Naturwiss* 53·442–443 (1966).

Klinge, H. C., *Report on Tropical Podzols*, FAO, Rome, 1968.

Richards, P. W., "Lowland tropical podzols and their vegetation," *Nature* 148:129–131 (1941).

Schwartz, D., "Some podzols on Bateke sands and their origins, People's Republic of Congo," *Geoderma* 43:229–247 (1988).

Schwartz, D., B. Guillet, G. Villemin, and F. Toutain, "Les alios humiques des podzols tropicaux du Congo: Constituants, micro- et ultra structure," *Pédologie* 36(2):179–198 (1986).

Soil Survey Staff, *Keys to Soil Taxonomy*. SMSS Technical Monograph #6, Department of Agronomy, Cornell University, Ithaca, NY, 1987.

Thompson, C. H. and G. D. Hubble, "Sub-tropical podzols (Spodosols and related soils) of coastal eastern Australia," *Proceedings Conf. Classification and Management of Tropical Soils*, Malaysian Soil Sci. Soc., Kuala-Lumpur, 1977, pp. 203–213.

13

Entisols

13.1 Definition

Entisols only have a definition by default. They include all soils that do not qualify for membership in any of the taxa that are recognized in the first 10 orders of the identification key of *Soil Taxonomy* (Soil Survey Staff, 1990). Hence they do not possess any of the diagnostic horizons that would justify their inclusion in one of the *mature* soils. However, that does not mean that Entisols are featureless or that they do not include strikingly different soils that characterize extremely large areas in the tropics.

The reason for the lack of development of Entisols is that they are either too young or the materials in which they occur are inert and cannot react to soil-forming factors. Nevertheless, some diagnostic properties are allowed in Entisols: sulfidic materials, mottling, albic horizons, durinodes, and many others. Albic horizons and E horizons are most common in deep, sandy soils with a udic soil moisture regime. The fact that they do not exhibit pedologically significant features does not imply that they cannot be highly productive. Some of them, for example those that occur in recent alluvial valleys, are excellent for agriculture.

It is not the purpose of this chapter to review in detail the various kinds of Entisols in the tropics. One of the suborders, the Psamments, has been retained for discussion because they cover extensive areas. The soils of one great group, the Sulfaquents, are described because they often develop into Sulfaquepts, a great group of Inceptisols (see Sec. 11.5.1) that may cause severe management problems.

13.2 Psamments

13.2.1 Definition

The Psamments are characterized by coarse textures that by definition should not be finer than loamy find sand. This texture is required in all subhorizons either to a depth of 1 m or to a lithic, paralithic, or petroferric contact, whichever is shallower (Soil Survey Staff, 1990). The sandy texture is the major attribute of Psamments; it has many accessory properties such as high water permeability, low water-holding capacity, low specific heat, and often minimal nutrient contents.

13.2.2 Great groups

Soil Taxonomy recognizes several great groups of Psamments. Their definitions in tropical areas result from the application of the following key:

1. Psamments that have an aridic moisture regime
   ...TORRIPSAMMENTS
2. Other Psamments that contain more than 90 percent silica minerals, primarily quartz, in the 0.02 to 2 mm fractions
   ...QUARTZIPSAMMENTS
3. Other Psamments that have a udic moisture regime
   ...TROPOPSAMMENTS
4. Other Psamments that have an ustic moisture regime
   ...USTIPSAMMENTS

The key is extracted from *Soil Taxonomy* (Soil Survey Staff, 1990) and is only valid for tropical regions. It separates first the Psamments of arid regions. It then recognizes the profiles that are essentially composed of quartz and consequently lack adequate amounts of weatherable minerals that could provide plant nutrients. The last steps in the key distinguish the remaining soils that have either an udic or an ustic moisture regime.

Some Psamments in the tropics have a spodic horizon below 200 cm depth. In this case this diagnostic horizon is not retained as a criterion in the key to the soil orders because it is considered too deep to be significant. Profiles of this kind are often referred to in the literature as "giant podzols." Their nutrient contents are among the poorest in the world.

The equivalent of Psamments in the FAO-Unesco classification is called Arenosols (FAO-Unesco, 1977). Many *Dior* soils developed on dune sands in north and west Africa are Psamments.

13.2.3 Origin

Most large areas of Psamments result from accumulations of sand particles that at one stage were sorted and transported by wind; the deflation zones from which they originate have or had dry seasons long enough to allow sand grains to be picked up by wind. The rainy seasons, on the other hand, were too short to allow the vegetation to protect the soils against wind erosion.

Most Psamments are or were located in arid or semiarid regions where the substrates contain adequate amounts of quartz grains to feed the supply of sand. Continental shields with granites and gneisses, sandstones, or quartzites are the ideal environments to create very extensive uniform sand mantles. Because of the marked variations in climate during geological times, sands now cover large surfaces in humid or even in perhumid regions.

Sandy parent materials may have different ages: The Kalahari sands of Africa date back to the Tertiary and Pleistocene periods and may be more than several million years old. They are among the poorest parent materials in the tropics. Their low water-retention characteristics make them very vulnerable to drought, and their low nutrient contents constantly prevent improvement.

The Quaternary period was affected by drastic changes in climate. Although there are several hypotheses regarding the timing and the intensities of the climatic sequences, some general trends emerge from geological studies. It is generally accepted that at the end of the Pleistocene an extremely arid climate prevailed in the southern Sahara and the Sahel; temporally, this climate occurred during the maximum ice advance (Hochwürm) in Europe (Volkel and Grunert, 1990) and was called *Kanemien* by Servant (1983). Its maximum aridity occurred about 18,000 years before present (BP). This dry phase was followed by a humid phase in the early Holocene and reached its climax between 9000 and 8000 years BP. There were aridic interruptions during this wet period between 7000 and 4000 years BP until the present climatic conditions were reached. Each dry phase is an era of dune generation, and each wet period is a time of stabilization, weathering, and leaching.

The accumulation of sands by wind is not limited to deserts or areas that border deserts but also occurs near broad valleys cut by rivers carrying overloads of sand particles that they deposit as river banks. When they are exposed during the dry season, winds pick up the sand grains from the banks and deposit them as longitudinal dunes on the uplands. In certain cases sands may originate from the uplands themselves when land use has removed the protective plant cover from the soils. The existence of dunes in the wet and dry tropics demonstrates the fragility of ecosystems that are established on sands. Figure 13.1 is an illustration of this in an area dominated by isohyperthermic families of Quartzipsamments with an ustic moisture regime.

The mineralogical composition of Psamments is diverse.

Figure 13.1   Recent longitudinal dune in a tropical Quartzipsamments area with ustic moisture regimes.

Tropopsamments generally lack weatherable materials and have clays that are dominated by kaolinite. Ustipsamments may be rich in feldspars and may contain 2:1 clays, although mineral compositions similar to Tropopsamments are the most frequent in the tropics.

13.2.4   Occurrence

Figure 13.2 shows the major sand areas of Africa as they are shown on the FAO-Unesco soil map of Africa.

According to the 1:5,000,000 scale FAO-Unesco soil map of the world the Arenosols (Psamment equivalent) would cover almost 10 percent of the land area between the tropical parallels or approximately 480 million hectares; 371 million hectares would be located in tropical Africa or 16 percent of the land between 23.5° N and S latitudes. There are approximately 70.9 million hectares of Quartzipsamments (Quartz Sands) in Brazil or 8.4 percent of its land area (Bowen and Lobato, 1988).

13.2.5   Management properties

Sands have the largest number of unfavorable land attributes for sustained agriculture in the tropics. They have a low specific heat and

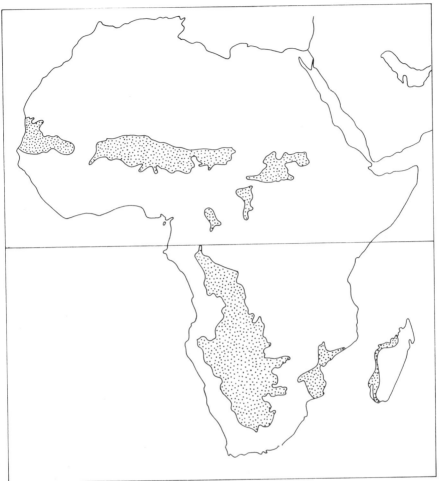

Figure 13.2   Extent of sandy areas in Africa. [*From FAO-Unesco (1977). Unesco, redrawn by permission.*]

their temperatures often exceed the critical levels for seed emergence and root growth. They are very permeable, only retaining little water and drying out very quickly. When they are dominated by low-activity clays, their nutrient-supplying power is minimal and even plants with large and deep rooting systems have difficulties in satisfying their requirements. Table 13.1 typifies the extreme poverty of aeolian acid sands in Niger (West et al., 1983).

The Quartzipsamments of the savannas of Guyana retain at field capacity approximately 75 mm available water per meter depth, equivalent to a volume percentage of 7.5. Their average saturated hydraulic conductivity is about 35 cm/h (Simpson, 1988). The particle size distribution of the materials, more specifically their fine and very

TABLE 13.1   Chemical Properties of a Quartzipsamment in Niger

Depth, cm	Organic C,%	Exchangeable cation, cmol(+)/kg				pH (H₂O)	CEC*
		Ca	Mg	K	Al		
15	0.25	0.2	0.1	0.1	0.6	5.1	1.5
27	0.15	0.1	—†	0.1	0.7	4.9	1.5
44	0.11	0.1	—	—	0.6	4.8	1.3
62	0.08	0.0	—	—	0.6	4.8	1.1
80	0.08	0.1	—	—	0.6	4.8	1.0
103	0.07	0.1	—	—	0.6	4.8	1.0
126	0.09	0.1	—	—	0.5	4.9	0.9
150	0.06	0.1	—	—	0.4	4.8	0.8

*CEC, cation exchange capacity; determined using NaOAc at pH 7.
†Trace levels found.
SOURCE: West et al. (1983).

fine sand contents, may change these physical characteristics. The water-holding capacity of the rooting zone in sandy soils of Senegal is approximately 80 mm (Chopart and Nicou, 1989).

The only land quality that in certain cases makes the use of sandy soils attractive to farmers is their easy workability. This is especially true for soils with a udic moisture regime, the Tropopsamments, that seldom become extremely hard when dry. However, the risks of failures both from an economic and soil conservation perspective are greater on sands than on any other soil that is richer in clay; the same limitations affect many *psammentic* subgroups in other orders of *Soil Taxonomy*.

Quartzipsamments dry out rapidly and when not covered by vegetation produce a single-grained, noncohesive, structureless topsoil at a surface that is easily eroded by wind and runoff water. In spite of the high permeability of the sandy materials, tropical rain intensities at times exceed the water uptake capacities of the soils. Van Caillie (1990) reported that in savanna areas of Zaïre rill erosion starts on sands with 5 to 9 percent slopes, and gullies develop as soon as the gradients are steeper than 12.5 percent. When runoff water flows on slopes of more than 20 percent, headward erosion gullies develop and undermine the surrounding plateaus. The average annual lowering of the topography in the area is estimated at 7.4 mm/year or 118 Mg/ha/year.

13.2.6   Management practices

Whenever possible the best land use decision is to avoid Psamments for annual crops and leave them under their natural vegetation if it is judged to be adequate to protect the soils against erosion. At the rain forest–savanna boundary bush fires must be avoided or properly managed so that tree regrowth rather than grass invasion occurs, being well aware that this is extremely difficult to achieve. The forests on sands are better left untouched to protect the environment, or their composition can be improved to make them economically more productive. There are situations in which communities have access only to sands as land resources. Range management and nomadic livestock approaches are traditional methods to harvest the small amounts of produce that the sands can offer. However, the environments are extremely fragile and the carrying capacity of psammentic regions is easily exceeded. Once deteriorated, the sandy land transforms into badlands that cannot be recovered and may become the deflation zone from which sands invade neighboring areas and ruin their resources. There have been several attempts to increase the productivity of sandy soils in the tropics, and some of the approaches are discussed below.

Seed bed preparation. The sandy surface layers of Psamments, more specifically Ustipsamments, may suffer from sealing, crusting, and hard setting that harm root growth in the early stages of crop development. In this case Ahmad (1988) recommended periodic deep plowing to incorporate lime (if needed), plant residues, and fertilizers that alternates with seed bed preparation under no-tillage. Hand-clearing methods for the vegetation are reported to result in better root distribution with depth than mechanical clearing that includes trampling or excessive wheel pressures.

Deep plowing (15 to 18 cm) of the dry soil instead of scraping the 5-cm topsoil results in roots penetrating into deeper layers (25 cm) without necessarily increasing the length of the rooting system itself; however, plowing improves the *Rhizobium* nodulation of groundnuts. The deep plowing produces a 30 percent average increase in yields of unshelled groundnuts above the traditional seed bed preparation production of approximately 1000 kg/ha that was obtained in a 20-year continuous cropping experiment. The effects of the deep plowing treatments are more evident for groundnuts than for millets (Chopart and Nicou, 1989).

The 15- to 18-cm dry plowing for 20 years cropping using chemical fertilizers and lime without importing organic amendments does not cause any significant reduction in the organic matter contents of the topsoils below the "traditional" levels that are considered bottom line (approximately 1.5 percent carbon). Millets seem to maintain more soil organic matter in the topsoils than groundnuts (Chopart and Nicou, 1989).

Fertilization and liming. Quartzipsamments and Tropopsamments are usually very poor in nutrients, particularly Ca, Mg, and K, and for reasonable yield performances they need fertilization and liming. Applications between 1.5 and 2.0 Mg/ha of lime are recommended. Fletcher (1988) pointed out that in many sands the solubility of Mg contained in dolomitic lime is inadequate to satisfy the Mg demand of most crops, and that supplemental fertilization with more soluble sources of Mg is necessary (i.e., potassium magnesium sulphates). Liming intervals after 3 to 4 annual cropping cycles or every 1.5 to 2 years are suggested, but strict monitoring of soil pH is needed to avoid overliming. Deep incorporation of the lime to at least 30 cm gives the best results.

Continuous cropping with chemical fertilization leads to a rapid drop in soil pH and severe reductions of Ca and Mg contents in sandy *Dior* soils (Quartzipsamments) of Senegal. The declining trend can be prevented by small amounts of dolomitic lime (1000 kg/ha every 10 years) (Chopart and Nicou, 1989).

Strongly leached Quartzipsamments and Tropopsamments that have been recently cleared are characterized by an extreme spatial microvariability. The uneven distribution of the ashes produced by burning the vegetation results in nutrient concentrations in the soil that vary from zero to optimum to excessive over short distances, sometimes less than a few meters. Many soil fertility experiments that overlook this microvariability in defining the dimensions of their test plots cannot be interpreted for this reason.

The available P levels (Truog) in Psamments rarely exceed 6 mg/kg soil, with higher values being restricted to the topsoils. The recommended fertilizer sources are triple superphosphates and diammonium phosphates. They build a labile soil P pool that reaches the point of adequacy after a few years. Rock phosphates in sands release P too slowly to satisfy the needs of annual crops.

The P fertilization of Psamments in Senegal for continuous groundnuts or a millet and groundnut rotation starts with a base application of 75 kg P_2O_5/ha that is continued by an annual maintenance application of 30 kg/ha (Chopart and Nicou, 1989). Zn and B are the most frequently reported micronutrient deficiencies (Fletcher, 1988).

Water conservation. The water-supplying power of Psamments is the most important land quality to affect crop yields. It causes a tremendous temporal variability from one year to another that in Senegal may range from 200 to 2700 kg/ha of unshelled groundnuts. It adds a large risk factor in the use of these soils in rainfed agriculture. Chopart and Nicou (1989) calculated a regression equation between yields and the degree of satisfaction of the evapotranspiration needs of groundnuts that reads as follows:

$$Y = -16,771\,X^2 + 24,930\,X - 7541$$

where Y is the yield in kg/ha of unshelled groundnuts and X is the ratio between the actual evapotranspiration and the crop water requirements (plowed and fertilizer plus lime treatments). The regression equation is not linear, and maximum yields are obtained at about 75 percent satisfaction, indicating that excessive rains or diminished insolation affects the performance of the groundnuts (Chopart and Nicou, 1989).

In semiarid regions that receive not more than 550 mm/year rainfall and where there are no other alternatives than sandy soils for annual crops, bare fallows are often considered to reduce the risks of crop failures caused by water shortages. *Bare fallows* are defined as practices that stop plant growth during one season to allow the storage of soil water to be used during a subsequent growing period. Positive results with bare fallows are generally obtained in temperate regions where winter rain is an effective source of soil water. The tropical climates with summer rains are less suitable for such practices, even more so in soils with low water-storage capacities such as the Psamments. Bare fallows in addition increase the risks of wind erosion in the tropics as well as elsewhere.

Payne et al. (1990) experimented with fallowing techniques on Psamments and psammentic Alfisols in Niger. Bare fallow plots were kept free of vegetation during the growing season by hoeing to a depth of approximately 5 cm; millet was grown in the cultivated plots that served as controls. Half of each plot in the experiments was covered with straw mulch at a rate of approximately 25 Mg/ha, equivalent to a thickness of 10 to 15 cm at harvest time. The water contents of the soils were monitored during the whole year at different depths down to 165 cm until the end of the dry season that followed the growing season.

The interpretation of the results indicates that the mulch had no effect on the "carry over" of soil moisture from one growing season to the next. Neither did the Psamments at the end of the dry season show any significant differences in moisture contents at less than 1 m depth between fallowed and cropped land. The gains by bare fallow could only be detected in the subsoil of a psammentic Alfisol that contains an argillic horizon in the lower part of the profile. In this case the difference for the whole profile (165 cm) was 40.6 ± 17.9 mm water at the end of the dry season—in other words, at the beginning of the following growing season.

The data show that bare fallow and dry season mulching are inefficient methods of water conservation in deep Psamments that have isotemperature regimes. This would be due to water losses during the dry season both by evaporation and drainage. There was a continuous downward unsaturated flow in the fallowed plots during the dry season. There was also a year-round evaporative loss of water from surface horizons (Payne et al., 1990).

Farming systems. Quartzipsamments and Tropopsamments are the least suitable for *intensive* annual crop production and should not be uocd for that purpoco whonovor other soils are available. *Extensive* rainfed production of annual crops is possible. Production has then to rely on correct spacing of plants that is adapted to the minimal available water content in the sands or on selection of crops such as millets and sorghums that can survive long periods of drought.

Groundnuts and cassava are plants that are often considered best adapted to the stresses imposed by the sandy nature of the soils. The most popular crop on white sands in Guyana is pineapples planted in newly cleared land; given the low fertility, only one harvest after clearing the forest is possible (Ahmad, 1988). On coastal sands in the Far East coconuts and cashews are grown, usually near or in villages.

Psamments in the Brazilian savannas are used for extensive grazing in cow-calf operations at 0.2 to 0.6 animals (Zebu) per hectare and per year. Life weight gains are particularly poor during the dry season. The severe P deficiencies can be alleviated by simple superphosphate applications of 22 kg/ha. Further improvements in productivity cannot be achieved, however, without complete fertilization. The most promising plants in improved pastures are *Stylosanthes* and *Andropogon* species that are tolerant to high Al saturations, low P, and drought (Bowen and Lobato, 1988).

Eucalyptus, Pinus, cashew (*Anacardium occidental* L.), and *Citrus* are tree species that do best in sandy soils. However, according to Brazilian experience, fertilization is necessary, either by chemical fertilizers, manure, or cultivation near homesteads. Cashews grow best in sands that contain a water table between 2 and 6 m depths.

Irrigation agriculture on sandy materials requires frequent applications of small amounts of water either by sprinkler or drip irrigation. These techniques can only be economically justified for high-value crops and cannot compete with soils that have higher water-retention capacities.

13.3 Sulfaquents

13.3.1 Definitions

The etymology of the word "Sulfaquent" leads to an identification of the major characteristics of this great soil group. The suffix "ent" indicates a young, poorly developed soil; the radical "aqu" implies that the soil has an aquic moisture regime, and in this case, is almost permanently saturated with water; and the prefix "sulf" denotes the presence of sulfidic material, more specifically pyrite (FeS_2). The equivalent in the FAO-Unesco legend of the soil map of the world is *Thionic Fluvisols*. Another name for these soils is *Potential Acid Sulfate Soils*.

Sulfidic material. Sulfidic materials are waterlogged mineral or organic materials that contain 0.75 percent or more sulfur, mostly as sulfides, and that have less than three times as much calcium carbonate equivalents as sulfur (Soil Survey Staff, 1990).

13.3.2 Origin

The optimum conditions for the formation of Sulfaquents and for the accumulation of pyrite (FeS_2) exist in the tidal flats of river deltas covered by mangrove forests. The rivers bring the iron as oxides or hydroxides in their solid load, the sea supplies the sulfates, and the swamp vegetation produces the organic matter that releases electrons to reduce both sulfates and iron into pyrite. Tidal movement that washes the sediments to remove the alkalinity released during pyrite formation is necessary to obtain a sulfidic material that does not contain large amounts of carbonates.

13.3.3 Management

Transformation of pyrite. The pyrite contained in sulfidic materials does not cause problems as long as it remains under anaerobic conditions. However, as soon as oxygen enters the sulfidic layers, for example, as a result of drainage, pyrite oxidizes and produces large quantities of acid according to the following equations:

$$FeS_2 + 7/2O_2 + H_2O \rightarrow Fe^{2+} + 2SO_4^{2-} + 2H^+$$

The ferrous iron oxidizes further to ferric hydroxides as follows:

$$6Fe^{2+} + 15H_2O + 3/2O_2 \rightarrow 6Fe(OH)_3 + 12H^+$$

If calcium carbonates are present, gypsum is formed and no extreme acidity occurs. In the absence of carbonates the pH drops to very low values, and the iron hydroxides may combine with potassium ions to form the mineral Jarosite that has characteristic straw-yellow colors and a very penetrating odor that resembles cat excrement. Sulfaquents that are drained develop into *actual* acid sulfate soils that are also called cat-clays. The acidity in these cat-clays is such that the soil environment becomes extremely toxic to the root systems of most cultivated plant species (see Sect. 11.5.1 on Sulfaquepts).

Management strategies. There are several approaches to manage Sulfaquents. All of them include the control of the water table. The irrigation and drainage system should be such that the acidity that is produced can be leached out of the soil without harm to the environment.

Drastic drainage programs that permanently lower the water table below the rooting zone are expensive and difficult to manage. The leaching of the acidity may also deplete the soil of its major nutrients. On the other hand, the quantity of lime that would be needed to neutralize the acidity is economically prohibitive.

The more gradual drainage schemes have more chances to be successful. They vary according to the depth and the thickness of the sulfidic materials. They only progressively drain the land so that less acidity is produced each year and are often combined with raised beds and furrow systems that leave a large portion of the pyrite-containing layers under the water table. The irrigation, leaching, and drainage programs have to take into account the water table fluctuations that follow the tides and seasons. The land utilization types used in the rotations usually include paddy rice, fish ponds, and dryland crops.

Sulfaquents that are drained, either artificially or naturally, develop into Sulfaquepts, an Inceptisol that is characterized by the presence of a *sulfuric* horizon. The sulfuric horizon is composed of materials that have both a pH of less than 3.5 and jarosite mottles. Sulfaquepts are seldom used for agriculture, except where population pressure is such that pioneer farmers have over many years developed inventive systems, taking advantage of tides, utilizing diversion of river waters in irrigation canals and flap gates, building beds and furrows, and growing a variety of crops to extract the maximum from such hostile environments. A prerequisite is the constant availability of sufficient irrigation water to control the water table, especially where a severe dry season would lower it too drastically and produce acidity that cannot be leached out by the rains. Monsoon climates are therefore at a disadvantage to climatic regimes with more uniform distributions of precipitation.

References

Ahmad, N., "Acid sandy soils of the tropics with particular reference to the Guyanas," in E. Walmsley (ed.), *Farming Systems for Low-Fertility Acid Sandy Soils,* CTA Seminar Proceedings, EDE-Wageningen, The Netherlands, 1988, pp. 12–31.

Bowen, W. T. and E. Lobato, "Possibilities and constraints for crop production on acid sandy soils (Quartz Sands) in Brazil," in E. Walmsley (ed.), *Farming Systems for*

Low-Fertility Acid Sandy Soils, CTA Seminar Proceedings, EDE-Wageningen, The Netherlands, 1988, pp. 75–85.

Chopart, J. L. and R. Nicou, "Vingt ans de culture continué avec ou sans labour au Sénégal," *Agronomie Tropicale* 44(4):269–281 (1989).

FAO-Unesco, *Soil Map of the World,* vol. VI, Africa, Unesco, Paris, 1977.

Fletcher, R. E., "Improvement of soil fertility on acid sandy soils of the intermediate savannahs of Guyana," in E. Walmsley (ed.), *Farming Systems for Low-Fertility Acid Sandy Soils,* CTA Seminar Proceedings, EDE-Wageningen, The Netherlands, 1988, pp. 38–51.

Payne, W. A., C. W. Wendt, and R. J. Lascano, "Bare fallowing on sandy fields of Niger, West Africa," *Soil Sci. Soc. Amer. J.* 54:1079–1084 (1990).

Pereira, H. C., R. A. Wood, H. W. Brzostowski, and P. H. Hosegood, "Water conservation by fallowing in semi-arid tropical East Africa," *Emp. J. Exp. Agric.* 26:213–228 (1958).

Servant, M., "Séquence continentales et variations climatiques: évolution du bassin du Tchad au Cénozoïque supérieur," *Travaux et documentation de l'O.R.S.T.O.M.*-159, Paris, 1983.

Simpson, L. A., "Some aspects of soil physical characteristics and their management in the intermediate savannahs of Guyana," in E. Walmsley (ed.), *Farming Systems for Low-Fertility Acid Sandy Soils,* CTA Seminar Proceedings, EDE-Wageningen, The Netherlands, 1988, pp. 52–68.

Soil Survey Staff, *Keys to Soil Taxonomy,* 4th ed., SMSS Technical Monograph #19, Blacksburg, VA., 1990.

Van Caillie, X. D., "Erodabilité des terrains sableux du Zaïre et contrôle de l'érosion," *Cahiers Orstom, série Pédologie* 25(1–2):197–208 (1990).

Volkel, J. and J. Grunert, "To the problem of dune formation and dune weathering during the late Pleistocene and Holocene in the southern Sahara and the Sahel," *Zeitschrift für Geomorphologie* 34:1–17 (1990).

West, L. T., L. P. Wilding, J. K. Landek, and F. G. Calhoun, *Soil Survey of the Icrisat Sahelian Centre, Niger, West Africa,* Texas A&M University, 1983.

A horizons Mineral horizons that formed at the surface or below an O horizon and (i) are characterized by an accumulation of humified organic matter intimately mixed with the mineral fraction and not dominated by properties characteristic of E or B horizons; or (ii) have properties resulting from cultivation, pasturing, or similar kinds of disturbance.

adsorption The process by which atoms, molecules, or ions are taken up and retained on the surfaces of solids by chemical or physical binding, e.g., the adsorption of cations by negatively charged minerals.

adsorption isotherm A graph of the quantity of a given chemical species bound to an adsorption complex, at fixed temperature, as a function of the concentration of the species that is in equilibrium with the complex.

aerobic (i) Having molecular oxygen as a part of the environment. (ii) Growing only in the presence of molecular oxygen, as aerobic organisms. (iii) Occurring only in the presence of molecular oxygen (said of certain chemical or biochemical processes such as aerobic decomposition).

aggregate A unit of soil structure, usually formed by natural processes in contrast with artificial processes, and generally < 10 mm in diameter.

agroforestry Any type of multiple cropping land use that entails complementary relations between tree and agricultural crops and produces some combination of food, fruit, fodder, fuel, wood, mulches, or other products.

air dry (i) The state of dryness at equilibrium with the water content in the surrounding atmosphere. The actual water content will depend upon the relative humidity and temperature of surrounding atmosphere. (ii) To allow to reach equilibrium in water content with the surrounding atmosphere.

albic horizon A mineral soil horizon from which clay and free iron oxides have been removed or in which the oxides have been segregated to the extent

---

*Extracted with permission from *Glossary of Soil Science Terms*, Soil Science Society of America, 1987.

that the color of the horizon is determined primarily by the color of the primary sand and silt particles rather than by coatings on these particles.

Alfisols   See page 165.

allophane   An aluminosilicate with primarily short-range structural order. Occurs as exceedingly small spherical particles especially in soils formed from volcanic ash. Also, it occurs in podzolic soils formed on weathered granite in a cool moist climate.

alluvial   Pertaining to processes or materials associated with transportation or deposition by running water.

ammonification   The biological process leading to a formation of ammoniacal nitrogen from nitrogen containing organic compounds.

amorphous material   Non-crystalline constituents that either do not fit the definition of allophane or it is not certain if the constituent meets allophane criteria.

anaerobic respiration   The metabolic process whereby electrons are transferred from an organic compound to an inorganic acceptor molecule other than oxygen. The most common acceptors are carbonate, sulfate, and nitrate. See denitrification.

Andepts   See page 207.

anion exchange capacity   The sum total of exchangeable anions that a soil can adsorb. Expressed as centimoles of charge per kilogram of soil (or of other adsorbing material such as clay).

aquic   A mostly reducing soil moisture regime nearly free of dissolved oxygen due to saturation by groundwater or its capillary fringe and occurring at periods when the soil temperature at 50 cm below the surface is >5°C.

argillan   A cutan composed dominantly of clay minerals.

argillic horizon   See page 166.

aridic   See page 127.

Aridisols   Mineral soils that have an aridic moisture regime, an ochric epipedon, and other pedogenic horizons but no oxic horizon. (An order in the U.S. system of *Soil Taxonomy.*)

available water   The portion of water in a soil that can be absorbed by plant roots. It is the amount of water released between in situ field capacity and the permanent wilting point (usually estimated by water content at soil matric potential of −1.5 MPa).

B horizons   Horizons that formed below an A, E, or O horizon and are dominated by (i) carbonates, gypsum, or silica, alone or in combination; (ii) evidence of removal of carbonates, (iii) concentrations of sesquioxides; (iv) alterations that form silicate clay; (v) formation of granular, blocky, or prismatic structure; or (vi) combination of these.

backslope   The slope component that is the steepest, straight then concave, or merely concave middle portion of an erosional slope.

base saturation percentage   The extent to which the adsorption complex of a soil is saturated with alkali or alkaline earth cations expressed as a percentage of the cation exchange capacity measured at pH 7.0, which may include acidic cations such as H^+ and Al^{3+}.

bedrock   The solid rock underlying soils and the regolith in depths ranging from zero (where exposed by erosion) to several hundred centimeters.

broadcast   The application of fertilizer on the soil surface. Usually done prior to planting and normally incorporated with tillage but may be unincorporated in no-till systems.

buffer power   The ability of ions associated with the solid phase to buffer changes in ion concentration in the solution phase. Cs/Cl where Cs represents the concentration of ions on the solid phase in equilibrium with Cl, the concentration of ions in the solution phase.

bulk density, soil (rb)   The mass of dry soil per unit bulk volume. The bulk volume is determined before drying to constant weight at 105°C. The value is expressed in grams per cubic centimeter.

buried soil   Soil covered by an alluvial, loessal, or other depositional surface mantle of new material, usually to a depth greater than the thickness of the solum.

C horizons   Horizons or layers, excluding hard bedrock, that are little affected by pedogenic processes and lack properties of O, A, E, or B horizons. Most are mineral layers, but limnic layers, whether organic or inorganic, are included.

calcareous soil   Soil containing sufficient free $CaCO_3$ and/or $MgCO_3$ to effervesce visibly when treated with cold 0.1M HCl. These soils usually contain from as little as 10 to as much as 200 g kg^{-1} $CaCO_3$ equivalent.

calcic horizon   A mineral soil horizon of secondary carbonate enrichment that is > 15 cm thick, has a $CaCO_3$ equivalent of > 150 g kg^{-1}, and has at least 50 g kg^{-1} more $CaCO_3$ equivalent than the underlying C horizon.

cambic horizon   See page 234.

cat clay   Wet clay soils containing ferrous sulfide which become highly acidic when drained.

category   Any one of the ranks of the system of soil classification in which soils are grouped on the basis of their characteristics.

catena   (as used in the U.S.A.) A sequence of soils of about the same age, derived from similar parent material, and occurring under similar climatic conditions, but having different characteristics due to variation in relief and in drainage. See toposequence.

cation exchange   The interchange between a cation in solution and another cation on the surface of any negatively charged material such as clay colloid or organic colloid.

cation exchange capacity (CEC)   The sum of exchangeable cations that a soil, soil constituent, or other material can adsorb at a specific pH. It is usually expressed in centimoles of charge per kilogram of exchanger ($cmol_c \ kg^{-1}$).

chemical potential   (i) The intensive variable that determines equilibrium with respect to matter flux. The chemical potential is the partial derivative of the Gibbs free energy with respect to the mass variable (e.g., the mole number) of a component of a thermodynamic system, calculated with temperature, applied pressure, mass variables for all the other components of the system, and any other independent, extensive variables, held fixed. (ii) Informally, it is the capacity of a solution or other substances to do work by virtue of its chemical composition.

chemical weathering   The breakdown of rocks and minerals due to chemical activity, primarily due to the presence of water and components of the atmosphere. See weathering.

chlorite   A layer-structured group of silicate materials of the 2:1 type that has the interlayer filled with a positively charged metal-hydroxide octahedral sheet. There are both trioctahedral (e.g., $M = Fe^{2+}$, Mg^{2+}) and dioctahedral ($M = Al^{3+}$) varieties.

chroma   The relative purity, strength, or saturation of a color; directly related to the dominance of the determining wavelength of the light and inversely related to grayness; one of the three variables of color. See Munsell color system, hue, and value.

chronosequence   A sequence of related soils that differ, one from the other, in certain properties primarily as a result of time as a soil-forming factor.

class, soil   A group of soils having a definite range in a particular property, such as acidity, degree of slope, texture, structure, land-use capability, degree of erosion, or drainage.

clay films   Coatings of clay on the surfaces of soil peds and mineral grains and in soil pores. (Also called clay skins, clay flows, illuviation cutans, argillans or tonhäutchen.)

clay mineral   (i) Any crystalline inorganic substance of clay size (i.e., <2 µm equivalent spherical diameter); (ii) Any phyllosilicate of clay size. See phyllosilicate mineral terminology.

colluvium   A general term applied to deposits on a slope or at the foot of a slope or cliff that were moved there chiefly by gravity. Talus and cliff debris are included in such deposits.

compressibility   The property of a soil pertaining to its susceptibility to decrease in bulk volume when subjected to a load.

concretion   A local concentration of a chemical compound, such as calcium carbonate or iron oxide, in the form of a grain or nodule of varying size, shape, hardness, and color.

constant-charge surface   A mineral surface carrying a net electrical charge whose magnitude depends only on the structure and chemical composition of the mineral itself. Constant-charge surfaces usually arise from isomorphous substitution in mineral structures.

constant-potential surface   A solid surface carrying a net electrical charge which may be positive, negative, or zero, depending on the activity of one or more species of ion (called a potential-determining ion) in a solution phase contacting the surface. For minerals common in soils, the potential-determining ion usually is H^+ or OH^-, but any ion that forms a complex with the surface may be potential-determining. See also pH-dependent charge.

crest   The slope component that is commonly at the top of an erosional ridge, hill, mountain, etc. See summit.

critical nutrient concentration   The nutrient concentration in the plant, or specified plant part, below which the nutrient becomes deficient for optimum growth rate.

crust   A soil-surface layer, ranging in thickness from a few millimeters to a few tens of millimeters, that is much more compact, hard, and brittle, when dry, than the material immediately beneath it.

crystalline rock   A rock consisting of various minerals that have crystallized in place from magma. See igneous rock and sedimentary rock.

cultivation   A tillage operation used in preparing land for seeding or transplanting or later for weed control and for loosening the soil.

cutan   A modification of the texture, structure, or fabric at natural surfaces in soil materials due to concentration of particular soil constituents or in situ modification of the plasma. See also clay films.

denitrification   Reduction of nitrate or nitrite to molecular nitrogen or nitrogen oxides by microbial activity (dissimilatory nitrate reduction) or by chemical reactions involving nitrite (chemical denitrification).

deposit   Material left in a new position by a natural transporting agent such as water, wind, ice, or gravity, or by the activity of man.

desert pavement   The layer of gravel or stones left on the land surface in desert regions after the removal of the fine material by wind erosion.

dinitrogen fixation   Conversion of molecular nitrogen (N_2) to ammonia and subsequently to organic combinations or to forms utilizable in biological processes.

dioctahedral   An octahedral sheet or a mineral containing such a sheet that has two-thirds of the octahedral sites filled by trivalent ions such as aluminum or ferric iron.

disperse   (i) To break up compound particles, such as aggregates, into the individual component particles. (ii) To distribute or suspend fine particles, such as clay, in or throughout a dispersion medium, such as water.

dissection   The partial erosional destruction of land surface or landform by gully, arroyo, canyon, or valley cutting leaving flattish remnants, or ridges, hills, or mountains separated by drainage ways.

dolomitic lime   A naturally occurring liming material composed chiefly of carbonates of Mg and Ca in approximately equimolar proportions.

drainage class (natural)   Has been used in humid areas; includes phrases such as well-drained, moderately well-drained, imperfect, or somewhat poorly and poorly drained.

duripan   See page 111.

E horizons   Mineral horizons in which the main feature is loss of silicate clay, iron, aluminum, or some combination of these, leaving a concentration of sand and silt particles of quartz or other resistant materials.

EC$_e$ The electrolytic conductivity of an extract from saturated soil, normally expressed in units of siemens per meter at 25°C.

Eh   The potential that is generated between an oxidation or reduction half-reaction and the H electrode in the standard state.

eluvial horizon   A soil horizon that has been formed by the process of eluviation. See illuvial horizon.

eluviation   The removal of soil material in suspension (or in solution) from a layer or layers of a soil. Usually, the loss of material in solution is described by the term "leaching." See illuvation and leaching.

eolian   A term applied to materials deposited by wind such as sand dunes, sandsheets, and loess.

erodibility   The state or condition of being erodible.

erodible   Susceptible to erosion. (Expressed by terms such as highly erodible, slightly erodible, etc.)

erosional surface   A land surface shaped by the erosive action of ice, wind, or water; but usually as the result of running water.

erosivity   The potential ability of water, wind, gravity, etc., to cause erosion.

eutrophic   Having concentrations of nutrients optimal, or nearly so, for plant or animal growth. (Said of nutrient or soil solutions and bodies of water.) The term literally means "self-feeding."

evapotranspiration   The combined loss of water from a given area, and during a specified period of time, by evaporation from the soil surface and by transpiration from plants.

exchange capacity The total ionic charge of the adsorption complex active in the adsorption of ions. See anion exchange capacity and cation exchange capacity.

exchangeable anion A negatively charged ion held on or near the surface of a solid particle by a positive surface charge and which may be replaced by other negatively charged ions.

exchangeable bases See base saturation percentage.

exchangeable cation A positively charged ion held on or near the surface of a solid particle by a negative surface charge of a colloid and which may be replaced by other positively charged ions in the soil solution. Often determined as the salt-extractable minus water-soluble cations in a saturation extract, and expressed in centimoles of charge per kilogram.

family, soil In soil classification one of the categories intermediate between the great soil group and the soil series. Families provide groupings of soils with ratings in texture, mineralogy, temperature, and thickness.

ferrihydrite See page 45.

ferrolysis Clay destruction process involving disintegration and solution in water by a process based upon the alternate reduction and oxidation of iron.

fertilizer, controlled-release A fertilizer term used interchangeably with delayed release, slow release, controlled availability, slow acting, and metered release to designate a controlled dissolution of fertilizer at a lower rate than conventional water-soluble fertilizers. Controlled-release properties may result from coatings on water-soluble fertilizers or from low dissolution and/or mineralization rates of fertilizer materials in soil.

fertilizer requirement The quantity of certain plant nutrients needed, in addition to the amount supplied by the soil, to increase plant growth to a designated level.

fertilizer, side-dressed Application made to the side of crop rows after plant emergence.

fertilizer, slow-release See fertilizers, controlled-release.

fertilizer, top-dressed A surface application of fertilizer to a soil after the crop has been established.

field capacity, in situ (field water capacity) The content of water, on a mass or volume basis, remaining in a soil 2 or 3 days after having been wetted with water and after free drainage is negligible. See available water.

flood plain The land bordering a stream, built up of sediments from overflow of the stream and subject to inundation when the stream is at flood stage.

Fluvents Entisols that form in recent loamy or clayey alluvial deposits, are usually stratified, and have an organic carbon content that decreases irregu-

larly with depth. Fluvents are not saturated with water for periods long enough to limit their use for most crops. (A suborder in the U.S. system of *Soil Taxonomy.*)

footslope The relatively gently sloping, slightly concave to straight slope component of an erosional slope that is at the base of the backslope component.

forest floor All dead vegetable or organic matter, including litter and unincorporated humus, on the mineral soil surface under forest vegetation.

fragile land Land that is sensitive to degradation when disturbed; such as with highly erodible soils, where salts can and do accumulate, and at high elevations.

fragipan See page 112.

free iron oxides A general term for those iron oxides that can be reduced and dissolved by a dithionite treatment. Generally includes goethite, hematite, ferrihydrite, lepidocrocite, and maghemite, but not magnetite. See iron oxides.

fulvic acid (i) The mixture of organic substances remaining in solution upon acidification of a dilute alkali extract from soil, in which case the expression fulvic acid fraction is often used, and (ii) the colored material that remains in solution after removal of humic acid by acidification.

geomorphic surface A portion of the landscape specifically defined in space and time that has determinable geographic boundaries and is formed by one or more agencies during a given time period.

geomorphology The science that studies the evolution of the earth's surface. The science of landforms. The systematic examination of landforms and their interpretation as records of geologic history.

gibbsite $Al(OH)_3$. A mineral with a platy habit that occurs in highly weathered soils and in laterite. Also, may be prominent in the subsoil and saprolite of soils formed on crystalline rock high in feldspar.

gilgai See page 194.

goethite $FeOOH$. A yellow-brown iron oxide mineral. Goethite is very common and is responsible for the brown color in many soils.

great soil group One of the categories in the system of soil classification that has been used in the United States for many years. Great groups place soils according to soil moisture and temperature, base saturation status, and expression of horizons.

green manure Plant material incorporated into soils while green or at maturity, for soil improvement.

ground water That portion of water below the surface of the ground at a pressure equal to or greater than atmospheric. See water table.

gullied land Areas where all diagnostic soil horizons have been removed by water, resulting in a network of V-shaped or U-shaped channels. Some areas resemble miniature badlands. Generally, gullies are so deep that extensive re-shaping is necessary for most uses.

gully A channel resulting from erosion and caused by the concentrated but intermittent flow of water usually during and immediately following heavy rains. Deep enough to interfere with, and not to be obliterated by, normal till-age operations.

gypsic horizon A mineral soil horizon of secondary $CaSO_4$ enrichment that is > 15 cm thick, has at least 50 g kg^{-1} more gypsum than the C horizon, and in which the product of the thickness in centimeters and the amount of $CaSO_4$ is equal to or greater than 1500 g kg^{-1}.

gypsum The common name for calcium sulphate ($CaSO_4 \cdot 2H_2O$), used to supply calcium and sulfur and to ameliorate sodic soils.

gypsum requirement The quantity of gypsum or its equivalent required to reduce the exchangeable sodium content of a given amount of soil to an ac-ceptable level.

halloysite A member of the kaolin subgroup of clay minerals. It is similar to kaolinite in structure and composition except that hydrated varieties occur that have interlayer water molecules and they are usually tubular or spheroi-dal particles. Halloysite is most common in soils formed from volcanic ash, but also occurs in subsoil and saprolite of soils formed on acid crystalline rock.

hematite Fe_2O_3. A red iron oxide material that contributes red color to many soils.

histic epipedon See page 98.

Histosols These organic soils have organic soil materials in more than half of the upper 80 cm, or of any thickness if overlying rock or fragmental mate-rials that have interstices filled with organic soil materials. (An order in the U.S. system of *Soil Taxonomy*.)

hue One of the three variables of color. It is caused by light of certain wave-lengths and changes with the wavelength. See Munsell color system, chroma, and value.

humic acid The dark-colored organic material that can be extracted from soil by various reagents (e.g., dilute alkali) and that is precipitated by acidi-fication to pH 1 to 2.

humic substances A series of relatively high-molecular-weight, brown to black colored substances formed by secondary synthesis reactions. The term is used in a generic sense to describe the colored material or its fractions ob-tained on the basis of solubility characteristics (e.g., humic acid, fulvic acid).

humification The process whereby the carbon of organic residues is trans-formed and converted to humic substances through biochemical and/or chem-ical processes.

humin   The fraction of the soil organic matter that is not dissolved upon extraction of the soil with dilute alkali.

hydrated lime   A liming material composed chiefly of calcium and magnesium hydroxides. It reacts quickly to neutralize acid soils.

hydrogen bond   The chemical bond between a hydrogen atom of one molecule and two unshared electrons of another molecule.

hydrophobic soils   Soils that are water repellent, often due to dense fungal mycelial mats or hydrophobic substances vaporized and reprecipitated during fire.

hydroxy-interlayered vermiculite   A vermiculite with partially filled interlayers of hydroxy-aluminum groups. It is normally dioctahedral in both the interlayer and the vermiculite layer. It is common in the coarse clay fraction of acid surface soil horizons. It has intermediate cation exchange properties between vermiculite and chlorite. Synonyms are "chlorite-vermiculite intergrade; vermiculite-chlorite intergrade."

igneous rock   Rock formed from the cooling and solidification of magma, and that has not been changed appreciably since its formation.

illite   The mica component of a structurally mixed fine grained mica and smectite or vermiculite. Sometimes the entire mixture is referred to as illite. Illites are dioctahedral.

illuvial horizon   A soil layer or horizon in which material carried from an overlying layer has been precipitated from solution or deposited from suspension. The layer of accumulation. See eluvial horizon.

illuviation   The process of deposition of soil material removed from one horizon to another in the soil; usually from an upper to a lower horizon in the soil profile. See eluviation.

imogolite   A poorly crystalline aluminosilicate mineral with an ideal composition $SiO_2 \, Al_2O_3 \, 2.5H_2O(+)$. It occurs as exceedingly small fibers mostly in soils formed from volcanic ash but also in podzolic soils formed on weathered granite in a cool moist climate.

Inceptisols   See page 233.

infiltration   The downward entry of water into the soil through the soil surface.

infiltration flux (or rate)   The volume of water infiltrated downward into the soil per unit cross-sectional soil area in unit time, with dimensions of velocity (i.e., $m^3 \, m^{-2} \, s^{-1} = m \, s^{-1}$).

inoculate   To treat, usually seeds, with microorganisms for the purpose of creating a favorable response. Most often refers to the treatment of legume seeds with *Rhizobium* to stimulate dinitrogen fixation.

ion activity   Informally, the effective concentration of an ion in solution. Numerically, it approaches the value of the ionic concentration at infinite dilution of the ion under consideration and otherwise satisfies the formal, thermodynamic relationship between activity and chemical potential as applied to a single ionic species.

ion selectivity   The relative adsorption of an ion by the solid phase in relation to the adsorption of other ions. The relative absorption of an ion by a root in relation to absorption of other ions.

ion activity coefficient   The ratio between ionic activity and an ionic concentration measured on an appropriate scale. In an aqueous solution at low ionic strength, the ionic activity coefficient may be defined by the Debye-Huckel equation.

ionic strength   A parameter that estimates the interaction between ions in solution. It is calculated as one-half the sum of the products of ionic concentration and the square of ionic charge for all the charged species in a solution.

iron oxides   Group name for the oxides and hydroxides of iron. Includes the minerals goethite, hematite, lepidocrocite, ferrihydrite, maghemite, and magnetite. Sometimes referred to as "free iron oxides," "sesquioxides," or "hydrous oxides."

ironstone   Hardened plinthite materials often occurring as nodules and concentrations.

isoelectric point   The pH value of a solution in equilibrium with a constant potential surface whose net electrical charge is zero and is dependent only on the presence of H^+, OH^-, or H_2O to form species on the surface.

isomorphous substitution   The replacement of one atom by another of similar size (but not necessarily of the same valence) in a crystal structure without disrupting or seriously changing the structure.

jarosite   A pale yellow potassium iron sulfate material.

kandic horizon   See page 170.

kaolin   A subgroup name of aluminum silicates with a 1:1 layer structure. Kaolinite is the most common clay mineral in the subgroup. Also, a soft, usually white rock composed largely of kaolinite.

kaolinite   A clay mineral of the kaolin subgroup. Its general formula is $Al_2Si_2O_5(OH)_4$. It has a 1:1 layer structure composed of shared sheets of Si-O tetrahedrons and Al-(O,OH) octahedrons.

lacustrine deposit   Material deposited in lake water and later exposed either by lowering of the water level or by the elevation of the land.

land classification   The arrangement of land units into various categories based upon the properties of the land or its suitability for some particular purpose.

landform   A three-dimensional part of the land surface, formed of soil, sediment, or rock that is distinctive because of its shape, that is significant for land use or to landscape genesis, that repeats in various landscapes, and that also has a fairly consistent position relative to surrounding landforms.

landscape   All the natural features such as fields, hills, forests, water, etc., which distinguish one part of the earth's surface from another part. Usually that portion of land or territory which the eye can comprehend in a single view, including all its natural characteristics.

lattice   A regular geometric arrangement of points in a plane or in space. Lattice is used to represent the distribution of repeating atoms or groups of atoms in a crystalline substance. A lattice is a mathematical concept. It has no mass.

lava flows   Areas that are covered by lava. Most have a sharp jagged surface, crevices, and angular blocks characteristic of lava. Others are relatively smooth and have a ropey glazed surface. Soil material may be in a few cracks and sheltered pockets, but the flows are virtually devoid of plants except for lichens.

layer silicate minerals   Minerals with the sheet silicate structures of the phyllosilicates.

leaching   The removal of materials in solution from the soil. See eluviation.

leaching requirement   The leaching fraction necessary to keep soil salinity, chloride, or sodium (the choice being that which is most demanding) from exceeding a tolerance level of a crop in question. It applies to steady-state or long-term average conditions.

lepidocrocite   FeOOH. An orange iron oxide mineral that is found in mottles and concretions of wet soils.

lime, agricultural   A soil amendment containing calcium carbonate, magnesium carbonate and other materials, used to neutralize soil acidity and furnish calcium and magnesium for plant growth. Classification including calcium carbonate equivalent and limits in lime particle size is usually prescribed by law or regulation.

lithic contact   A boundary between soil and continuous, coherent, underlying material. The underlying material must be sufficiently coherent to make hand-digging with a spade impractical. If it is a single mineral, it must have a hardness of 3 or more (Mohs scale); if not, gravel size chunks that can be broken out do not disperse with 15 hours shaking in water or sodium hexametaphosphate solution.

litter   The surface layer of the forest floor consisting of freshly fallen leaves, needles, twigs, stems, bark, and fruits.

loess   Material transported and deposited by wind and consisting of predominantly silt-sized particles.

maghemite Fe_2O_3. A dark reddish-brown, magnetic iron oxide mineral chemically similar to hematite, but structurally similar to magnetite. Often found in well-drained, highly weathered soils of tropical regions.

magnetite Fe_3O_4. A black, magnetic iron oxide mineral usually inherited from igneous rocks. Often found in soils as black magnetic sand grains.

manganese oxides   A group term for oxides of manganese. They are typically black and frequently occur in soils as nodules and coatings on ped faces usually in association with iron oxides. Birnessite and lithiophorite are common manganese oxide minerals in soils.

manure The excreta of animals, with or without an admixture of bedding or litter, fresh or at various stages of further decomposition or composting. In some countries may denote any fertilizer material.

marl Soft and unconsolidated calcium carbonate, usually mixed with varying amounts of clay or other impurities.

marsh Periodically wet or continually flooded areas with the surface not deeply submerged. Covered dominantly with sedges, cattails, rushes, or other hydrophytic plants. Subclasses include fresh-water and salt-water marshes. See swamp.

mass flow (nutrient) The movement of solutes associated with net movement of water.

mesofauna Nematodes, oligochaete worms, smaller insect larvae, and microarthropods.

metamorphic rock Rock derived from preexisting rocks but that differ from them in physical, chemical, and mineralogical properties as a result of natural geological processes, principally heat and pressure, originating within the earth. The preexisting rocks may have been igneous, sedimentary, or another form of metamorphic rock.

mica A layer-structured aluminosilicate mineral group of the 2:1 type that is characterized by its high layer charge, which is usually satisfied by potassium. The major types are moscovite, biotite, and phlogopite.

microfauna Protozoa, nematodes, and arthropods of microscopic size.

microflora Bacteria (including actinomycetes), fungi, algae, and viruses.

micronutrient A chemical element necessary for plant growth found in small amounts, usually < 100 mg kg^{-1} in the plant. The elements consist of B, Cl, Cu, Fe, Mn, Mo, and Zn.

microrelief (i) Small scale, local differences in topography, including mounds, swales, or pits that are usually < 1 m in diameter and with elevation differences of up to 2 m. (ii) Differences in topography altered by tillage operations, over an area of about 1 m^2 with elevation difference of a few centimeters or less.

mollic epipedon A surface horizon of mineral soil that is dark colored and relatively thick, contains at least 5.8 g kg^{-1} organic carbon, is not massive and hard or very hard when dry, has a base saturation of more than 50% when measured at pH 7, has less than 10 mg P kg^{-1} soluble in 0.05 M citric acid, and is dominantly saturated with bivalent cations.

Mollisols Mineral soils that have a mollic epipedon overlying mineral material with a base saturation of 50% or more when measured at pH 7. Mollisols may have an argillic, natric, albic, cambic, gypsic, calcic, or petrocalcic horizon, a histic epipedon, or a duripan, but not an oxic or spodic horizon. (An order in the U.S. system of *Soil Taxonomy.*)

montmorillonite   An aluminum silicate (smectite) with a layer structure composed of two silica tetrahedral sheets and a shared aluminum and magnesium octahedral sheet. Montmorillonite has a permanent negative charge that attracts interlayer cations that exist in various degrees of hydration thus causing expansion and collapse of the structure (i.e., shrink-swell). Its general formula is $Si_4Al_{1.5}Mg_{0.5}O_{10}(OH)_2Ca_{0.25}$. The calcium is exchangeable.

mor   A type of forest humus in which the Oa horizon is present and in which there is practically no mixing of surface organic matter with mineral soil; that is, the transition from the Oa to A horizon is abrupt. (Sometimes differentiated into thick mor, thin mor, granular mor, greasy mor, or felty mor.)

mottled zone   A layer that is marked with spots or blotches of different color or shades of color. The pattern of mottling and the size, and color contrast of the mottles may vary considerably and should be specified in soil description.

mottles   Spots or blotches of different color or shades of color interspersed with the dominant color.

muck   Organic soils in which the original plant parts are not recognizable. Contains more mineral matter and is usually darker in color than peat.

mulch   (i) Any material such as straw, sawdust, leaves, plastic film, loose soil, etc., that is spread upon the surface of the soil to protect the soil and plant roots from the effects of raindrops, soil crusting, freezing, evaporation, etc. (ii) To apply mulch to the soil surface.

mull   A type of forest humus in which the Oe horizon may or may not be present and in which there is no Oa horizon. The A horizon consists of an intimate mixture of organic matter and mineral soil with gradual transition between the A horizon and the horizon beneath. (Sometimes differentiated into firm mull, sand mull, coarse mull, medium mull, and fine mull.)

Munsell color system   A color designation system that specifies the relative degrees of the three simple variables of color: hue, value, and chroma. For example: 10YR 6/4 is a color (of soil) with a hue = 10YR, value = 6, and chroma = 4. These notations can be translated into several different systems of color names as desired. See chroma, hue, and value.

mycorrhiza   Literally "fungus root." The association, usually symbiotic, of specific fungi with roots of higher plants.

natric horizon   A mineral soil horizon that satisfies the requirements of an argillic horizon, but that also has prismatic, columnar, or blocky structure and a subhorizon having > 15% saturation with exchangeable Na^+.

nitrification   Biological oxidation of ammonium to nitrite and nitrate, or a biologically induced increase in the oxidation state of nitrogen.

nodule bacteria   The bacteria that fix dinitrogen (N_2) within organized structures (nodules) on the roots, stems, or leaves of plants. Sometimes used as a synonym for "rhizobia."

nodules   (i) Specialized tissue enlargements, or swellings, on the roots or leaves of plants, such as are caused by nitrogen-fixing microorganisms, (ii) glaebules with an undifferentiated fabric; in the context undifferentiated fabric includes recognizable rock and soil fabrics.

Oa horizon (H layer)   A layer occurring in mor humus consisting of well-decomposed organic matter of unrecognizable origin. The O2 horizon.

Oe horizon (F layer)   A layer of partially decomposed litter with portions of plant structures still recognizable. Occurs below the L layer (O11 horizon) on the forest floor in forest soils. It is the fermentation layer or the O12 layer. See litter.

Oi horizon [L layer (litter)]   The surface layer of the forest floor consisting of freshly fallen leaves, needles, twigs, stems, bark, and fruits. This layer may be very thin or absent during the growing season. The O1 horizon.

organic soil   A soil which contains a high percentage (>200 g kg^{-1}, or $>120–180$ g kg^{-1} if saturated with water) of organic carbon throughout the solum.

organic soil materials   Soil materials that are saturated with water and have 174 g kg^{-1} or more organic carbon if the mineral fraction has 500 g kg^{-1} or more clay, or 166 g kg^{-1} organic carbon if the mineral fraction has no clay, or has proportional intermediate contents, or if never saturated with water, have 203 g kg^{-1} or more organic carbon.

oxic horizon   See page 141.

paleosol, buried   A soil formed on a landscape during the geological past and subsequently buried by sedimentation.

paleosol, exhumed   A formerly buried paleosol that has been exposed on the landscape by the erosive stripping of an overlying mantle of sediment.

pans   Horizons or layers, in soils, that are strongly compacted, indurated, or very high in clay content.

paralithic contact   Similar to a lithic contact. It differs from a lithic contact in that the underlying material, if it is a single mineral, has a hardness of less than 3 (Mohs scale). If not a single mineral, gravel size chunks that can be broken out partially disperse within 15 hours shaking in water or sodium hexametaphosphate solution. In addition when moist, the material can be dug with difficulty with a spade.

parent material   The unconsolidated and more or less chemically weathered mineral or organic matter from which the solum of soils is developed by pedogenic processes.

pascal   A unit of pressure equal to one Newton per square meter.

peat   Unconsolidated soil material consisting largely of undecomposed, or only slightly decomposed, organic matter accumulated under conditions of excessive moisture.

ped   A unit of soil structure such as an aggregate, crumb, prism, block, or granule, formed by natural processes (in contrast with a clod, which is formed artificially).

pediment   The footslope component of an erosional slope; geomorphologically an erosional surface that lies at the foot of a receded slope, with underlying rocks or sediments that also underlie the upland, which is barren of, or mantled with sediment, and which normally has a concave upward profile.

pediplain   A geomorphic term for an outwash plain landform.

peneplain   A once high, rugged area which has been reduced by erosion to a low, gently rolling surface resembling a plain.

percolation, soil water   The downward movement of water through soil. Especially, the downward flow of water in saturated or nearly saturated soil at hydraulic gradients of the order of 1.0 or less.

permanent wilting point   The largest water content of a soil at which indicator plants, growing in that soil, wilt and fail to recover when placed in a humid chamber. Often estimated by the water content at -1500 kPa soil matric potential.

permeability, soil   (i) The ease with which gases, liquids, or plant roots penetrate or pass through a bulk mass of soil or a layer of soil. Since different soil horizons vary in permeability, the particular horizon under question should be designated. (ii) The property of a porous medium itself that expresses the ease with which gases, liquids, or other substances can flow through it, and is the same as intrinsic permeability k.

petrocalcic horizon   A continuous, indurated calcic horizon that is cemented by calcium carbonate and, in some places, with magnesium carbonate. It cannot be penetrated with a spade or auger when dry, dry fragments do not slake in water, and it is impenetrable to roots.

petrogypsic horizon   A continuous, strongly cemented, massive, gypsic horizon that is cemented by calcium sulfate. It can be chipped with a spade when dry. Dry fragments do not slake in water and it is impenetrable to roots.

pH-dependent charge   The portion of the cation or anion exchange capacity which varies with pH.

pH, soil   The negative logarithm of the hydrogen ion activity of a soil. The degree of acidity (or alkalinity) of a soil as determined by means of a glass, quinhydrone, or other suitable electrode or indicator at a specified moisture content or soil-water ratio, and expressed in terms of the pH scale.

phosphate   In fertilizer trade terminology, phosphate is used to express the sum of the water-soluble and the citrate-soluble phosphoric acid (P_2O_5); also referred to as the available phosphoric acid (P_2O_5).

phosphate rock   A porous, lower-density, microcrystalline, calcium fluorophosphate of sedimentary or igneous origin. It is usually concentrated and solubilized to be used directly or concentrated in manufacture of commercial phosphate fertilizers.

phyllosilicate mineral terminology   Phyllosilicate minerals have layer structures composed of shared octahedral and tetrahedral sheets. (See table on page 289.)

interlayer—Materials between structural layers of minerals, including cations, hydrated cations, organic molecules, and hydroxide groups or sheets.

layer—A combination of sheets in a 1:1 or 2:1 assemblage.

plane of atoms—A flat (planar) array of one atomic thickness. Example: plane of basal oxygen atoms within a tetrahedral sheet.

sheet of polyhedra—Flat array of more than one atomic thickness and composed of one level of linked coordination polyhedra. A sheet is thicker than a plane and thinner than a layer. Example: tetrahedral sheet, octahedral sheet.

physical properties (of soils)   Those characteristics, processes, or reactions of a soil which are caused by physical forces and which can be described by, or expressed in, physical terms or equations. Sometimes confused with and difficult to separate from chemical properties; hence, the terms "physical-chemical" or "physico-chemical." Examples of physical properties are bulk density, water-holding capacity, hydraulic conductivity, porosity, pore-size distribution, etc.

physical weathering   The breakdown of rock and mineral particles into smaller particles by physical forces such as frost action. See weathering.

placic horizon   A black to dark reddish mineral soil horizon that is usually thin but that may range from 1 mm to 25 mm in thickness. The placic horizon is commonly cemented with iron and is slowly permeable or impenetrable to water and roots.

plinthite   See page 100.

pore-size distribution   The volume fractions of the various size ranges of pores in a soil, expressed as percentages of soil bulk volume (soil particles plus pores.)

pore space   The portion of soil bulk volume occupied by soil pores.

porosity   The volume of pores in a soil sample (nonsolid volume) divided by the bulk volume of the sample.

primary mineral   A mineral that has not been altered chemically since deposition and crystallization from molten lava. See secondary mineral.

profile, soil   A vertical section of the soil through all its horizons and extending into the C horizon.

pyroclastics   A general term applied to detrital volcanic materials that have been explosively or aerially ejected from a volcanic vent.

R layers   Hard bedrock including granite, basalt, quartzite and indurated limestone or sandstone that is sufficiently coherent to make hand digging impractical.

rainfall erosion index   A measure of the erosive potential of a specific rainfall event. In the Universal Soil Loss Equation it is defined as the product of two rainstorm characteristics: total kinetic energy of the storm × its maximum 30-minute intensity.

regolith   The unconsolidated mantle of weathered rock and soil material on the earth's surface; loose earth materials above solid rock. (Approximately equivalent to the term "soil" as used by many engineers.)

residual material   Unconsolidated and partly weathered mineral materials accumulated by disintegration of consolidated rock in place.

residual shrinkage   In a compressible, water-saturated soil, the decrease in bulk volume of soil in addition to the decrease caused by the loss of water.

reticulate mottling   A network of streaks of different color; most commonly found in the deeper profiles of Lateritic soils containing plinthite.

rhizobia   Bacteria capable of living symbiotically in roots of leguminous plants, from which they receive energy and often utilize molecular nitrogen. Collective common name for the genus *Rhizobium*.

rill   A small, intermittent water course with steep sides; usually only several centimeters deep and, hence, no obstacle to tillage operations.

runoff   That portion of the precipitation on an area which is discharged from the area through stream channels. That which is lost without entering the soil is called surface runoff and that which enters the soil before reaching the stream is called ground water runoff or seepage flow from ground water. (In soil science "runoff" usually refers to the water lost by surface flow; in geology and hydraulics "runoff" usually includes both surface and subsurface flow.)

saline soil   A nonsodic soil containing sufficient soluble salt to adversely affect the growth of most crop plants. The lower limit of saturation extract electrical conductivity of such soils is conventionally set at 0.4 siemens per meter. Actually, sensitive plants are affected at half this salinity and highly tolerant ones at about twice this salinity.

saline-sodic soil   A soil containing both sufficient soluble salt and exchangeable Na^+ to adversely affect crop production under most soil and crop conditions. The electrical conductivity and sodium adsorption ratio of the saturation extract are at least 0.4 siemens per meter and 13, respectively.

salt-affected soil   Soil that has been adversely modified for the growth of most crop plants by the presence of soluble salts, exchangeable sodium, or both. See saline soil, saline-sodic soil, and sodic soil.

saprolite   Weathered rock material that retains the original rock structure. It may be soil parent material.

saturated soil paste   A particular mixture of soil and water. At saturation, the soil paste glistens as it reflects light, flows slightly when container is tipped, and the paste slides freely and cleanly from a spatula.

secondary mineral   A mineral resulting from the decomposition of a primary mineral or from the reprecipitation of the products of decomposition of a primary mineral. See primary mineral.

sediment   Transported and deposited particles derived from rocks, soil, or biological material.

sedimentary rock   A rock formed from materials deposited from suspension or precipitated from solution and usually being more or less consolidated. The principal sedimentary rocks are sandstones, shales, limestones, and conglomerates.

sedimentation   The process of sediment deposition.

self-mulching soil   A soil in which the surface layer becomes so well aggregated that it does not crust and seal under the impact of rain but instead serves as a surface mulch upon drying.

sesquioxides   A general term for oxides and hydroxides of iron and aluminum.

shear   Force, as with a tillage tool, acting at a right angle to the direction of movement of the tillage implement.

shear strength   The maximum resistance of a soil to shearing stresses.

shoulder   The convex slope component at the top of an erosional sideslope.

silica-alumina ratio   The molecules of silicon dioxide (SiO_2) per molecule of aluminum oxide (Al_2O_3) in clay minerals or in soils.

silica-sesquioxide ratio   The molecules of silicon dioxide (SiO_2) per molecule of aluminum oxide (Al_2O_3) plus ferric oxide (Fe_2O_3) in clay minerals or in soils.

skeleton grains   Individual grains that are relatively stable and not readily translocated, concentrated, or reorganized by soil-forming processes; they include mineral grains and resistant siliceous and organic bodies larger than colloidal size.

slickensides   Polished and grooved surfaces produced by one mass sliding past another. Slickensides are common in Vertisols.

smectite   A group of 2:1 layer structured silicates with a high cation exchange capacity and variable interlayer spacing. Formerly called the montmorillonite group. The group includes di- and trioctahedral members.

sodic soil   A nonsaline soil containing sufficient exchangeable sodium to adversely affect crop production and soil structure under most conditions of soil and plant type. The lower limit of the saturation extract SAR of such soils is conventionally set at 13.

sodium adsorption ratio   A relation between soluble divalent cations which can be used to predict the exchangeable sodium percentage of soil equilibrated with a given solution. It is defined as follows:

$$SAR = \frac{(Na)}{(Ca + Mg)^{1/2}}$$

where concentrations, denoted by parentheses, are expressed in moles per liter.

solum (plural: sola)   The upper and most weathered part of the soil profile; the A, E, and B horizons.

sombric horizon   See page 108.

specific surface   The solid-particle surface area (of a soil or porous medium) divided by the solid-particle mass or volume, expressed in m^2 kg^{-1} or $m^2/m^3 = m^{-1}$, respectively.

spodic horizon   See page 246.

substrate   (i) That which is laid or spread under; an underlying layer, such as the subsoil. (ii) The substance, base, or nutrient on which an organism grows. (iii) Compounds or substances that are acted upon by enzymes or catalysts and changed to other compounds in the chemical reaction.

substratum   Any layer lying beneath the soil solum, either conforming or unconforming.

sulfidic material   See page 263.

sulfuric horizon   See page 236.

summit   The highest point of any landform remanant, hill, or mountain.

surface sealing   The orientation and packing of dispersed soil particles in the immediate surface layer of the soil, thus greatly reducing its permeability.

swamp   An area saturated with water throughout much of the year but with the surface of the soil usually not deeply submerged. Usually characterized by tree or shrub vegetation. See marsh.

symbiosis   To obligate living together in intimate association of two dissimilar organisms, the cohabitation being mutually beneficial.

talc   A magnesium silicate mineral with a 2:1 type layer structure but without isomorphous substitution. May occur in soils as an inherited mineral. It is trioctahedral.

terrace   (i) A level, usually narrow, plain bordering a river, lake, or the sea. Rivers sometimes are bordered by terraces at different levels. (ii) A raised, more or less level or horizontal strip of earth usually constructed on or nearly on a contour and supported on the downslope side by rocks or other similar barrier and designed to make the land suitable for tillage and to prevent accelerated erosion. For example, the ancient terraces built by the Incas in the Andes. (iii) An embankment with the uphill side sloping toward and into a channel for conducting water, and the downhill side having relatively sharp decline; constructed across the direction of the slope for the purpose of conducting water from the area above the terrace at a regulated rate of flow and to prevent the accumulation of large volumes of water on the downslope side of cultivated fields. The depth of the channel, the width of the terrace ridge, and the spacings of the terraces on a field are varied with soil types, cropping systems, climatic conditions, and other factors.

tidal flats   Areas of nearly flat, barren mud periodically covered by tidal waters. Normally these materials have an excess of soluble salt.

tilth   The physical condition of soil as related to its ease of tillage, fitness as a seedbed, and its impedance to seedling emergence and root penetration.

top dressing   An application of fertilizer to a soil surface, without incorporation, after the crop stand has been established.

toposequence   A sequence of related soils that differ, one from the other, primarily because of topography as a soil-formation factor.

trioctahedral   An octahedral sheet that has all of the sites filled, usually by divalent ions such as magnesium or ferrous iron.

Tropepts   See page 240.

tuff   Volcanic ash usually more or less stratified and in various states of consolidation.

umbric epipedon   A surface layer of mineral soil that has the same requirements as the mollic epipedon with respect to color, thickness, organic carbon content, consistence, structure, and phosphorus content, but that has a base saturation of less than 50% when measured at pH 7.

unsaturated flow   The movement of water in soils in which the pores are not completely filled with water.

vadose water   See water table, perched.

value, color   The relative lightness or intensity of color and approximately a function of the square root of the total amount of light. One of the three variables of color. See Munsell color system, hue, and chroma.

vermiculite   A highly charged layer-structured silicate of the 2:1 type that is formed from mica. It is characterized by adsorption preference for potassium and cesium over smaller exchange cations. It may be di- or trioctahedral.

volumetric water content   The soil-water content expressed as the volume of water per unit bulk volume of soil.

water (or matric) suction   Same as water tension, but avoids overworking the term "tension," which also appears in other soil-water-related contexts and terms (e.g., surface tension, vapor tension, and aqueous tension).

water table   The upper surface of ground water or that level in the ground where the water is at atmospheric pressure.

water table, perched   The water table of a saturated layer of soil which is separated from an underlying saturated layer by an unsaturated layer (vadose water).

water tension (or pressure)   The equivalent negative pressure in the soil water. It is equal to the equivalent pressure that must be applied to the soil water to bring it to hydraulic equilibrium, through a porous permeable wall or membrane, with a pool of water of the same composition.

weathering   All physical and chemical changes produced in rocks, at or near the earth's surface, by atmospheric agents. See chemical and physical weathering.

yield   The amount of a specified substance produced (e.g., grain, straw, total dry matter) per unit area.

yield, sustained   A continual annual, or periodic, yield of plants or plant material from an area; implies management practices which will maintain the productive capacity of the land.

zero point of charge   The pH value of a solution in equilibrium with a particle whose net charge from all sources is zero.

Classification Scheme for Phyllosilicates Related to Clay Minerals

Type	Group (x = charge per formula unit)	Subgroup	Species*
1:1	Kaolin-serpentine	Kaolins	Kaolinite, halloysite (7 Å or 0.7 nm) halloysite (10 Å or 1.0 nm)
	$x \sim 0$	Serpentines	Chrysotile, lizardite, antigorite
2:1	Pyrophyllite-talc	Pyrophyllites	Pyrophyllite
	$x \sim 0$	Talcs	Talc
	Smectite	Dioctahedral smectites	Montmorillonite, beidellite, nontronite
	$x \sim 0.25–0.6$	Trioctahedral smectites	Saponite, hectorite, sauconite
	Vermiculite	Dioctahedral vermiculite	Dioctahedral vermiculite
	$x \sim 0.6–0.9$	Trioctahedral vermiculite	Trioctahedral vermiculite
	Mica	Dioctahedral micas	Moscovite, paragonite
	$x \sim 1$	Trioctahedral micas	Biotite, phlogopite
	Brittle mica	Dioctahedral brittle micas	Margarite
	$x \sim 2$	Trioctahedral brittle micas	Clintonite
	Chlorite	Dioctahedral chlorites (4–5 octahedral cations per formula unit)	
	x variable	Trioctahedral chlorites (5–6 octahedral cations per formula unit)	Pennine, chlinochlore, prochlorite

*Only a few examples are given.

Conversion Factors for SI and non-SI Units

To convert column 1 into column 2, multiply by	Column 1 SI unit	Column 2 non-SI unit	To convert column 2 into column 1 multiply by
		Length	
0.621	Kilometer, km (10^3 m)	Mile, mi	1.609
1.094	Meter, m	Yard, yd	0.914
3.28	Meter, m	Foot, ft	0.304
1.0	Micrometer, μm (10^{-6} m)	Micron, μ	1.0
3.94×10^{-2}	Millimeter, mm (10^{-3} m)	Inch, in	25.4
10	Nanometer, nm (10^{-9} m)	Angstrom, Å	0.1
		Area	
2.47	Hectare, ha	Acre	0.405
247	Square kilometer, km^2 (10^3 m)^2	Acre	4.05×10^{-3}
0.386	Square kilometer, km^2 (10^3 m)^2	Square mile, mi^2	2.590
2.47×10^{-4}	Square meter, m^2	Acre	4.05×10^3
10.76	Square meter, m^2	Square foot, ft^2	9.29×10^{-2}
		Volume	
35.3	Cubic meter, m^3	Cubic foot, ft^3	2.83×10^{-2}
1.057	Liter, L (10^{-3} m^3)	Quart (liquid), qt	0.946
3.53×10^{-2}	Liter, L (10^{-3} m^3)	Cubic foot, ft^3	28.3
0.265	Liter, L (10^{-3} m^3)	Gallon	3.78
33.78	Liter, L (10^{-3} m^3)	Ounce (fluid), oz	2.96×10^{-2}
2.11	Liter, L (10^{-3} m^3)	Pint (fluid), pt	0.473
		Mass	
2.20×10^{-3}	Gram, g (10^{-3} kg)	Pound, lb	454
3.52×10^{-2}	Gram, g (10^{-3} kg)	Ounce (avdp), oz	28.4
2.205	Kilogram, kg	Pound, lb	0.454
10^{-2}	Kilogram, kg	Quintal (metric), q	10^2
1.10×10^{-3}	Kilogram, kg	Ton (2000 lb), ton	907
1.102	Megagram, Mg (tonne)	Ton (US), ton	0.907
1.102	Tonne, t	Ton (US), ton	0.907
		Yield and Rate	
0.893	Kilogram per hectare, kg ha^{-1}	Pound per acre, lb acre^{-1}	1.12
1.49×10^{-2}	Kilogram per hectare, kg ha^{-1}	Bushel per acre, 60 lb	67.19
1.59×10^{-2}	Kilogram per hectare, kg ha^{-1}	Bushel per acre, 56 lb	62.71
1.86×10^{-2}	Kilogram per hectare, kg ha^{-1}	Bushel per acre, 48 lb	53.75
893	Tonnes per hectare, t ha^{-1}	Pound per acre, lb acre^{-1}	1.12×10^{-3}
893	Megagram per hectare, Mg ha^{-1}	Pound per acre, lb acre^{-1}	1.12×10^{-3}
0.446	Megagram per hectare, Mg ha^{-1}	Ton (2000 lb) per acre, ton acre^{-1}	2.24
2.24	Meter per second, m s^{-1}	Mile per hour	0.447

(Continued)

Conversion Factors for SI and non-SI Units (*Continued*)

To convert column 1 into column 2, multiply by	Column 1 SI unit	Column 2 non-SI unit	To convert column 2 into column 1 multiply by
		Pressure	
9.90	Megapascal, MPa (10^6 Pa)	Atmosphere	0.101
10	Megapascal, MPa (10^6 Pa)	Bar	0.1
1.00	Megagram per cubic meter, Mg m^{-3}	Gram per cubic centimeter, g cm^{-3}	1.00
2.09×10^{-2}	Pascal, Pa	Pound per square foot, lb ft^{-2}	47.9
1.45×10^{-4}	Pascal, Pa	Pound per square inch, lb in^{-2}	6.90×10^3
		Temperature	
1.00 (K−273)	Kelvin, K	Celsius, °C	1.00 (°C+273)
(9/5°C) + 32	Celsius, °C	Fahrenheit, °F	5/9 (°F−32)
		Energy, Work, and Quality of Heat	
0.239	Joule, J	Calorie, cal	4.19
0.735	Joule, J	Foot-pound	1.36
2.387×10^{-5}	Joule per square meter, J m^{-2}	Calorie per square centimeter (langley)	4.19×10^4
10^5	Newton, N	Dyne	10^{-5}
		Angle	
57.3	Radian, rad	Degrees (angle), °	1.75×10^{-2}
		Electrical Conductivity	
10	Siemen per meter, S m^{-1}	Millimho per centimeter, mmho cm^{-1}	0.1
		Concentrations	
1	Centimole per kilogram, cmol kg^{-1} (ion exchange capacity)	Milliequivalents per 100 grams, meq 100 g^{-1}	1
0.1	Gram per kilogram, g kg^{-1}	Percent,	10
1	Megagram per cubic meter, Mg m^{-3}	Gram per cubic centimeter, g cm^{-3}	1
1	Milligram per kilogram, mg kg^{-1}	Parts per million, ppm	1
		Plant Nutrient Conversion	
	Elemental	Oxide	
2.29	P	P_2O_5	0.437
1.20	K	K_2O	0.830
1.39	Ca	CaO	0.715
1.66	Mg	MgO	0.602

Profile Descriptions
and Analyses

B.1 Analytical Methods

This appendix includes the descriptions and analyses of soil profiles that belong to the major soil orders. The laboratories that analyzed the soil samples did not use uniform methods. A brief reference to the methodology that was followed is given for each column of the analytical tables.

B.1.1 (Column 3): Horizon nomenclature

The horizon nomenclature is the symbol given to a particular horizon or layer in the soil by the persons describing the profile. It corresponds either to the horizon designations used by the Soil Conservation Service of the U.S. Department of Agriculture (Soil Survey Staff, 1988) or by the Food and Agriculture Organization of the United Nations (FAO, 1977). The symbols are an expression of the qualitative judgment of the author of the description on the kinds of changes that have taken place in the profile during its development.

B.1.2 (Columns 4, 5, 6, and 9):
Particle size distribution

The particle size distribution is given in percentages of the less than 2-mm fraction. When analyzed by NSSL[1] the fine earth is dispersed in water with sodium hexametaphosphate plus sodium carbonate and overnight reciprocal shaking. Sands are separated by sieving; clay

---

[1]National Soil Survey Laboratories of the Soil Conservation Service in Lincoln, Nebraska.

and fine silt by sedimentation, pipetting, drying, and weighing; and the silt content is calculated by difference. The samples are treated with H_2O_2 and candlestick filtering before dispersion.

In the analyses by SNLCS[2] the less than 2-mm fraction is dispersed in water with NaOH or occasionally with calgon, high-speed stirring, and sedimentation. The clay is measured in the supernatant by a modified hydrometer method; the sand is separated by sieving; and silt is calculated by difference. There is no pretreatment to destroy the organic matter.

The SNLCS procedure produces clay percentages that correlate well with several other soil properties. The NSSL method, when applied to strongly weathered soils that contain dominantly low-activity clays, leaves intact many of the iron cemented, very fine aggregates of clay that are erroneously included in the silt percentage that is calculated by difference (Kimble, 1986).

The samples of the Andisol profiles that were analyzed by ISRIC[3] were dispersed by ultrasonic treatment at pH 4 after organic matter and carbonate removal. The clay was collected by repeated sedimentation and siphoning.

B.1.3 (Column 7): Fine clay

Method 3A1b of Soil Survey Investigation Report (Soil Conservation Service, 1972) using a centrifuge at 1500 rpm on clay plus silt suspensions separated during the NSSL particle size analysis (columns 4, 5, 6, and 9); time is adjusted according to the temperature.

B.1.4 (Column 8): Water dispersible clay

Particle size analysis without dispersing agents and no peroxide treatment to destroy the organic matter. The "clay" aliquot that is collected during the analysis is actually a mixture of clay and finely divided organic matter.

B.1.5 (Column 10): Coarse fraction

Air-dried samples crushed by rolling manually with wood or hard rubber and passed through a 2-mm sieve; the >2-mm fraction is washed, dried, and weighed, and the percentage is expressed on the basis of the whole soil.

---

[2]Servico Nacional de Levantamento e Conservacao de Solos, Empresa Brasileira de Pesquisa Agropecuaria, Brazil.

[3]International Soil Reference and Information Center, Wangeningen, The Netherlands.

B.1.6   (Column 11): Organic carbon

Wet digestion with acid dichromate and automatic titration with $FeSO_4$. A recovery factor of 0.77 is used to adjust the results [NSSL method 6A1c (Soil Conservation Service, 1972)].

B.1.7   (Column 12):
Total nitrogen—Kjeldahl

Acid digestion, steam distillation and automatic titration with 0.1 or 0.05 M HCl to pH 4.6 [NSSL method 6B1b (Soil Conservation Service, 1972)].

B.1.8   (Column 13): pH in KCl

pH measured in a 1 M KCl solution mixed 1:1 with soil [NSSL method 8C1d (Soil Conservation Service, 1972)]; for the SNLCS method the soil to 1 M KCl ratio is 1:2.5.

B.1.9   (Column 14): pH in H₂O

Same procedure as for column 13, but water is used instead of 1 M KCl.

B.1.10   (Columns 15, 16, and 17):
Extractable Ca, Mg, and K

Cations extracted with NH_4OAc at pH 7 when analyzed by NSSL. The SNLCS laboratories extract Ca and Mg with 1 M KCl and titrate with EDTA (ethylenediaminetetraacetic acid). K is extracted with 0.05 M HCl plus 0.025 M H_2SO_4 and determined by flame photometry.

B.1.11   (Column 18): Sodium or aluminum

Sodium extracted with NH_4OAc at pH 7 when analyzed by NSSL. The SNLCS laboratories extract Na with 0.05 M HCl plus 0.025 M H_2SO_4 and measure its concentration by flame photometry.

The SNLCS extracts the aluminum with 1 M KCl and titrates the acidity with 0.025 M NaOH and bromothymol blue as the indicator. The Al data from NSSL extract Al with 1 M KCl and determine Al by Aluminon colorimetry.

B.1.12   (Column 19): Cation
exchange capacity, pH 7

NSSL: The soil is equilibrated and washed overnight with 1 M NH_4OAc at pH 7; the excess NH_4OAc is removed with dropwise ethanol leaching. The NH_4^+ in the soil is extracted by modified Kjeldahl distillation and autotitrated with 0.1 M HCl.

SNLCS: Sum of the extractable bases plus the acidity that is extracted by 1 M Ca(OAc)$_2$, pH 7, and titrated with NaOH and phenolphthalein as indicator.

B.1.13   (Column 20): Base saturation, sum

NSSL: The percentage extractable bases using their sum plus the extractable acidity determined by the BaCl$_2$-triethanolamine method at pH 8.2 as the denominator in the computation.

B.1.14   (Column 21): Base saturation, pH 7

NSSL: The sum of bases extracted by 1 M NH$_4$OAc at pH 7 expressed as a percentage of the cation exchange capacity (column 19).
SNLCS: Derived value based on the extractable bases according to SNLCS method, reported as a percentage of the cation exchange capacity (column 19).

B.1.15   (Column 22): Extractable Fe

Fe extracted by dithionite-citrate expressed as the percentage Fe in the fraction of less than 2 mm diameter.

B.1.16   (Column 23): Bulk density

Plastic-coated prewetted clods, desorbed to 33 kPa, weighed in air and in water.

B.1.17   (Column 24): Water content at 33 kPa

The water retained at 33 kPa is measured by desorption of prewet, plastic-coated clods.

B.1.18   (Column 25): Water content at 1500 kPa

The water retained at 1500 kPa is measured by desorption of prewet bulk samples on a pressure membrane.

B.1.19   (Column 26): WRD, water retention difference

The value computed from water retentions at 33- and 1500-kPa suction and converted to cm of water per cm of soil by using the bulk density of the less than 2-mm fraction, the percentage of coarse fragments, and their particle density.

B.2 Descriptions and Analyses

B.2.1 Oxisol, Anionic Acrustox

Description

PEDON NO: DPFS 83 GO
LOCATION: 7.1 km toward Planaltina from Brasilia stadium, Brasilia, DF. Brazil; 15°46'S. and
    47°56'W.Gr.
POSITION: top of plateau, 1 to 3% slope, under savannah (cerrado) vegetation.
ELEVATION: 1,100 meters.
GEOLOGICAL FORMATION: Paranoá formation. Bambuí Group. Pre-cambrian A.
PARENT MATERIAL: Pelitic mantle.
EROSION: Sheet erosion, slight.
LAND USE: None.
DESCRIBED BY: F.G. deFreitas et alii. (24 Jan. 1967)

A - 0 - 10 cm, dark reddish brown (2.5YR 3/4, moist), dark reddish brown (2.5YR 3/4,
    moist crushed), red (2.5YR 4/6, dry), red (2.5YR 5/6, dry crushed); clay (very
    fine); weak very fine to medium granular; very friable, plastic and sticky; clear and
    smooth boundary.

AB - 10-30 cm, dry red (2.5YR 3.5/6, moist), red (2.5YR 3.5/6, moist crushed), reddish
    brown (2.5YR 4/4, dry), yellowish red (5 YR 4.5/6, dry crushed); clay (very fine);
    some parts are weak fine to medium granular and other parts pseudostructureless
    formed by strong very fine granular; firm, plastic and sticky; gradual and smooth
    boundary.

BA - 30-65 cm, red (10 R 4/6); clay (very fine); pseudostructureless formed by strong
    very fine granular; friable, plastic and sticky; diffuse and smooth boundary.

Bo1 - 65-140 cm, red (10 R 4/7), same as above though very friable.

Bo2 - 140-235 cm; same as above; boundary not seen.

Bo3 -235-310 cm, red (10 R 4.5/6); presumably same as above as estimated from core
    sample of bucket auger.

Analyses

ANIONIC ACRUSTOX

PEDON DPFS 83 GO

ANALYSIS: Beltsville Soil Survey Laboratory
(Soil Conservation Service USDA) and
EMBRAPA-SNLCS (Brazil)

[1] HZN NO.	[2] DEPTH (CM)	[3] HORIZON	[4] SAND %	[5] SILT %	[6] CLAY %	[7] FINE CLAY %	[8] DISP CLAY %	[9] FINE SILT %	[10] COAR FRAC *	[11] O.C. %	[12] TOTAL N %
1	10	A	13	9	78	35			0	3.24	0.24
2	30	AB	11	8	81	25			0	2.10	0.14
3	65	BA	10	7	83	1			0	1.35	0.05
4	140	Bo1	9	7	84	0			0	0.92	0.03
5	235	Bo2	11	8	81	16			0	0.67	0.03
6	310	Bo3	8	6	86	53			0	0.53	0.02

[1] HZN NO.	[13] KCl	[14] H2O	[15] Ca	[16] Mg	[17] K	[18] Al	[19] pH7	[20] SUM	[21] pH7	[22] Fe	[23] DENS	[24] 33 kPa	[25] 1500 kPa	[26] WRD
	pH		NH40AC EXTRACTABLE				CEC	BASE SAT.	EXTR	BULK	WATER CONTENT		
			----cmol(+)/kg----						%				
1	4.2	5.0	0.42	0.32	0.15	1.1	15.0	4	6	-		24.0	
2	4.6	5.1	0.01	0.04	0.05	0.7	8.4	1	1		0.93	23.9	
3	4.7	5.0	0.02	0.02	0.01	0.2	6.0	1	1		0.90	24.1	
4	5.4	5.1	0.01	0.01	-	0.1	4.5		1		0.90	24.8	
5	6.2	5.5	0.01	TR	-	TR	3.1		1		0.90	24.4	
6	6.4	5.8	0.01	-	TR	TR	3.1		1			25.5	

_____

See page 1 of appendix for analytical methods.

B.2.2 Oxisol, Humic Rhodic Kandiudox

Description

PEDON 85PO725 (NSSL-SCS-USDA): 85PO725 (EMBRAPA-SNLCS): ISCW 5
LOCATION - Highway SP-127, Piracicaba-Rio Claro, Km 12, 22°34'S., 47°35'W.Gr.
       (Brazil)
POSITION - 8% slope in plowed land, with former application of limestone.
ELEVATION - 630 meters.
GEOLOGICAL FORMATION - Argillites and shales, occasionally calcareous. Corumbataí
       Formation. Passa Dois Group. Permian.
PARENT MATERIAL - Reworked pelitic colluvium from given sediments.
EROSION - Sheet erosion, slight.
PRIMARY VEGETATION - Semi-evergreen tropical forest.
LAND USE - Sugar cane crop.
SOURCE - SMSS-EMBRAPA, 1986

Ap1 - 0 - 11 cm, dark reddish brown (5 YR 3/3, moist), mixed with reddish brown (5 YR
      4/4, moist), common coatings dark reddish brown (5 YR 2.5/2, moist), dark reddish
      brown (4 YR 3/2, moist crushed), dark reddish brown (5 YR 3/3, dry); clay (very
      fine); moderate fine and very fine granular; variably friable to firm, very plastic and
      very sticky; abrupt and smooth boundary.

Ap2*- 11 - 19 cm, dark red (3.5 YR 3/5, moist), mixed with colors similar to those of upper
      horizon, dark reddish brown (3.5 YR 3/4, moist crushed, reddish brown (4 YR 4/4,
      dry); clay (very fine); moderate medium to fine subangular blocky (somewhat
      compacted); variably friable to very firm, very plastic and very sticky; abrupt and
      wavy boundary (4-13 cm).

Bto1 - 19 - 43 cm, dark red (2.5 YR 3/6, moist), red (2.5 YR 4.5/6, dry); clay (very fine);
      moderate very fine and fine subangular blocky; moderate and common clay skins; firm,
      plastic and sticky; gradual and smooth boundary.

Bto2 - 43 - 78 cm, red (2.5 YR 4/6, moist), red (2.5 YR 5/6, dry); clay (very fine); weak
      very fine subangular blocky breaking to strong very fine granular; very firm clayey
      bodies 2 to 4 cm in diameter; friable with parts very firm, slightly plastic and slightly
      sticky (not rubbed), plastic and sticky (well rubbed); diffuse and smooth boundary.

Bo1 - 78 - 190 cm, red (1.5 YR 4/6, moist), red (1.5 YR 4/6, dry), reddish yellow (4 YR
      6/6, dry ground); clay (very fine); pseudostructureless formed by strong very fine
      granular; very friable, slightly plastic and nonsticky (not rubbed), plastic and sticky
      (well rubbed).

Bo2 - 190 - 290 cm, red (9 R 4/6, moist), red (1.5 YR 4.5/6, dry); clay (very fine);
      nonplastic and nonsticky (not rubbed), plastic and sticky (well rubbed).

BC - 290 - 320 cm+, red (10 R 4.5/6, moist), red (9 YR 5/6, dry); same as above.
_____
* Actually, if admitted, would correspond to Bp/Ap

Analyses

HUMIC RHODIC KANDIUDOX

PEDON 85PO725 (NSSL-SCS-USDA)
ISCW 5 (EMBRAPA-SNLCS)
ANALYSIS*:   National Soil Survey Laboratory
Soil Conservation Service (USDA)

[1]	[2]	[3]	[4]	[5]	[6]	[7]	[8]	[9]	[10]	[11]	[12]
HZN	DEPTH	HORIZON	SAND	SILT	CLAY	FINE	DISP	FiNE	COAR	O.C.	TOTAL N
NO.	(CM)		%	%	%	CLAY	CLAY	SILT	FRAC	%	%
						%	%	%			
1	11	Ap1	17.6	21.4	61.0	20.1	52	31.1		4.54	0.346
2	19	Ap2	17.3	15.8	66.9	34.0	53	8.0		2.01	0.150
3	43	Bto1	11.0	12.2	76.8	44.1	0	6.2		1.15	0.085
4	78	Bto2	11.0	12.5	76.5	40.9	0	6.3		0.82	
5	190	Bo1	11.0	15.3	72.8	35.3	0	7.8		0.47	0.036
6	290	Bo2	14.7	23.4	61.9	30.8	0	13.0		0.25	
7	320	BC	14.2	34.4	51.4		0	21.2		0.18	0.021

[1]	[13]	[14]	[15]	[16]	[17]	[18]	[19]	[20]	[21]	[22]	[23]	[24]	[25]	[26]
HZN	pH		NH40AC EXTRACTABLE			A I	CEC	BASE SAT.		EXTR	BULK	WATER CONTENT		
NO.	KCl	H2O	Ca	Mg	K		pH7	SUM	pH7	Fe	DENS	33	1500	WRD
										%		kPa	kPa	
1	5.0	6.1	18.1	2.23	1.00	0.1	17.7	67	100	4.6			22.2	
2	4.7	5.3	6.80	1.37	0.51	TR	11.6	43	75	5.2	1.32	29	21.2	0.10
3	4.3	4.6	2.07	0.79	0.12	1.1	7.4	21	41	5.5	1.25	32.2	25.0	0.09
4	4.3	4.5	0.89	0.35	0.03	1.2	6.8	11	19	5.6	1.07	33.7	26.0	0.08
5	4.5	5.2	0.18	0.24	0.03	0.4	5.9	5	8	5.7	1.11	33.1	26.3	0.08
6	4.4	5.3	0.15	0.12	0.03	0.7	5.7	4	6	5.8			26.4	
7	4.0	5.2		0.16		2.6	6.8	2	3	5.5			26.5	

B.2.3 Oxisol, Humic Rhodic Eutrustox

Description

PEDON NO: 85P 736 (NSSL-SCS-USDA)
LOCATION: Highway BR-153, Goiania-Anápolis, 200 m before km 140, 15°24'S, 49°00'W.GR.
POSITION: lower third of hillside, with 5% slope, under grasses.
ELEVATION: 1,070 meters
GEOLOGICAL FORMATION: metamorphic rocks, Pre-cambrian..
PARENT MATERIAL: Reworked sandy-clay.
TOPOGRAPHY: undulating.
PRIMARY VEGETATION: Semi-deciduous tropical forest.
LAND USE: Pasture (Hyparrhenia sp.).
DESCRIBED BY: P.K.T. Jacomine, C.S. Holzhey, J. Olmos I.L., M.N. Camargo, P.C. de Lima and A.L.
Lemos (SMSS-EMPRAPA, 1986).

Ap - 0 - 23 cm, dusky red (1.5 YR 3/2, moist and moist crushed), reddish brown (4 YR
4/4, dry); clay; moderate fine and medium granular and strong fine granular; firm,
slightly plastic and slightly sticky (not rubbed), plastic and sticky (well rubbed); clear
and smooth boundary.

AB - 23-40 cm, dark reddish brown (1.5 YR; 3/3, moist and moist crushed), reddish brown
(3.5 YR 4/5, dry); clay; weak fine and medium subangular blocky and strong very fine
granular; consistence as above; gradual and smooth boundary.

BA - 40-58 cm, dark reddish brown (1.5 YR 3/4, moist), red (2.5 YR 4/5, dry); same as
above though very friable.

Bo1 - 58-95 cm, dark red (1.5 YR 3/5, moist), red (2.5 YR 4/6, dry) and red (3.5 YR 5/7,
dry ground); clay; pseudostructureless formed by strong very fine granular;
consistence as above; diffuse and smooth boundary.

Bo2 - 95-130 cm, dark red (1.5 YR 3/6, moist), red (2.5 YR 4.5/6, dry); same as above.

Bo3 - 130-200 cm, same as above.

Bo4 - 200-275 cm, dark red (9 R 3/6, moist), red (1.5 YR 5/7, dry); clay; weak fine
subangular blocky breaking to strong very fine granular; friable to very friable, wet
consistence and transition as above.

Bo5 - 275-410 cm, dark red (10 R 3/6, moist), red (1.5 YR 5/7, dry); same as above
though friable with parts firm and very firm; gradual? or clear? boundary.

2BC - 410-430 cm+ sandy clay- skeletal, similar to upper horizon interspersed with some
rounded and subrounded gravels and few stones consisting of quartz, quartzite, gneissic
rock fragments and lateritic nodules.

Analyses

HUMIC RHODIC EUTRUSTOX

PEDON NO: 85P 736 (NSSL-SCS-USDA)

ANALYSIS: National Soil Survey Laboratory
Soil Conservation Service (USDA)

[1]	[2]	[3]	[4]	[5]	[6]	[7]	[8]	[9]	[10]	[11]	[12]
HZN	DEPTH	HORIZON	SAND	SILT	CLAY	FINE	DISP	FINE	COAR	O.C.	TOTAL N
NO.	(CM)		%	%	%	CLAY	CLAY	SILT	FRAC	%	%
						%	%	%	•		
1	23	Ap	41	11	48		30	9.8		3.97	0.268
2	40	AB	41	9	52		42	6.4		1.99	0.124
3	58	BA	41	7	54		47	5.1		1.29	0.069
4	95	Bo1	41	6	55		47	4.8		0.89	0.047
5	130	Bo2	38	7	55		47	3.2		0.68	
6	200	Bo3	38	8	54		0	4.7		0.53	0.024
7	275	Bo4	35	8	57		0	4.0		0.42	

[1]	[13]	[14]	[15]	[16]	[17]	[18]	[19]	[20]	[21]	[22]	[23]	[24]	[25]	[26]
HZN	pH		NH40AC EXTRACTABLE				CEC	BASE SAT.		EXTR	BULK	WATER CONTENT		
NO.	KCl	H2O	Ca	Mg	K	Al	pH7	SUM	pH7	Fe	DENS	33	1500	WRD
			----cmol(+)/kg----							%		kPa	kPa	
1	5.0	5.6	6.46	0.67	0.10	0.1	11.6	33	62	6.4	1.13	26.8	17.7	1.10
2	5.2	6.0	4.01	0.21	0.04	0.1	6.9	32	62	7.4	1.14	24.2	16.5	0.09
3	5.4	6.0	2.50	0.12	0.03	0.1	3.9	29	68	7.4	1.19	21.7	16.2	0.07
4	5.7	6.2	1.52	0.12	0.03		2.7	23	62	7.7	1.21	22.9	16.8	0.07
5	6.1	6.3	1.01	0.12	0.04		1.9	18	61	7.6	1.17	21.5	17.1	0.05
6	6.5	6.7	0.56	0.12	0.05		0.9	14	79	7.7	1.21	21.4	17.2	0.05
7	6.5	6.2		0.21	0.04		0.9	6	27	7.7	1.23	23.3	16.9	0.08

B.2.4   Ultisol, Acrustoxic Kandiustult

Description

PEDON NO: 84PO508 (NSSL-SCS-USDA)
LOCATION: Miasmfu Reg. Res. St., 8 km N. of Kasama, Zambia (Pedon 3) 10°10'S, 031°10'E
PHYSIOGRAPHY: Upland slope in plateaus of tablelands.
POSITION: summit interfluve.
ELEVATION: 1384 m MSL
PARENT MATERIAL: residuum from metamorphic-acidic material.
DESCRIBED BY: D. Hallbick, O. Spaargaren and L. Bustness (Soil Survey Unit, 1985.
SAMPLE DATE: 11/83

A   - 0 - 16 cm; reddish brown (5 YR 4/4) sandy loam; dark reddish brown (5 YR 3/3)
        moist; weak fine and medium crumb structure; slightly hard, nonsticky, plastic; many
        fine roots throughout and few medium roots throughout; many very fine and fine
        interstitial and tubular and common fine to coarse continuous tubular pores; clear
        smooth boundary.

BA   -16-37 cm; red (2.5 YR 4/8) clay; red (2.5 YR 4/6) moist; massive parting to strong
        very fine granular; hard, very sticky, plastic; common fine roots throughout; common
        to many very fine and fine interstitial and tubular pores and few to common fine and
        medium continuous tubular pores; clear wavy boundary.

Bw1 - 37 - 63 cm; red (2.5 YR 4/8) clay; dark red (2.5 YR 3/6) moist; massive parting to
        strong very fine granular; slightly hard, very sticky, plastic; common fine roots
        throughout and few medium and coarse roots throughout; common to many very fine and
        fine interstitial and tubular pores and few fine to coarse continuous tubular pores;
        diffuse smooth boundary.
        In Bw1 and Bw2 horizons few termite chambers occur inter-connected by channels of 2
        to 3 mm wide and connected with the surface by channels 5 to 10 mm wide. Some of
        these vertical channels are partly filled and lined.

Bw2 - 63-124 cm; red (2.5 YR 4/8) sandy clay; dark red (2.5 YR 3/6) moist; massive
        parting to strong very fine granular; soft, very sticky, plastic; few fine and medium
        roots throughout; many very fine and fine interstitial and tubular pores and few fine to
        coarse continuous tubular pores; gradual smooth boundary.
        Horizon split for sampling 63 to 95 cm and 95 to 124 cm.

Bw3 - 124 - 148 cm; red (2.5 YR 5/8) sandy clay; dark red (2.5 YR 3/6) moist; massive
        parting to strong very fine granular; soft, very sticky, plastic; very few fine and
        medium roots throughout; many very fine and fine interstitial and tubular pores and
        very few and medium continuous tubular pores.

Analyses

ACRUSTOXIC KANDIUSTULT

PEDON NO: 84PO508 (NSSL-SCS-USDA)

ANALYSIS: National Soil Survey Laboratory
Soil Conservation Service (USDA)

[1] HZN NO.	[2] DEPTH (CM)	[3] HORIZON	[4] SAND %	[5] SILT %	[6] CLAY %	[7] FINE CLAY %	[8] DISP CLAY %	[9] FINE SILT %	[10] COAR FRAC *	[11] O.C. %	[12] TOTAL N %
1	16	A	77.0	6.0	17.0			2.5	-	1.22	0.067
2	37	BA	68.0	5.5	26.5			2.2	-	0.48	0.030
3	63	Bw1	62.3	6.0	31.7			2.4	-	0.30	0.023
4	95	Bw2	61.4	5.9	32.7			2.2	-	0.19	
5	124	Bw2	65.4	5.7	28.9			2.2	-	0.15	
6	148	Bw3	67.7	5.7	26.6			2.3	-	0.12	

[1] HZN NO.	[13] pH KCl	[14] pH H2O	[15] NH40AC EXTRACTABLE Ca	[16] Mg	[17] K	[18] Al	[19] CEC pH7	[20] BASE SAT. SUM	[21] pH7	[22] EXTR Fe	[23] BULK DENS	[24] WATER CONTENT 33 kPa	[25] 1500 kPa	[26] WRD
			----cmol(+)/kg----							%		kPa	kPa	
1	4.6	5.7	1.5	0.8	0.1		4.5	36	53	2.6	1.33	10.4	6.7	0.05
2	4.7	5.4	0.5	0.5	TR	0.3	2.6	26	38	3.0	1.42	12.2	8.7	0.05
3	4.7	5.7	TR	0.6	0.1		2.3	21	30	3.1	1.39	13.2	10.1	0.04
4	4.4	5.6	-	0.3	TR		2.3	9	13	3.2	1.34	12.7	10.2	0.03
5	4.3	5.7	-	0.2	TR		1.7	7	12	3.0	1.34	12.8	9.4	0.05
6	4.3	5.6	-	0.1	-		1.5	4	7	3.0	1.35	12.2	8.9	0.04

B.2.5 Ultisol, Acrustoxic Kandiustult

Description

PEDON NO: 85P0734 (NSSL-SCS-USDA)
LOCATION: Highway BR-153, Itumbiara-Goiania, 300m after km 1,347, Brazil, 17°20'S and 49°15'W.Gr.
POSITION: top of hill, 1 to 2% slope, at the border of disturbed forest remnant.
ELEVATION: 800 meters.
GEOLOGICAL FORMATION: Mica-schists and quartzites. Arixá Group. Pre-cambrian.
EROSION: Sheet erosion, slight.
LAND USE: Pasture.
DESCRIBED BY: P.K.T. Jacomine, J.Olmos I.L., P.C. de Lima, C.S. Holzhey (SMSS-EMBRAPA, 1986).

Ap   - 0 - 13 cm, dark reddish brown (4 YR 3/2.5, moist), pinkish gray (6 YR 5.5/2, dry); sandy clay loam; moderate fine and medium granular; friable, plastic and sticky; abrupt and smooth boundary.

AB   - 13-23 cm, dark reddish brown (4 YR 3/5, moist), light reddish brown (4 YR 5.5/4, dry); sandy clay loam slightly gravelly; moderate very fine and fine subangular and angular blocky; common weak and few moderate clay skins; firm, plastic and sticky; clear and smooth boundary.

BA   - 23-33 cm, dark red (3.5 YR 3/5, moist), light reddish brown (4 YR 5.5/4, dry); sandy clay loam slightly gravelly; moderate very fine and fine subangular and angular blocky; common weak and few moderate clay skins; firm, plastic and sticky; clear and smooth boundary.

Bto1 - 33-70 cm, dark red (3.5 YR 3/6, moist), reddish yellow (4 YR 6/6, dry), reddish yellow (6 YR 6/6, dry, ground); slightly gravelly clay; structure as above; few weak and discontinuous clay skins; consistence as above; gradual and smooth boundary.

Bto2 - 70-100 cm, dark red (3.5 YR 3.5/6, moist), yellowish red (4 YR 5.5/6, dry); texture as above; weak very fine and fine subangular and angular blocky; friable with firm parts, plastic and sticky; abrupt and wavy boundary (20-60 cm).

2BC - 100-175 cm, clay loam-skeletal fine earth, red (3.5 YR 4/5, moist) interspersed with great amount of subangular and angular gravel and stones consisting mostly of quartz and quartzites.

Analyses

ACRUSTOXIC KANDIUSTULT

PEDON NO: 85PO734 (NSSL-SCS-USDA)

ANALYSIS: National Soil Survey Laboratory
Soil Conservation Service (USDA)

[1]	[2]	[3]	[4]	[5]	[6]	[7]	[8]	[9]	[10]	[11]	[12]
HZN	DEPTH	HORIZON	SAND	SILT	CLAY	FINE	DISP	FINE	COAR	O.C.	TOTAL N
NO.	(CM)		%	%	%	CLAY	CLAY	SILT	FRAC	%	%
						%	%	%	*		
1	13	Ap	58.3	14.3	27.4	7.4	24	6.7	5	2.56	0.198
2	23	AB	57.0	14.5	28.5	8.3	25	7.3	5	1.59	0.124
3	33	BA	53.8	14.6	31.6	9.4	29	7.0	10	1.00	0.074
4	70	Bto1	47.2	14.6	38.2	11.2	36	7.3	8	0.71	0.058
5	100	Bto2	46.4	15.8	37.8	11.6	0	7.7	11	0.50	0.035
6	175	2BC	54.5	15.6	29.9	8.3	0	8.2	57	0.30	0.019

[1]	[13]	[14]	[15]	[16]	[17]	[18]	[19]	[20]	[21]	[22]	[23]	[24]	[25]	[26]
HZN	ph		NH40AC EXTRACTABLE				CEC	BASE SAT.		EXTR	BULK	WATER CONTENT		
NO.	KCl	H2O	Ca	Mg	K	Al	pH7	SUM	pH7	Fe	DENS	33	1500	WRD
			----cmol(+)/kg----							%		kPa	kPa	
1	4.7	5.2	4.19	0.83	0.27	0.1	8.8	35	60	2.9			11.9	
2	4.7	5.4	2.89	0.74	0.18	0.1	6.2	36	62	2.7	1.35	17.2	11.5	0.08
3	4.7	5.5	1.92	0.57	0.18	TR	4.0	35	67	3.0	1.39		11.6	
4	4.9	5.5	1.21	0.41	0.21	TR	3.3	28	56	3.6	1.48	18.0	13.7	0.06
5	5.0	5.4	0.61	0.35	0.14	TR	2.5	22	45	3.5	1.33	19.9	14.2	0.08
6	5.9	5.8	-	0.43	-	TR	1.5	11	31	3.6			12.3	

B.2.6   Ultisol, Plinthic Kandiustult

Description

PEDON NO: 7008-7014 (Chalong Series)
LOCATION: 2 km West of Ban Mab Ta Put (100 m. South of Sukumvit Rd.), Amphoe Muang;
          Rayong Province, Thailand.
ELEVATION: Approx. 34 m.
RELIEF: Undulating.
PARENT MATERIAL: Transported materials from granite and other contact metamorphic rocks.
PHYSIOGRAPHY: Buried erosion surface.
SLOPE: 3 percent.
DRAINAGE: well drained.
CLIMATE: annual rainfall 1493 mm. mean annual temperature 27°C
VEGETATION: Cassava.
SOURCE: Proceedings Second International Soil Classification Workshop, Part II. Soil Survey
         Division, 1979.

A1   - 0 - 10 cm, brown loamy sand; moderate medium subangular blocky structure; very
       friable, slightly sticky, slightly plastic; common fine roots; clear, smooth boundary.

B1   - 10-40 cm, brown yellow (10 YR 6/6) medium sandy clay loam; moderate medium and
       coarse subangular blocky structure; friable, sticky, slightly plastic; patchy thin clay
       and organic matter coatings on ped faces and along root channels; few fine roots; gradual
       smooth boundary.

B21t- 40-63 cm, reddish yellow (7.5 YR 6/6) medium sandy clay loam; common medium
       prominent yellowish red (5 YR 5/8) mottles; moderate subangular blocky structure;
       friable, sticky, plastic; broken thin clay coatings on ped faces; moderately thick organic
       matter coatings along root channels; few fine roots; clear, wavy boundary.

B22tcn-63-96 cm, reddish yellow (7.5 YR 6/8) gravelly medium sandy clay; common
       medium prominent red (10 R 4/8) mottles; weak medium subangular blocky structure;
       friable, sticky, plastic; broken moderately thick clay coatings on ped faces; 30% by
       volume of ironstone concretions of 0.5 to 1.0 cm. in diameter, few quartz; clear,
       irregular boundary.

B23tcn-96-130 cm, mixed brownish yellow (10 YR 6/8) and yellowish brown (10 YR 5/8)
       gravelly clay; many medium prominent red (10 R 4/8) and strongly brown (7.5 YR
       5/8) mottles; weak medium subangular blocky structure, friable, sticky, plastic;
       broken moderately thick clay coatings on ped faces, 20% by volume of ironstone
       concretions of 0.5 to 1.0 cm. in diameter; gradual, smooth boundary.

B24t- 130-180 cm, mixed brownish yellow (10 YR 6/6), yellowish brown (10 YR 5/6) and
       light gray (10 YR 7/1) clay; many coarse prominent yellowish red (5 YR 4/6), dark
       red and red (10 R 3-4/6) mottles; firm, sticky, plastic; broken moderately thick clay
       coatings on ped faces; few ironstone concretions, few fine and medium roots.

Analyses

PLINTHIC KANDIUSTULT

PEDON NO: 7008-7014 (Chalong Series)

ANALYSIS: Soil Analysis Division
Dept. of Land Development
Bangkok, Thailand

[1]	[2]	[3]	[4]	[5]	[6]	[7]	[8]	[9]	[10]	[11]	[12]
HZN	DEPTH	HORIZON	SAND	SILT	CLAY	FINE	DISP	FINE	COAR	O.C.	TOTAL N
NO.	(CM)		%	%	%	CLAY	CLAY	SILT	FRAC	%	%
						%	%	%	*		
1	10	A1	71.8	18.0	10.2				2.40	1.07	0.06
2	40	B1	74.8	7.0	18.2				4.35	0.31	0.03
3	63	B21t	62.5	6.8	30.7				13.74	0.40	0.04
4	96	B22tcn	48.5	11.3	40.2				35.96	0.31	0.02
5	130	B23tcn	39.4	17.2	43.4				40.89	0.25	0.03
6	150	B24t	35.5	20.5	44.0				34.11	0.25	0.03
7	180		31.0	17.7	51.3					0.27	0.03

[1]	[13]	[14]	[15]	[16]	[17]	[18]	[19]	[20]	[21]	[22]	[23]	[24]	[25]	[26]
HZN	pH		NH4OAC EXTRACTABLE				CEC	BASE SAT.		EXTR	BULK	WATER CONTENT		
NO.	KCl	H2O	Ca	Mg	K	Al	pH7	SUM	pH7	Fe	DENS	33	1500	WRD
			----cmol(+)/kg----							%		kPa	kPa	
1	4.6	5.1	1.66	0.74	0.14	TR	3.64	36.1	70.6	0.30	1.48	7.4	6.8	
2	3.9	4.5	0.65	0.32	0.06	0.20	1.42	26.0	74.6	0.45	1.90	7.4	6.0	
3	3.6	4.0	0.41	0.22	0.10	0.07	2.15	14.3	35.3	0.78	1.67	13.1	10.5	
4	3.7	4.2	0.61	0.16	0.09	0.91	2.23	15.4	43.5	1.17	1.75	18.2	15.4	
5	3.7	4.2	0.26	0.25	0.06	1.09	3.42	8.8	17.2	1.99	1.77	30.5	25.4	
6	4.0	4.0	0.29	0.31	0.08	1.07	5.32	17.6	13.2	2.71	1.69	30.5	25.4	
7	3.9	4.2	0.26	0.33	0.08	1.39	3.35	10.4	21.5	2.41		3.6	1.4	

See page 1 of appendix for analytical methods.

B.2.7   Alfisol, Typic Kandiustalf

Description

PEDON NO:84PO522 (NSSL-SCS-USDA)
LOCATION: 10 km SE of Choma, Zambia (Pedon 13).
PHYSIOGRAPHY: level or undulating uplands
ELEVATION: 1075 m MSL.
PRECIPITATION: 830 mm
DRAINAGE: Moderately well drained
LAND USE: Pasture land and native pasture
PARENT MATERIAL: residuum from schist and phyllite material.
DESCRIBED BY: D. Hallbick, O. Spaargaren and C. Kalima (Soil Survey Unit, 1985).
SAMPLE DATE: 11/83

Ap1 - 0 - 19 cm; loamy sand; brown to dark brown (7.5 YR 4/4) and strong brown (7.5 YR 5/6) moist; massive; very friable, nonsticky, nonplastic; common fine and medium roots throughout; very fine and fine total porosity and few fine and medium continuous tubular pores; clear smooth boundary.

Ap2 - 19 - 31 cm; loamy sand; brown to dark brown (7.5 YR 4/4) and strong brown (7.5 YR 5/6) moist; massive; very friable, nonsticky, nonplastic; common fine and medium roots throughout; very fine and fine total porosity and few fine and medium continuous tubular pores; clear smooth boundary.

Bt1 - 31 - 72 cm; yellowish red (5 YR 5/8) sandy clay loam; yellowish red (5 YR 4/6) moist; weak coarse and very coarse angular blocky structure; slightly hard, slightly sticky, plastic; few patchy faint-thin clay films on faces of peds; common patchy faint-thin clay films between sand grains; few to common fine roots throughout; common very fine and fine interstitial pores and few fine and medium continuous tubular pores; gradual smooth boundary.

Bt2 - 72 - 125 cm; reddish yellow (5 YR 6/8) sandy clay; yellowish red (5 YR 5/8) moist; weak coarse and very coarse angular blocky structure; slightly hard, very sticky, plastic; few patchy faint-thin clay films on faces of peds; common patchy faint-thin clay films between sand grains; few fine roots throughout; common very fine and fine interstitial pores and few fine and medium continuous tubular pores; clear smooth boundary.

Bt3 - 125 - 185 cm; reddish yellow (5 YR 6/8) clay; yellowish red (5 YR 5/8) moist; common medium faint red (2.5 YR 5/8) and common medium prominent yellow (10 YR 7/6) mottles; moderate medium and coarse subangular blocky structure; hard, very sticky, very plastic; few discontinuous distinct-thin clay films on faces of peds; common patchy faint-thin clay films between sand grains; very few fine roots throughout, common very fine and fine interstitial pores, abrupt wavy boundary.

2BC - 185 - 198 cm; yellowish red (5 YR 5/6) gravelly clay; yellowish red (5 YR 5/6) moist; common medium prominent dark red (2.5 YR 3/6) and common medium faint strong brown (7.5 YR 5/6) mottles; massive; hard, slightly sticky, very plastic; few discontinuous prominent-thin dark red (10 R 3/6) clay films on sand and gravel; very few fine roots throughout; few very fine and fine interstitial pores; common medium and coarse irregular soft masses of iron; 25% pebbles, 5% stones from siltstone.

Analyses

TYPIC KANDIUSTALF

PEDON NO:84PO522 (NSSL-SCS-USDA)

ANALYSIS: National Soil Survey Laboratory
Soil Conservation Service (USDA)

[1]	[2]	[3]	[4]	[5]	[6]	[7]	[8]	[9]	[10]	[11]	[12]
HZN	DEPTH	HORIZON	SAND	SILT	CLAY	FINE	DISP	FINE	COAR	O.C.	TOTAL N
NO.	(CM)		%	%	%	CLAY	CLAY	SILT	FRAC	%	%
						%	%	%	*		
1	19	Ap1	87.0	7.6	5.4			3.4	TR	0.46	0.036
2	31	Ap2	85.8	8.2	6.0			3.6	TR	0.24	0.019
3	72	Bt1	74.8	6.8	18.4			3.1	1	0.15	0.014
4	125	Bt2	69.1	6.1	24.8			2.7	1	0.14	
5	185	Bt3	61.0	9.7	29.3			4.7	2	0.09	
6	198	2BC	60.0	16.4	23.6			9.5	31	0.08	

[1]	[13]	[14]	[15]	[16]	[17]	[18]	[19]	[20]	[21]	[22]	[23]	[24]	[25]	[26]
HZN	pH		NH40AC EXTRACTABLE				CEC	BASE SAT.	EXTR	BULK	WATER CONTENT			
NO.	KCl	H2O	Ca	Mg	K	Al	pH7	SUM	pH7	Fe	DENS	33	1500	WRD
			----cmol(+)/kg----								%	kPa	kPa	
1	4.3	4.8	0.7	0.3	0.2	TR	1.2	50	100	0.3	1.63	7.0	2.0	0.08
2	4.2	4.7	0.2	0.1	0.2	0.2	0.9	25	56	0.3	1.61	8.3	2.2	0.10
3	4.7	5.7	1.3	0.2	0.2		2.1	44	90	0.7	1.66	7.4	6.0	0.02
4	5.0	5.8	1.2	0.5	0.2		2.4	53	79	0.9	1.54	10.3	7.5	0.04
5	4.8	5.8	0.7	0.7	0.1		2.7	38	56	1.2	1.64	12.6	9.1	0.06
6	5.0	5.6	0.7	0.9	0.2		3.9	35	46	2.2	1.55	19.8	9.6	0.13

B.2.8 Alfisol, Typic Kandiustalf

Description

PEDON NO: 84P0514 (NSSL-SCS-USDA)
LOCATION: 10 km SE of Mkushi, Zambia (Pedon 6) 13°40'S, 029°29'E
PHYSIOGRAPHY: Upland slope in plateaus or tablelands.
ELEVATION: 1330 m MSL
PARENT MATERIAL: residuum from metamorphic-acidic material.
DESCRIBED BY: D. Hallbick, O. Spaargaren C. English and P. Woode.
SAMPLE DATE: 11/83

A   - 0 - 11 cm; grayish brown (10 YR 5/2) sandy loam; very dark grayish brown (10 YR
       3/2) moist; weak fine and medium granular structure; soft, nonsticky, nonplastic;
       many fine roots throughout and few medium roots throughout; many fine and medium
       continuous tubular pores and many fine and medium void between rock fragments pores;
       clear smooth boundary.

EA   - 11 - 34 cm; very pale brown (10 YR 7/3) sandy loam; brown to dark brown (10 YR
       4/3) moist; massive parting to strong very fine granular; slightly hard, nonsticky,
       nonplastic; common fine roots throughout and very few medium roots throughout;
       common fine and medium continuous tubular pores and many fine interstitial pores;
       gradual smooth boundary.

EB   - 34-61 cm; pink (7.5 YR 7/4) sandy clay loam; brown (7.5 YR 5/4) moist; massive
       parting to strong very fine granular; slightly hard, slightly sticky, plastic; very few
       fine roots throughout; common medium and coarse continuous tubular pores and many
       fine interstitial pores; clear smooth boundary.
       Cracks of about 0.5 to 1 cm wide occur from the base of the EA down to Bw3. Distance
       between cracks is 60 to 150 cm.

Bw1 - 61 - 99 cm; reddish yellow (7.5 YR 7/6) clay; strong brown (7.5 YR 5/6) moist;
       massive parting to strong very fine granular structure; slightly hard, nonsticky,
       nonplastic; very few fine roots throughout; few medium and coarse continuous tubular
       pores; gradual smooth boundary.

Bw2 - 99 - 151 cm; reddish yellow (7.5 YR 7/6) clay; strong brown (7.5 YR 5/6) moist;
       massive parting to strong very fine granular; hard, very sticky, plastic; very few fine
       roots throughout; very few medium and coarse continuous tubular pores; gradual smooth
       boundary.

Bw3 -151 - 174 cm; reddish yellow (7.5 YR 7/6) clay; yellowish red (5 YR 5/6) moist;
       massive parting to strong very fine granular; slightly hard, very sticky, plastic; very
       few medium and coarse continuous tubular pores; very few very fine and fine irregular
       gibbsite concretions. White crystals probably gibbsite occur locally near cracks.

Analyses

TYPIC KANDIUSTALF

PEDON NO: 84P0514 (NSSL-SCS-USDA)

ANALYSIS: National Soil Survey Laboratory
Soil Conservation Service (USDA)

[1]	[2]	[3]	[4]	[5]	[6]	[7]	[8]	[9]	[10]	[11]	[12]
HZN	DEPTH	HORIZON	SAND	SILT	CLAY	FINE	DISP	FINE	COAR	O.C.	TOTAL N
NO.	(CM)		%	%	%	CLAY	CLAY	SILT	FRAC	%	%
						%	%	%	*		
1	11	A	85.3	7.4	7.3			2.5	TR	1.60	0.082
2	34	EA	87.0	7.1	5.9			2.1	TR	0.26	0.016
3	61	EB	73.0	7.1	19.9			2.1	TR	0.23	0.017
4	99	Bw1	57.8	5.2	37.0			1.5	TR	0.13	
5	151	Bw2	58.5	5.6	35.9			1.7	TR	0.10	
6	174	Bw3	57.0	6.5	36.5			1.6	TR	0.09	

[1]	[13]	[14]	[15]	[16]	[17]	[18]	[19]	[20]	[21]	[22]	[23]	[24]	[25]	[26]
HZN	pH		NH40AC EXTRACTABLE				CEC	BASE SAT.	EXTR	BULK	WATER CONTENT			
NO.	KCl	H2O	Ca	Mg	K	Al	pH7	SUM	pH7	Fe	DENS	33	1500	WRD
			----cmol(+)/kg----							%		kPa	kPa	
1	5.7	6.6	4.0	0.7	0.5		5.1	78	100	0.3	1.56	6.7	4.2	0.04
2	5.2	6.3	0.6	0.2	0.2		0.9	100	100	0.2	1.55	5.0	2.3	0.04
3	5.2	6.2	0.5	0.6	0.5		1.6	73	100	0.4	1.60	8.2	6.1	0.03
4	4.4	5.2	0.6	0.7	0.3	0.3	2.7	46	59	0.7	1.60	13.6	11.0	0.04
5	4.4	5.6	0.5	0.5	0.3		2.5	41	52	0.7	1.51	13.1	10.7	0.04
6	4.7	5.8	0.7	0.6	0.2		2.9	44	52	0.7	1.44	13.2	10.6	0.04

B.2.9 Alfisol, Kandic Paleustalf

Description

PEDON NO: 84P0521 (NSSL-SCS-USDA)
LOCATION: Magoye Reg. Res. St. 15 km W. of Mazabukai, Zambia (Pedon 11) 15°49'S, 27°45'E
ELEVATION: 1040 m MSL
PARENT MATERIAL: residuum from metamorphic material.
DESCRIBED BY: D. Hallbick, O. Spaargaren and C. Kaliman (Soil Survey Unit, 1985).

A   - 0 - 8 cm; reddish brown (5 YR 5/3) sandy loam; dark reddish brown (5YR 3/3) moist;
      weak fine and medium subangular blocky structure parting to massive; soft, very
      friable, nonsticky, nonplastic; many fine and medium roots throughout; common to many
      very fine and fine interstitial and tubular pores; clear smooth boundary.

EA   - 8 - 21 cm; yellowish red (5 YR 5/6) sandy clay loam; dark reddish brown (5 YR 3/4)
      moist; massive; slightly hard, slightly sticky, nonplastic; common to many fine and
      medium roots throughout; common fine and medium continuous tubular pores and many
      very fine and fine interstitial pores; clear smooth boundary.

Bt1 - 21 - 43 cm; yellowish red (5 YR 4/6) clay; dark red (2.5 YR 3/6) moist; moderate
      fine and medium subangular blocky structure; slightly hard, very sticky, plastic; few
      patchy faint-thin clay films on vertical faces of peds; common to many fine and medium
      roots throughout; common fine and medium continuous tubular pores and common to
      many fine and medium void between rock fragments pores; gradual smooth boundary.

Bt2 - 43 - 112 cm; red (2.5 YR 4/6) clay; red (2.5 YR 4/6) moist; moderate fine and
      medium subangular blocky structure; slightly hard, very sticky, plastic; common
      discontinuous distinct-thin clay films on vertical and horizontal faces of peds; common
      fine and medium roots throughout; common fine and medium continuous tubular pores
      and common to many fine and medium void between rock fragments pores; clear smooth
      boundary.
      Clod 43 to 112 cms bulk samples split. Horizon split for sampling 43 to 75 cm and 75
      to 112 cm.

Bt3 - 112 - 157 cm; red (2.5 YR 5/8) clay; red (2.5 YR 4/6) moist; moderate fine and
      medium subangular blocky structure; slightly hard, very sticky, plastic; common
      patchy distinct-thin clay films on vertical and horizontal faces of peds; few fine and
      medium roots throughout; few medium continuous tubular pores and common fine and
      medium interstitial pores; gradual smooth boundary.

Bt4 - 157 - 186 cm; red (2.5 YR 5/8) clay; red (2.5 YR 4/8) moist; weak fine and medium
      subangular blocky structure; soft, very sticky, plastic; few patchy distinct-thin clay
      films on vertical and horizontal faces of peds; few fine and medium roots throughout; few
      medium continuous tubular and common fine and medium interstitial pores.

Analyses

KANDIC PALEUSTALF

PEDON NO: 84P0521 (NSSL-SCS-USDA)

ANALYSIS: National Soil Survey Laboratory
Soil Conservation Service (USDA)

[1]	[2]	[3]	[4]	[5]	[6]	[7]	[8]	[9]	[10]	[11]	[12]
HZN	DEPTH	HORIZON	SAND	SILT	CLAY	FINE	DISP	FINE	COAR	O.C.	TOTAL N
NO.	(CM)		%	%	%	CLAY	CLAY	SILT	FRAC	%	%
						%	%	%	*		
1	8	A	75.5	12.8	11.7			5.0	-	0.88	0.072
2	21	EA	71.7	11.7	16.6			4.4	-	0.74	0.057
3	43	Bt1	61.8	12.1	26.1			5.1	-	0.57	0.022
4	75	Bt2	51.2	10.7	38.1			4.7	-	0.52	
5	112	Bt2	46.2	10.4	43.4			4.0	-	0.44	
6	157	Bt3	47.8	11.4	40.8			4.5	-	0.27	
7	186	Bt4	47.4	12.1	40.5			4.6	-	0.22	

[1]	[13]	[14]	[15]	[16]	[17]	[18]	[19]	[20]	[21]	[22]	[23]	[24]	[25]	[26]
HZN	pH		NH40AC EXTRACTABLE				CEC	BASE SAT.		EXTR	BULK	WATER CONTENT		
NO.	KCl	H2O	Ca	Mg	K	Al	pH7	SUM	pH7	Fe	DENS	33	1500	WRD
			----cmol(+)/kg----							%		kPa	kPa	
1	5.0	5.8	1.4	1.4	0.6		4.1	56	83	1.1	1.43	11.6	4.3	0.10
2	4.8	5.8	1.1	1.3	0.4		4.4	44	64	1.4	1.46	11.0	5.3	0.08
3	4.0	4.9	0.9	1.4	0.2	0.8	5.7	34	53	2.1	1.45	15.9	8.5	0.11
4	4.1	5.4	1.2	2.1	0.3	1.1	7.8	32	47	2.8	1.44	20.2	12.0	0.12
5	4.2	5.4	1.4	2.6	0.3	0.9	8.4	36	51	3.3	1.44	20.2	13.4	0.10
6	4.5	5.5	1.6	2.9	0.4	0.1	7.8	46	64	3.1	1.33	21.2	12.2	0.12
7	4.7	5.6	1.6	3.1	0.3	0.1	8.1	48	63	3.1	1.36	20.2	11.9	0.11

B.2.10 Alfisol, Rhodic Paleudalf

Description

PEDON NO: ISCW-BR 6 (77P1256-61)
LOCATION: Londrina, PR. 34 km from Londrina on the highway to Ponta Grossa.
           23°40'S, 51°10'W
TOPOGRAPHY: Middle of 15% slope.
ELEVATION: 480 meters.
PARENT MATERIAL: Diabase.
VEGETATION: Semi-evergreen tropical forest.
SOURCE: Camargo, M. and F. Beinroth (1978).

Ap - 0 - 15 cm, dusky red (2.5 YR 3/2, moist), dusky red (2.5 YR 3/3, dry); clay; strong
      fine to medium granular and subangular blocky; many very fine to medium and some
      coarse pores; hard, firm, plastic to very plastic and very sticky; gradual and smooth
      boundary.

B1t - 15-32 cm, dark reddish brown (1.5 YR 3/4, moist), dark reddish brown (1.5 YR 3/4,
      dry); clay; moderate medium prismatic breaking easily to strong fine to medium
      subangular and angular blocky; common moderate clay films; common very fine and
      some coarse pores; hard, firm, plastic and sticky; diffuse and smooth boundary.

B21t- 32-74 cm, dark red (1.5 YR 3/5); clay; moderate coarse prismatic breaking easily to
      strong medium to coarse subangular and angular blocky; continuous strong clay films;
      common very fine to fine pores; hard, firm, plastic and sticky; diffuse and smooth
      boundary.

B22t- 74-154 cm, dark red (10 R 3/6); clay; moderate coarse prismatic breaking easily to
      strong medium to coarse subangular and angular blocky; continuous strong clay films;
      common very fine to fine pores; hard, firm, plastic and sticky; diffuse and smooth
      boundary.

B3t - 154-227 cm, weak red (10 R 4/4), few fine and diffuse mottles of red (10 R 4/8),
      few fine and prominent mottles of light yellowish brown (10 YR 4/4) and black (N 2/);
      clay; moderate medium to coarse subangular and angular blocky; common strong clay
      films; many very fine and fine pores; slightly hard to hard, friable, plastic and sticky;
      diffuse and smooth boundary.

C1 - 227-317 cm, variegated color of black (N 2/), yellowish brown (10 YR 5/6), reddish
      brown (5 YR 5/4), red (2.5 YR 4/5); clay; saprolite mixed by the bucket auger;
      plastic and sticky.

C2 - 317-370 cm, variegated color of black (N 2/), yellowish brown (10 YR 5/6), reddish
      brown (5 YR 5/4), red (2.5 YR 4/6); clay; consisting in saprolite mixed by bucket
      auger; plastic and sticky.

Analyses

RHODIC PALEUDALF

PEDON NO: ISCW-BR 6 (77P1256-61)

ANALYSIS: SNLCS, EMBRAPA, Brazil and
NSSL, US Dept. of Agriculture

[1]	[2]	[3]	[4]	[5]	[6]	[7]	[8]	[9]	[10]	[11]	[12]
HZN	DEPTH	HORIZON	SAND	SILT	CLAY	FINE	DISP	FINE	COAR	O.C.	TOTAL N
NO.	(CM)		%	%	%	CLAY	CLAY	SILT	FRAC	%	%
						%	%	%	*		
1	15	Ap	7	34	59		52			2.49	0.32
2	32	B1t	5	25	70		63			1.08	0.16
3	74	B21t	3	14	83		-			0.78	0.14
4	154	B22t	3	18	79		-			0.41	0.08
5	227	B3t	3	26	71		-			0.28	0.05
6	317	C1	5	29	66		1			0.22	0.04
7	370	C2	10	31	59		2			0.12	0.04

[1]	[13]	[14]	[15]	[16]	[17]	[18]	[19]	[20]	[21]	[22]	[23]	[24]	[25]	[26]
HZN	ph		NH4OAC EXTRACTABLE				CEC	BASE SAT.	EXTR	BULK	WATER CONTENT			
NO.	KCl	H2O	Ca	Mg	K	A l	pH7	SUM	pH7	Fe	DENS	33	1500	WRD
			----cmol(+)/kg----							%		kPa	kPa	
1	5.1	6.1	12.8	2.6	0.5		23.2		69	8.2	1.19		25.2	
2	5.2	6.0	9.1	2.5	0.5		18.9		64	8.5	1.26		26.7	
3	4.5	5.5	7.3	2.3	0.2	0.2	19.6		50	8.9	1.20		31.1	
4	4.9	5.6	6.8	2.2	0.6	TR	17.9		54	. 8.9	1.25		32.1	
5	4.2	5.4	7.7	4.0	0.9	0.8	21.8		58	7.1	1.22		30.8	
6	3.9	5.2	6.9	5.5	1.0	2.1	23.9		56	6.6			30.3	
7	3.8	5.2												

B.2.11   Vertisol, Entic Chromustert

Description

PEDON NO: 82P0243 (NSSL-SCS-USDA)
LOCATION: Broad plain in river valley. Terrace number 2 of Blue Nile.
POSITION: interfluve summit.
SLOPE: 0 percent, plane.
PARENT MATERIAL: alluvium from basalt.
DESCRIBED BY: H. Fadul, W.D. Nettleton (SMSS-Soil Survey Administration, 1982).

Ap1 - 0 - 5 cm.   Brown to dark brown (10YR 4/3) clay, brown to dark brown (10YR 4/3
          moist); moderate fine subangular blocky structure; hard, firm, sticky; plastic; many
          fine roots; many fine tubular, discontinuous pores; a few rounded calcium carbonate
          nodules; weakly effervescent; pH = 8.4, moderately alkaline; clear smooth boundary.

Ap2 - 5-23 cm.   Brown to dark brown (10YR 4/3) clay, strong coarse primatic parting to
          moderate fine subangular blocky structure; hard, firm, sticky; plastic; common fine
          roots; common rounded calcium carbonate nodules; weakly effervescent; pH = 8.6
          strongly alkaline; gradual smooth boundary.

Bwk1-23-45 cm.   Brown to dark brown (10YR 4/3, moist) clay; moderate fine subangular
          blocky parting to moderate medium subangular blocky structure; hard, firm, sticky;
          plastic; many slickensides on faces of peds; a few fine roots; common rounded calcium
          carbonate nodules; weakly effervescent; pH - 8.8, strongly alkaline; clear wavy
          boundary.

Bwk2-45-70 cm.   Brown to dark brown (10YR 4/3, moist) clay; weak medium subangular
          blocky structure; hard, firm, sticky; plastic; a few slickensides on faces of peds; very
          few fine roots; common rounded calcium carbonate nodules; weakly effervescent; pH =
          8.8, strongly alkaline; clear wavy boundary.

BCk - 70-90 cm.   Very dark gray (10YR 3/1, moist) clay; weak medium subangular blocky
          structure; hard, firm , sticky; plastic; very few fine roots; weakly effervescent; pH =
          8.8, strongly alkaline; irregular boundary.

Cky1 - 90-122 cm.   Yellowish brown (10YR 5/4, moist) clay; weak medium platy structure;
          hard, firm, sticky; plastic; a few gypsum crystals; weakly effervescent; pH = 8.8,
          strongly alkaline; clear wavy boundary.

Cky2 - 122-165 cm.   Very dark gray (10YR 3/1, moist) clay; weak medium platy structure;
          hard, firm, sticky; plastic; a few gypsum crystals; weakly effervescent; pH = 8.8,
          strongly alkaline.

Analyses

ENTIC CHROMUSTERT

PEDON NO: 82P0243 (NSSL-SCS-USDA)

ANALYSIS: National Soil Survey Laboratory
Soil Conservation Service (USDA)

[1]	[2]	[3]	[4]	[5]	[6]	[7]	[8]	[9]	[10]	[11]	[12]
HZN	DEPTH	HORIZON	SAND	SILT	CLAY	FINE	DISP	FINE	COAR	O.C.	TOTAL N
NO.	(CM)		%	%	%	CLAY	CLAY	SILT	FRAC	%	%
						%	%	%	*		
1	5	Ap1	16.9	29.0	54.1			18.7	2	0.52	0.040
2	23	Ap2	15.6	28.6	55.8			19.2	2	0.41	0.028
3	45	Bwk1	15.0	27.5	57.5			17.9	-	0.37	0.023
4	70	Bwk2	14.8	26.5	58.7			16.6	-	0.36	
5	90	BCk	15.0	26.8	58.2			17.0	-	0.45	
6	122	Cky1	11.5	30.2	58.3			19.7	-	0.56	
7	165	Cky2	12.1	31.5	56.4			24.0	-	0.42	

[1]	[13]	[14]	[15]	[16]	[17]	[18]	[19]	[20]	[21]	[22]	[23]	[24]	[25]	[26]
HZN	pH		NH40AC EXTRACTABLE				CEC	BASE SAT.	EXTR	BULK	WATER CONTENT			
NO.	KCl	H2O	Ca	Mg	K	Na	pH7	SUM	pH7	Fe	DENS	33	1500	WRD
			------cmol(+)/kg-------							%		kPa	kPa	
1		8.2	10.1	1.6	1.5	47.7		100	1.6	1.20	31.9	20.3	0.14
2		8.5	10.4	1.2	4.2	48.0		100	1.6		22.5		
3		8.6	10.6	1.0	9.1	48.2		100	1.6	1.24	35.8	23.6	0.15
4		8.6	11.3	0.8	13.1	48.4		100	1.6		23.6		
5		8.3	11.7	0.7	13.8	51.2		100	1.4	1.26	33.8	24.1	0.12
6		7.9	14.3	0.8	17.6	52.6		100	1.2		25.5		
7		7.9	15.9	0.9	17.7	49.3		100	1.2	1.17	41.4	26.1	0.18

B.2.12 Vertisol, Typic Chromustert

Description

PEDON: 84P0520 (NSSL-SCS-USDA)
LOCATION: Harthoorn's farm 10 km W of Kafue, Zambia. Pit located about 250 m SE of fish farm
    offices. 15°46' S; 27°55' E
PHYSIOGRAPHY: Broad plain in lake plains.
MICRORELIEF: gilgai less than 20 cm on middle third.
ELEVATION: 990 m.
LAND USE: Pasture land and native pasture.
PARENT MATERIAL: Lacustrine from sedimentary material
DESCRIBED BY: D. Hallbick, O. Spaargaren and C. Kalima (Soil Survey Unit, 1985)
SOURCE: Tour guide Zambia Forum

A   - 0 to 8 cm,; dark grayish brown (2.5Y 4/2) clay; very dark grayish brown (2.5Y 3/2)
         moist; few to common fine prominent yellowish red (5YR 5/8) mottles; strong medium
         and coarse granular structure; hard, very sticky, very plastic; many fine and medium
         roots throughout; many very fine and fine void between rock fragments pores; common
         fine rounded ironstone nodules; clear smooth boundary.

Bw   - 8 to 24 cm; olive (5Y 4/3) clay; dark olive gray (5Y 3/2) moist; few to common fine
         prominent yellowish red (5YR 5/8) mottles; strong medium and coarse subangular
         blocky structure; hard, very sticky, very plastic; few continuous thin intersecting
         slickensides on faces of peds; many fine and medium roots throughout; many fine and
         medium discontinuous tubular pores; common fine and medium rounded ironstone
         nodules and few fine and medium rounded lime nodules; clear smooth boundary.

Bk1   - 24 to 74 cm; olive gray (5Y 4/2) clay; dark olive gray (5Y 3/2) moist; strong medium
         and coarse angular blocky structure; hard; many thin intersecting slickensides on faces
         of peds; common to many fine roots throughout; few to common very fine and fine
         discontinuous tubular pores; common fine to coarse rounded lime nodules and common
         fine and medium rounded ironstone nodules; gradual wavy boundary.

Bk2   - 74 to 133 cm; olive gray (5Y 4/2) clay; dark olive gray (5Y 3/2) moist; strong
         medium and coarse angular blocky structure; hard, very firm; many thin intersecting
         slickensides on faces of peds; common fine roots throughout; few to common very fine
         and fine discontinuous tubular pores; many fine to coarse rounded lime nodules and
         common fine and medium rounded ironstone nodules; gradual smooth boundary.

BCk   - 133 to 154 cm; clay; olive (5Y 4/3) moist; moderate medium and coarse angular
         blocky structure; very firm; common thin intersecting slickensides on faces of peds; few
         fine roots throughout; few fine discontinuous tubular pores; many fine to coarse rounded
         lime nodules and many fine rounded ironstone nodules.

Analyses

TYPIC CHROMUSTERT

PEDON 84PO520 (NSSL-SCS-USDA):

ANALYSIS: National Soil Survey Laboratory
Soil Conservation Service (USDA)

[1]	[2]	[3]	[4]	[5]	[6]	[7]	[8]	[9]	[10]	[11]	[12]
HZN	DEPTH	HORIZON	SAND	SILT	CLAY	FINE	DISP	FINE	COAR	O.C.	TOTAL N
NO.	(CM)		%	%	%	CLAY	CLAY	SILT	FRAC	%	%
						%	%	%	*		
1	8	A	30.0	19.4	50.6	38.3		9.3	1	2.45	0.186
2	24	Bw	26.1	21.0	52.9	34.9		11.7	-	0.81	0.067
3	74	Bk1	26.9	20.7	52.4	29.6		11.9	3	0.57	0.045
4	105	Bk2	25.3	20.9	53.8	29.1		12.1	1	0.56	
5	133	Bk2	25.0	19.0	56.0	27.7		12.8	3	0.47	
6	154	BCk	25.4	18.6	56.0	25.1		12.8	4	0.36	

[1]	[13]	[14]	[15]	[16]	[17]	[18]	[19]	[20]	[21]	[22]	[23]	[24]	[25]	[26]
HZN	pH		NH40AC EXTRACTABLE				CEC	BASE SAT.		EXTR	BULK	WATER CONTENT		
NO.	KCl	H2O	Ca	Mg	K	Na	pH7	SUM	pH7	Fe	DENS	33	1500	WRD
			------cmol(+)/kg------							%		kPa	kPa	
1		6.7	18.1	11.0	0.7	2.3	30.9	83	97	2.3	1.40	26.5	17.4	0.13
2		7.4	17.7	10.0	0.2	2.7	29.3	89	97	2.7	1.55	23.1	16.2	0.11
3		8.4	19.2	12.1	0.2	2.6	28.6		100	2.6	1.57	23.3	17.2	0.09
4		8.5		15.7	0.2	2.6	28.7			2.6	1.57	24.3	19.1	0.08
5		8.1		16.0	0.2	2.5	27.9			2.5	1.53	26.1	19.3	0.10
6		8.3		14.5	0.2	2.5	28.5			2.5	1.53	27.0	20.0	0.11

B.2.13 Andisol, Typic Melanudand

Description

PEDON NO: INS 18 (ISRIC)
LOCATION: N. Sumatra, Indonesia, 5 km WSW of Takengon, 4°36'18"N, 96°47'24"W.
ALTITUDE: 1395 m
PHYSIOGRAPHY: lower volcanic slope of Salak Nama volcano.
SLOPE: 2-6%
PARENT MATERIAL: Andesitic ash.
LAND USE: Coffee plantation
CLIMATE: MAR 1700 mm, MAT 18.8°C.
SOIL CLIMATE: Udic.
DESCRIBED BY: P. Buurman (Mizota and Van Reenwyk, 1989).

Ah1 - 0 - 30 cm; black (10YR 2/1, moist) loam; moderate fine subangular blocky; friable,
thixotropic; common roots; few coarse and medium, common fine and many very fine
pores; gradual wavy boundary.

Ah2 - 30-37 cm; very dark brown (10YR 2/2 moist) loam; moderate fine angular blocky;
friable, thixotropic; few coarse and medium, common fine roots; few coarse, common
medium and many fine and very fine pores; clear smooth boundary.

A/B - 37-48 cm; dark yellowish brown (10YR 4/4 moist); loam to clay loam; moderate
medium-coarse subangular blocky; friable to firm, thixotropic; few roots; few medium
and many fine and very fine pores; clear and smooth boundary.

B   - 48-110 cm; yellowish brown (10YR 5/6 moist) clay loam to silty clay loam, 2-5%
stones; weak medium to coarse subangular blocky; firm, thixotropic; few roots; few
coarse and medium, many fine and very fine pores; clear and smooth boundary.

B/Cg - 110-120/130 cm; brownish yellow (10YR 6/8 moist) loam to silt loam, 15-20%
stones increasing with depth; firm, thixotropic; coarse and medium yellow (7.5YR 6/8)
mottles; gradual smooth boundary.

C   - 130+ cm; andesite, 15-20% saprolite.

Analyses

TYPIC MELANUDAND

PEDON NO: INS 18 (ISRIC)

ANALYSIS· International Soil Reference and
Information Centre (ISRIC) Wageningen, The Netherlands

[1]	[2]	[3]	[4]	[5]	[6]	[7]	[8]	[9]	[10]	[11]	[12]
HZN	DEPTH	HORIZON	SAND	SILT	CLAY	FINE	DISP	FINE	COAR	O.C.	TOTAL N
NO.	(CM)	.	%	%	%	CLAY	CLAY	SILT	FRAC	%	%
						%	%	%	*		
1	30	Ah1	38.5	43.5	18.0					7.8	
2	37	Ah2	37.3	50.4	12.3					5.7	
3	48	2AB	20.7	62.2	17.1					4.8	
4	110	2Bw	20.2	60.0	19.8					2.9	
5	110+	3BC	34.9	45.6	19.5					1.4	

[1]	[13]	[14]	[15]	[16]	[17]	[18]	[19]	[20]	[21]	[22]	[23]	[24]	[25]	[26]
HZN	pH		NH40AC EXTRACTABLE				CEC	BASE SAT.	EXTR	BULK	WATER CONTENT			
NO.	KCl	H2O	Ca	Mg	K	NA	pH7	SUM	pH7	Fe	DENS	33	1500	WRD
			------cmol(+)/kg------							%		kPa	kPa	
1	4.9	5.3	11.3	1.1	0.3	0.4	27.3		48	1.2	0.60		28	
2	5.2	5.8	11.5	1.2	0.1	0.2	23.5		55	0.6				
3	5.3	5.8	10.0	1.2	0.1	0.2	20.2		57	2.2	0.50		30	
4	5.0	5.5	3.2	0.5	0.1	0.2	13.9		28	2.5	0.50		33	
5	5.4	5.5	0.6	0.1	0.1	0.1	6.4		14	3.0				

B.2.14   Andisol, Hydric Pachic Fulvudand

Description

PEDON NO: CO 12 (ISRIC)
LOCATION: N. 2.5 km N. of Narino, Colombia, 1°20'N, 7°420'W.
ALTITUDE: 2350 m
PHYSIOGRAPHY: dissected footslopes of Galeras volcano.
SLOPE: 5%
PARENT MATERIAL: Andesitic ash and tuff.
VEGETATION: original forest cleared for cultivation.
LAND USE: sisal; in immediate vicinity also maize, potatoes, onions, Eucalyptus and Pinus
SOIL CLIMATE: Udic, isomesic.
DESCRIBED BY: W. Siderius (Mizota and Van Reenwyk, 1989).

Ah1 - 0 - 30 cm; dark brown (7.5YR 3/2) moist, loam, weak fine to medium granular; slightly hard (dry), friable (moist), non sticky and non plastic (wet); few volcanic glass; many fine, medium and coarse tubular pores; many fine medium and coarse roots; gradual smooth boundary.

Ah2 - 30-65 cm; black (N 2.5/0), with few fine faint diffuse dark reddish brown (5YR 3/4) mottles along root channels; sandy loam; weak fine to very friable (moist), slightly sticky and non plastic (wet); much volcanic glass; roots and pores as Ah1; gradual smooth boundary.

AC - 65-88 cm; very dark grey (10YR 3/1) moist, sandy loam; weak medium to fine subangular blocky; slightly hard dry, very friable (moist), slightly sticky and non plastic (wet); common volcanic glass; many fine, medium and coarse pores; few fine and medium roots; gradual smooth boundary.

C1 - 88-108 cm; yellowish brown (10YR 5/6) moist, clay loam; porous, massive; slightly hard (dry), very friable (moist), slightly sticky and non plastic (wet); few volcanic glass; common pores; very few roots; gradual smooth boundary.

C2 - 108-150 cm; yellowish brown (10YR 5/8) moist, sandy loam; porous, massive; soft (dry), very friable (moist), non sticky and non plastic (wet); few volcanic glass, some rock fragments; pores and roots as in C1.

Analyses

HYDRIC PACHIC FULVUDAND

PEDON NO: CO 12 (ISRIC)

ANALYSIS: International Soil Reference and
Information Centre (ISRIC) Wageningen, The Netherlands

[1]	[2]	[3]	[4]	[5]	[6]	[7]	[8]	[9]	[10]	[11]	[12]
HZN	DEPTH	HORIZON	SAND	SILT	CLAY	FINE	DISP	FINE	COAR	O.C.	TOTAL N
NO.	(CM)		%	%	%	CLAY	CLAY	SILT	FRAC	%	%
						%	%	%	*		
1	30	Ah1	53.4	27.6	19.0			19.3		15.9	
2	65	Ah2	58.9	22.0	19.1			16.2		9.3	
3	88	AC	24.6	32.8	42.6			27.1		11.2	
4	108	C1	12.5	47.6	39.9			33.0		1.9	
5	150	C2	20.2	44.8	35.0			35.5		1.3	

[1]	[13]	[14]	[15]	[16]	[17]	[18]	[19]	[20]	[21]	[22]	[23]	[24]	[25]	[26]
HZN	pH		NH40AC EXTRACTABLE					CEC	BASE SAT.	EXTR	BULK	WATER CONTENT		
NO.	KCl	H2O	Ca	Mg	K	Al	pH7	SUM	pH7	Fe	DENS	33	1500	WRD
			----cmol(+)/kg----								%	kPa	kPa	
1	4.2	4.3	0.6	0.5	0.1	3.2	61.7					0.48		54
2	4.7	5.1	1.3	0.5	0.0	0.6	46.9					0.72		48
3	4.9	5.3	0.8	0.7	0.2	0.3	76.0					0.52		91
4	5.4	5.6	0.8	0.7	0.0	0.1	47.2					0.48		85
5	5.4	5.6	1.6	1.3	0.1	0.1	45.7							

B.2.15   Andisol, Acrudoxic Hydrudand

Description

PEDON NO: Hilo series (ISRIC-USA 7)
LOCATION: Wainaku, Island of Hawaii, sugarcane field of Mauna Kea Sugar Company.
    19°44'39"N,   155°6'2"W
PHYSIOGRAPHY: Gently sloping to steep uplands.
SLOPE: 5%
PARENT MATERIAL: Andesitic volcanic ash.
VEGETATION: Sugarcane, hilograss, california grass.
CLIMATE: MAR 3800 mm, MAT ca. 22.5°C.
SOIL CLIMATE: Perudic.
DESCRIBED BY: S. Nakamura (Mizota and Van Reenwyk, 1989)

Ap   - 0 - 23 cm; dark brown (7.5YR 3/4) silty clay loam (gritty due to irreversible
      drying); moderate very fine and fine subangular blocky structure; friable, slightly
      sticky, plastic; common roots; many pores; there is mixture of redder material from
      below; at the base of this horizon is a 1 to 5 cm thick discontinuous dark red ash band
      that rubs down to a sandy clay loam; clear smooth boundary.

B2i - 23-56 cm; dark brown (7.5YR 3/4) silty clay loam; weak medium prismatic structure
      parting to moderate fine subangular blocky; friable, sticky, plastic, moderately smeary;
      common roots; many very fine and few fine pores; horizon appears stratified due to ash
      deposits; the lower 5 cm is dark reddish brown (5YR 3/4); gradual smooth boundary.

B22 - 56-91 cm; dark reddish brown (5YR 3/4) and yellowish red (5YR 4/6) silty clay
      loam; moderate medium prismatic structure parting to strong fine and very fine
      subangular blocky; friable, sticky, plastic, moderately smeary; few roots; many very
      fine and common fine pores; horizon appears stratified due to ash deposits; clear smooth
      boundary.

B23 - 91-102 cm; dark brown (7.5YR 3/2) silty clay loam; moderate medium prismatic
      structure parting to strong fine subangular blocky; friable, sticky, plastic, moderately
      smeary; few roots; many very fine and common fine pores; clear smooth boundary.

B24 - 102-107 cm; dark reddish brown (5YR 3/4) silty clay loam: moderate medium
      prismatic structure parting to strong, fine subangular blocky; friable, sticky, plastic,
      moderately smeary; few roots; many very fine and common fine pores; clear smooth
      boundary.

B25 - 107-165 cm; dark brown (7.5YR 3/4) silty clay loam; strong medium prismatic
      structure parting to strong fine subangular blocky; friable, sticky, plastic, moderately
      smeary; few roots; many very fine and common fine pores.

Analyses

ACRUDOXIC HYDRUDAND

PEDON NO: Hilo series (ISRIC-USA 7)

ANALYSIS: International Soil Reference and
Information Centre (ISRIC) Wageningen, The Netherlands

[1]	[2]	[3]	[4]	[5]	[6]	[7]	[8]	[9]	[10]	[11]	[12]
HZN	DEPTH	HORIZON	SAND	SILT	CLAY	FINE	DISP	FINE	COAR	O.C.	TOTAL N
NO.	(CM)		%	%	%	CLAY	CLAY	SILT	FRAC	%	%
						%	%	%	*		
1	23	Ap	20.0	40.8	39.2			30.2		6.5	
2	56	B21	16.5	37.8	45.7			33.4		3.5	
3	91	B22	14.5	37.3	48.2			32.4		3.1	
4	102	B23	13.3	32.7	54.0			27.5		3.3	
5	107	B24	20.2	42.3	37.5			34.1		2.2	
6	165	B25	14.8	35.3	49.9			30.4		2.8	

[1]	[13]	[14]	[15]	[16]	[17]	[18]	[19]	[20]	[21]	[22]	[23]	[24]	[25]	[26]
HZN	pH		NH40AC EXTRACTABLE				CEC		BASE SAT.	EXTR	BULK	WATER CONTENT		
NO.	KCl	H2O	Ca	Mg	K	Al	pH7	SUM	pH7	Fe	DENS	33	1500	WRD
			----cmol(+)/kg----							%		kPa	kPa	
1	5.5	5.8	4.1	0.7	0.1	0.0	46.4			5.0	0.74		74	
2	6.0	6.1	1.2	0.0	0.0	0.0	42.6			11.5	0.29		148	
3	6.1	6.3	0.7	0.0	0.0	0.0	47.1			13.0	0.30		160	
4	6.1	6.3	0.7	0.0	0.0	0.1	53.6			8.6	0.42		154	
5	6.1	6.1	0.0	0.0	0.0	0.0	35.1			11.9				
6	6.0	6.2	0.0	0.0	0.0	0.1	55.0			14.3	0.29		185	

B.2.16 Inceptisol, Fluventic Ustropept

Description

PEDON 82P0375 (NSSL-SCS-USDA)
LOCATION: Wang Salavillage, Thailand.
PHYSIOGRAPHY: Terrace in river valley or delta.
SLOPE: 1 percent
LAND USE: Cropland
PARENT MATERIAL: Strongly weathered alluvium from acidic rock
DESCRIBED BY: W. Silicheuchu, C. Changprai, C. Manotham, C. Niamskul (Soil Survey Division, 1983)

Ap - 0 - 17 cm Pale brown (10YR 6/3) clay loam, brown to dark brown (10YR 4/3, moist): moderate medium to coarse subangular blocky parting to moderate fine to medium granular structure; hard, friable, slightly sticky; slightly plastic; common fine roots and common coarse roots; common worm casts; clear wavy boundary.

AB - 17-32 cm Pale brown (10YR 6/3) clay loam, brown to dark brown (10YR 4/3, moist); moderate fine to medium subangular blocky structure; hard, friable, slightly sticky; slightly plastic; common fine roots and common coarse roots; common worm casts; clear smooth boundary.

Bt1 - 32-65 cm Light yellowish brown (10YR 6/4) clay loam, yellowish brown (10YR 5/4, moist); moderate fine to medium subangular blocky structure; hard, friable, slightly sticky; slightly plastic; some thin patchy clay skins on faces of peds and some thick clay skins in channels and pores; common medium roots; a few iron concretions; gradual smooth boundary.

Bt2 - 65-90 cm Pale brown (10YR 6/3) light clay loam, yellowish brown (10YR 5/4, moist); moderate fine to medium subangular blocky structure; hard, friable, slightly sticky; slightly plastic; some thin patchy clay skins on faces of peds and some thick clay skins in channels and pores; common medium roots; a few iron concretions; gradual smooth boundary.

Bt3 - 90-122 cm Light yellowish brown (10YR 6/4) light clay loam, yellowish brown (10YR 5/5, moist); weak fine to medium subangular blocky structure; friable, slightly sticky; slightly plastic; some thin patchy clay skins on faces of peds and some thick clay skins in channels and pores; a few iron concretions; diffuse smooth boundary.

Bc1 - 122-138 cm Yellow (10YR 7/6) light clay loam, yellowish brown (10YR 5/6, moist); weak fine to medium subangular blocky structure; some thin patchy clay skins on faces of peds; a few iron concretions; clear smooth boundary.

Bc2 - 138-160 cm, Yellowish brown (10YR 5/8, moist) sandy clay loam; weak fine to medium subangular blocky structure; some thin patchy clay skins in channels and pores.

Analyses

FLUVENTIC USTROPEPT

PEDON 82P0375 (NSSL-SCS-USDA)

ANALYSIS: National Soil Survey Laboratory
Soil Conservation Service (USDA)

[1]	[2]	[3]	[4]	[5]	[6]	[7]	[8]	[9]	[10]	[11]	[12]
HZN	DEPTH	HORIZON	SAND	SILT	CLAY	FINE	DISP	FINE	COAR	O.C.	TOTAL N
NO.	(CM)		%	%	%	CLAY	CLAY	SILT	FRAC	%	%
						%	%	%	*		
1	17	Ap	38.9	35.1	26.0	8.7		19.4	1	0.83	0.072
2	32	AB	39.1	34.2	26.7	9.5		18.1	1	0.56	0.057
3	65	Bt1	38.1	34.9	27.0	12.3		17.2	TR	0.44	0.049
4	90	Bt2	41.2	33.3	25.5	8.6		17.7	2	0.41	0.043
5	122	Bt3	46.5	32.1	21.4	7.4		15.6	3	0.34	
6	138	BC1	58.6	25.9	15.5	6.5		12.2	27	0.22	
7	160	BC2	55.2	28.0	16.7	7.3		12.6	3	0.24	

[1]	[13]	[14]	[15]	[16]	[17]	[18]	[19]	[20]	[21]	[22]	[23]	[24]	[25]	[26]
HZN	pH		NH40AC EXTRACTABLE				CEC	BASE SAT.	EXTR	BULK	WATER CONTENT			
NO.	KCl	H2O	Ca	Mg	K	Na	pH7	SUM	pH7	Fe	DENS	33	1500	WRD
			------ cmol(+)/kg ------							%		kPa	kPa	
1		8.1		1.9	0.3	-	11.0		100	1.3	1.47	19.8	10.5	0.14
2		8.2		1.9	0.3	-	11.1		100	1.3	1.45	19.3	10.6	0.13
3		8.2		1.1	0.3	-	10.9		100	1.4	1.46	19.4	9.9	0.14
4		8.2		1.4	0.3	TR	11.3		100	1.2	1.39	19.3	9.9	0.13
5		8.2		1.7	0.3	-	10.3		100	1.1	1.45	15.6	8.7	0.10
6		8.3		1.8	0.2	TR	7.4		100	1.1	1.53	15.6	6.5	0.12
		8.3		2.2	0.2	TR	7.9		100	1.0	1.49	13.7	6.9	0.10

B.2.17 Spodosol, Tropohumod

Description

PEDON NO: S82FN-875-001 (NSSL-SCS-USDA)
LOCATION: 232 Km E of Bangkok. 100 m right of Rayong Ban Pae road (Thailand).
PHYSIOGRAPHY: Beach in coastal plains.
ELEVATION: 4 m M.S.L.
PRECIPITATION: 2400 mm. per year.
WATER TABLE: 140 cm Apparent.
DRAINAGE: Moderately well drained.
LAND USE: Cropland.
PARENT MATERIAL: Beach sand.
DESCRIBED BY: L. Moncharoen, W. Silicheuchu, M. Mausbach (August 1982)
SOURCE: 4th International Forum on Soil Taxonomy. Tour Guide. (Soil Survey Division, 1983).

Ap   - 0 - 17 cm Dark gray (10YR 4/1, dry) sand; weak fine granular structure; soft,
       friable, non-sticky; non-plastic; common fine roots; clear smooth boundary.

E1   - 17-37 cm Light brownish gray (10YR 6/2, dry) sand; structureless; loose, loose, non-
       sticky; non-plastic; a few fine roots; clear smooth boundary.

E2   - 37-80 cm White (10YR 8/1, moist) sand; structureless; loose, loose, non sticky; non-
       plastic; very few fine roots; clear broken boundary.

Bh1 - 80-90 cm Very dark grayish brown (10YR 3/2, moist) sand; structureless; firm
       moderately smeary, very moist or wet, weakly cemented; clear smooth boundary.

Bh2 - 90-130 cm Very dark grayish brown (2.5YR 3/2, moist) sand; weak medium blocky
       structure; friable, moderately smeary, very moist or wet; clear irregular boundary,
       lower parts sometimes strongly cemented.

Bh3 - 130-147 cm Very dark grayish brown (2.5YR 3/2, moist) sand; structureless; weakly
       cemented.

NOTE: The analysis of the Bh1 horizon shows that the ratio of aluminum plus carbon
       extractable by pyrophosphate at pH 10 to percentage clay is 0.7; the ratio of (Fe + Al)
       extractable by pyrophosphate to the (Fe + Al) extractable by dithionite-citrate is 0.6;
       the index of accumulation is 451. The chemical requirements for the identification of
       the spodic horizon are met.

Analyses

TROPOHUMOD

PEDON S82FN-85-001 (NSSL-SCS-USDA)

ANALYSIS: National Soil Survey Laboratory
             Soil Conservation Service (USDA)

[1]	[2]	[3]	[4]	[5]	[6]	[7]	[8]	[9]	[10]	[11]	[12]
HZN	DEPTH	HORIZON	SAND	SILT	CLAY	FINE	DISP	FINE	COAR	O.C.	TOTAL N
NO.	(CM)		%	%	%	CLAY	CLAY	SILT	FRAC	%	%
						%	%	%	*		
1	17	Ap	96.4	2.0	1.6			0.8	-	1.08	0.030
2	37	E1	97.6	0.2	2.2			0.1	-	0.20	0.004
3	80	E2	98.3	0.9	0.8			0.8	-	0.12	-
4	90	Bh1	88.5	5.3	6.2			3.7	-	4.32	0.101
5	130	Bh2	98.0	1.2	0.8			1.1	-	0.43	0.009
6	147	Bh3	96.4	0.7	2.9			-	TR	1.59	0.034

[1]	[13]	[14]	[15]	[16]	[17]	[18]	[19]	[20]	[21]	[22]	[23]	[24]	[25]	[26]
HZN	pH		NH40AC EXTRACTABLE				CEC		BASE SAT.	EXTR	BULK	WATER CONTENT		
NO.	KCl	H2O	Ca	Mg	K	Al	pH7	SUM	pH7	Fe	DENS	33	1500	WRD
			----cmol(+)/kg----							%		kPa	kPa	
1		4.2	0.4	0.1		0.5	2.9		17	-	1.63	5	2.1	0.05
2		4.7	-	TR		0.1	0.3		7	-			0.9	
3		5.0	-	-		-	0.2		10	-			0.8	
4		4.4	0.2	-		6.0	24.9		1	TR	1.37	17.3	9.0	0.11
5		4.5	TR	-		0.7	1.3		2	-			1.2	
6		4.4	TR	-		2.2	8.6		TR	-	1.58	10.9	3.1	0.12

References

FAO, Guidelines for Soil Profile Descriptions. FAO, Rome, 1977.

Kimble, J., "Introduction to data section," in *Tour Guide, Eighth International Soil Classification Workshop, Brazil,* Serviço Nacional de Levantamento e Conservação de Solos, Soil Management Support Services, Soil Conservation Service, U.S.D.A., Washington, D.C., 1986.

Soil Conservation Service, *Soil Survey Laboratory Methods and Procedures for Collecting Soil Samples,* Soil Survey Investigations Report No. 1, U.S. Government Printing Office, Washington, D.C., 1972.

Soil Survey Division, *Tour Guide, Fourth International Forum on Soil Taxonomy and Agrotechnology Transfer, Thailand,* Soil Survey Division, Land Development Department, Bangkok, Thailand, 1983.

Soil Survey Unit, *Field Guide of the 11th International Forum on Soil Taxonomy,* Department of Agriculture, Republic of Zambia, Lusaka, Zambia, 1985.

Soil Survey Staff, *Keys to Soil Taxonomy,* SMSS Technical Monograph #6, International Soils, Agronomy Department, Cornell University, Ithaca, New York, 1988.

SMSS-EMBRAPA, *Tour Guide, Eighth International Soil Classification Workshop, Brazil,* Serviço Nacional de Levantamento e Conservação de Solos (Embrapa) and Soil Management Support Services (SMSS), Soil Conservation Service, U.S.D.A., Washington, D.C., 1986.

SMSS-Soil Survey Administration, *Tour Guide, Fifth International Soil Classification Workshop, Sudan,* Soil Management Support Services (SMSS), Soil Conservation Service, U.S.D.A. and Soil Survey Administration of Sudan, Washington, D.C., 1982.

Index

ABOUT THE AUTHOR

Armand Van Wambeke is a professor of soil science at the Department of Soil, Crop, and Atmospheric Sciences, College of Agriculture and Life Science, Cornell University, Ithaca, New York.